BIBLIOGRAPHIC RECORD TARGET

AAM8263
UL FMT B RT a BL m T/C DT 09/12/88 R/DT 03/21/89 CC STAT mm E/L 1

010: : |a 02020693
035/1: : |a (RLIN)MIUG83-B73244
035/2: : |a (CaOTULAS)160088626
040: : |c MiU |d MiU
050/1:0 : |a QA453 |b .H19
100:1 : |a Halsted, George Bruce, |d 1853-1922.
245:04: |a The elements of geometry. |c By George Bruce Halsted ...
260: : |a New York, |b J. Wiley & sons, |c 1885.
300/1: : |a xvi, 366 p. |b diagr. |c 24 cm.
650/1: 0: |a Geometry.
998/1: : |c EM |s 9124

Alexander Ziwet

THE

ELEMENTS OF GEOMETRY.

BY

GEORGE BRUCE HALSTED,

A.B., A.M., AND EX-FELLOW OF PRINCETON COLLEGE; PH.D. AND EX-FELLOW OF JOHNS HOPKINS UNIVERSITY; INSTRUCTOR IN POST-GRADUATE MATHEMATICS, PRINCETON COLLEGE; PROFESSOR OF PURE AND APPLIED MATHEMATICS, UNIVERSITY OF TEXAS.

NEW YORK:

JOHN WILEY & SONS,

15 ASTOR PLACE.

1885.

ELECTROTYPED AND PRINTED
BY RAND, AVERY, AND COMPANY,
BOSTON.

Dedicated

(BY PERMISSION)

TO

J. J. SYLVESTER,

A.M., CAM.; F.R.S., L. AND E.; CORRESPONDING MEMBER INSTITUTE OF FRANCE; MEMBER ACADEMY OF SCIENCES IN BERLIN, GÖTTINGEN, NAPLES, MILAN, ST. PETERSBURG, ETC.; LL.D., UNIV. OF DUBLIN, AND U. OF E.; D.C.L., OXFORD; HON. FELLOW OF ST. JOHN'S COL., CAM.; SAVILIAN PROFESSOR OF GEOMETRY IN THE UNIVERSITY OF OXFORD;

In Grateful Remembrance

OF BENEFITS CONFERRED THROUGHOUT TWO FORMATIVE YEARS.

PREFACE.

In America the geometries most in vogue at present are vitiated by the immediate assumption and misuse of that subtile term "direction;" and teachers who know something of the Non-Euclidian, or even the modern synthetic geometries, are seeing the evils of this superficial "directional" method.

Moreover the attempt, in these books, to take away by definition from the familiar word "distance" its abstract character and connection with length-units, only confuses the ordinary student. A reference to the article *Measurement* in the "Encyclopædia Britannica" will show that around the word "distance" centers the most abstruse advance in pure science and philosophy. An elementary geometry has no need of the words direction and distance.

The present work, composed with special reference to use in teaching, yet strives to present the Elements of Geometry in a way so absolutely logical and compact, that they may be ready as rock-foundation for more advanced study.

Besides the acquirement of facts, there properly belongs to Geometry an educational value beyond any other elementary subject. In it the mind first finds logic a practical instrument of real power.

The method published in my *Mensuration* for the treatment of solid angles, with my words *steregon* and *steradian*, having been adopted by such eminent authorities, may I venture to recommend the use of the word *sect* suggested in the same volume?

From 1877 I regularly gave my classes the method of Book IX. In 1883 my pupil, H. B. Fine, at my suggestion, wrote out a Syllabus of Spherical Geometry on the lines of my teaching, which I have followed in Book IX.

The figures, which I think give this geometry a special advantage, owe all their beauty to my colleague, Professor A. V. Lane, who has given them the benefit of his artistic skill and mastery of graphics.

The whole work is greatly indebted to my pupil and friend, Dr. F. A. C. Perrine. We have striven after accuracy. Any corrections or suggestions relating to the book will be thankfully received.

GEORGE BRUCE HALSTED.

2004 Matilde Street,
Austin, Texas.

CONTENTS.

THE ELEMENTS OF GEOMETRY.

BOOK I.

CHAPTER I.

ON LOGIC.

CHAPTER II.

THE PRIMARY CONCEPTS OF GEOMETRY.

I. *Definitions of Geometric Magnitudes.*

II. *Properties of Distinct Things.*

III. *Some Geometrical Assumptions about Euclid's Space.*

IV. *The Assumed Constructions.*

CHAPTER III.

PRIMARY RELATIONS OF LINES, ANGLES, AND TRIANGLES.

I. *Angles about a Point.*

II. *Angles about Two Points.*

III. *Triangles.*

CHAPTER IV.

PROBLEMS.

CHAPTER V.

INEQUALITIES.

CHAPTER VI.

PARALLELS.

CHAPTER VII.

TRIANGLES.

CHAPTER VIII.

POLYGONS.

I. *Definitions.*

BOOK II.

RECTANGLES.

BOOK III.

THE CIRCLE.

BOOK IV.

REGULAR POLYGONS.

BOOK V.

RATIO AND PROPORTION.

BOOK VI.

RATIO APPLIED.

GEOMETRY OF THREE DIMENSIONS.

BOOK VII.

OF PLANES AND LINES.

BOOK VIII.

TRI-DIMENSIONAL SPHERICS.

BOOK IX.

TWO-DIMENSIONAL SPHERICS.

BOOK X.

POLYHEDRONS.

BOOK XI.

MENSURATION, OR METRICAL GEOMETRY.

CHAPTER I.

THE METRIC SYSTEM. — LENGTH, AREA.

CHAPTER II.

RATIO OF ANY CIRCLE TO ITS DIAMETER.

CHAPTER III.

MEASUREMENT OF SURFACES.

CHAPTER IV.

SPACE ANGLES.

CHAPTER V.

THE MEASUREMENT OF VOLUMES.

THE ELEMENTS OF GEOMETRY.

BOOK I.

CHAPTER I.

ON LOGIC.

I. Definitions. — Statements.

1. A statement is any combination of words that is either true or false; e.g. (*exempli gratia*, "for example"),

All x's are y's.

2. To pass from one statement to another, with a consciousness that belief in the first warrants belief in the second, is to infer.

3. Two statements are equivalent when one asserts just as much as the other, neither more nor less; e.g., *A* equals *B*, and *B* equals *A*.

4. A statement is implied in a previous statement when its truth follows of necessity from the truth of the previous statement; e.g.,

All x's are y's implies *some y's are x's.*

5. Any declarative sentence can be reduced to one or more simple statements, each consisting of three parts, namely, two terms or classes, and a copula, or relation, connecting them.

A equals *B*, *x* is *y*, are simple examples of this typical form of all statements. Here *x* and *y* each stand for any word or group of words that may have the force of a substantive in naming a class; e.g., *x* is *y* may mean all the *x*'s are *y*'s; man is mortal, means, all men are mortals.

6. A sentence containing only one such statement is a logically simple sentence; e.g., Man is the only picture-making animal.

7. A sentence that contains more than one such statement is a logically composite sentence; e.g., *A* and *B* are respectively equal to *C* and *D*, is composite, containing the two statements *A* equals *C*, and *B* equals *D*.

8. Statements that are expressly conditional, such as, if *A* is *B*, then *C* is *D*, reduce to the typical form as soon as we see that they mean

(*The consequence of M*) *is N*.

9. In the typical form, bird is biped, *x* is *y*, we call *x* and *y* the terms of the statement; the first being called the subject, and the last the predicate.

II. Definitions.—Classes.

10. Terms that name a single object are called individual terms; e.g., *Newton*.

11. Terms that name any one of a group of objects are called class terms; e.g., *man*, *crystal*. The name stands for any object that has certain properties, and these properties are possessed in common by the whole group for which the name stands.

12. A class is defined by stating enough properties to decide whether a thing belongs to it or not.

Thus, "rational animal" was given as a definition of "man."

13. If we denote by x the class possessing any given property, all things not possessing this property form another class, which is called the contradictory of the first, and is denoted by non-x, meaning "not x;" e.g., the contradictory of animate is inanimate.

14. Any one thing belongs either to the class x or to the class non-x, but no thing belongs to both.

It follows that x is just as much the contradictory of non-x as non-x is of x. So any class y and the class non-y are mutually the contradictories of each other, and both together include all things in the universe; e.g., *unconscious* and *conscious.*

III. The Universe of Discourse.

15. In most investigations, we are not really considering all things in the world, but only the collection of all objects which are contemplated as objects about which assertion or denial may take place in the particular discourse. This collection we call our universe of discourse, leaving out of consideration, for the time, every thing not belonging to it.

Thus, in talking of geometry, our terms have no reference to perfumes.

16. Within the universe of discourse, whether large or small, the classes x and non-x are still mutually contradictory, and every thing is in one or the other; e.g., within the universe mammals, every thing is man or brute.

17. The exhaustive division into x and non-x is applicable to any universe, and so is of particular importance in logic. But a special universe of discourse may be capable of some entirely different division into contradictories, equally exhaustive. Thus, with reference to any particular magnitude, all

magnitudes of that kind may be exhaustively divided into the contradictories,

Greater than, equal to, less than.

IV. Contranominal, Converse, Inverse, Obverse.

18. If x and y are classes, our typical statement x is y means, if a thing belongs to class x, then it also belongs to class y; e.g., Man is mortal, means, to be in the class men, is to be in the class mortals.

If the typical statement is true, then every individual x belongs to the class y: hence no x belongs to the class non-y, or no thing not y is a thing x; that is, every non-y is non-x: e.g., the immortals are not-human.

The statements x is y, and non-y is non-x, are called each the contranominal form of the other.

Though both forms express the same fact, it is, nevertheless, often of importance to consider both. One form may more naturally connect the fact with others already in our mind, and so show us an unexpected depth and importance of meaning.

19. Since x is y means all the x's are y's, the class y thus contains all the individuals of the class x, and may contain others, besides. Some of the y's, then, must be x's. Thus, from "a crystal is solid" we infer "some solids are crystals."

This guarded statement, some y is x, is called the logical converse of x is y. It is of no importance in geometry.

20. If, in the true statement x is y, we simply interchange the subject and predicate, without any restriction, we get the inverse statement y is x, which may be false.

In geometry it often happens that inverses are true and important. When the inverse is not true, this arises from the circumstance that the subject of the direct statement has been more closely limited than was requisite for the truth of the statement.

21. The contranominal of the inverse, namely, non-x is non-y, is called the obverse of the original proposition.

Of course, if the inverse is true, the obverse is true, and *vice versa.* To prove the obverse, amounts to the same thing as proving the inverse. They are the same statement, but may put the meaning expressed, in a different light to our minds.

22. If the original statement is *x is y*, its contranominal is *non-y is non-x*, its inverse is *y is x*, its obverse is *non-x is non-y*. The first two are equivalent, and the last two are equivalent.

Thus, of four such associated theorems it will never be necessary to demonstrate more than two, care being taken that the two selected are not contranominal.

23. From the truth of either of two inverses, that of the other cannot be inferred. If, however, we can prove them both, then the classes x and y are identical.

A perfect definition is always invertable.

V. On Theorems.

24. A theorem is a statement usually capable of being inferred from other statements previously accepted as true.

25. The process by which we show that it may be so inferred is called the proof or the demonstration of the theorem.

26. A corollary to a theorem is a statement whose truth follows at once from that of the theorem, or from what has been given in the demonstration of the theorem.

27. A theorem consists of two parts,—the hypothesis, or that which is assumed, and the conclusion, or that which is asserted to follow therefrom.

28. A geometric theorem usually relates to some figure, and says that a figure which has a certain property has of necessity also another property; or, stating it in our typical form, x is y, "a figure which has a certain property" is "a figure having another specified property."

The first part names or defines the figure to which the theorem relates; the last part contains an additional property. The theorem is first stated in general terms, but in the proof we usually help the mind by a particular figure actually drawn on the page; so that, before beginning the demonstration, the theorem is restated with special reference to the figure to be used.

29. **Type.** — Beginners in geometry sometimes find it difficult to distinguish clearly between what is assumed and what has to be proved in a theorem.

It has been found to help them here, if the special enunciation of what is given is printed in one kind of type; the special statement of what is required, in another sort of type; and the demonstration, in still another. In the course of the proof, the reason for any step may be indicated in smaller type between that step and the next.

30. When the hypothesis of a theorem is composite, that is, consists of several distinct hypotheses, every theorem formed by interchanging the conclusion and one of the hypotheses is an inverse of the original theorem.

VI. On Proving Inverses.

31. Often in geometry when the inverse, or its equivalent, the obverse, of a theorem, is true, it has to be proved geometrically quite apart from the original theorem. But if we have proved that every x is y, and also that there is but one individual in the class y, then we infer that y is x. The extra-logical proof required to establish an inverse is here contained in the proof that there is but one y.

Rule of Identity.

32. If it has been proved that x is y, that no two x's are the same y, and that there are as many individuals in class x as in class y, then we infer y is x.

Rule of Inversion.

33. If the hypotheses of a group of demonstrated theorems exhaustively divide the universe of discourse into contradictories, so that one must be true, though we do not know which, and the conclusions are also contradictories, then the inverse of every theorem of the group will necessarily be true.

Examples occur in geometry of the following type:—

If *a* is greater than *b*, then *c* is greater than *d*.

If *a* is equal to *b*, then *c* is equal to *d*.

If *a* is less than *b*, then *c* is less than *d*.

Three such theorems having been demonstrated geometrically, the inverse of each is always and necessarily true.

Take, for instance, the inverse of the first; namely, when *c* is greater than *d*, then *a* is greater than *b*.

This must be true; for the second and third theorems imply that if *a* is not greater than *b*, then *c* is not greater than *d*.

CHAPTER II.

THE PRIMARY CONCEPTS OF GEOMETRY.

I. Definitions of Geometric Magnitudes.

34. *Geometry* is the science which treats of the properties of space.

35. A part of space occupied by a physical body, or marked out in any other way, is called a *Solid.*

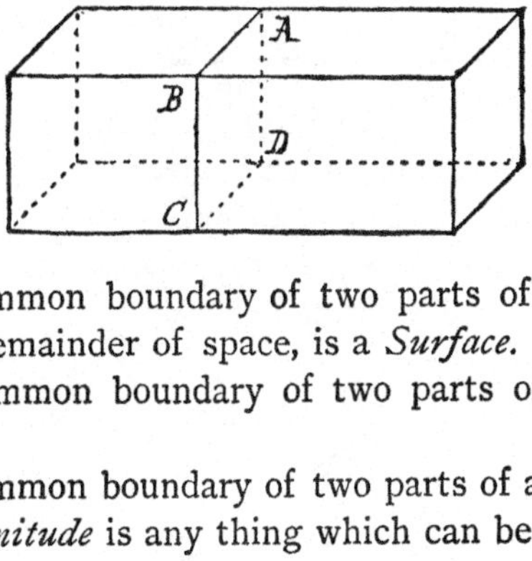

36. The common boundary of two parts of a solid, or of a solid and the remainder of space, is a *Surface.*

37. The common boundary of two parts of a surface is a *Line.*

38. The common boundary of two parts of a line is a *Point.*

39. A *Magnitude* is any thing which can be added to itself so as to double.

40. A point has position without magnitude.

41. A line may be conceived of as traced or generated by a point on a moving body. The intersection of two lines is a point.

42. A line on a moving body may generate a surface. The intersection of two surfaces is a line.

43. A surface on a moving body may generate a solid.

44. We cannot picture any motion of a solid which will generate any thing else than a solid.

Thus, in our space experience, we have three steps down from a solid to a point which has no magnitude, or three steps up from a point to a solid; so our space is said to have three dimensions.

45. A *Straight Line* is a line which pierces space evenly, so that a piece of space from along one side of it will fit any side of any other portion.

A line.

46. A *Curve* is a line no part of which is straight.

A curve.

47. Take notice: the word "line," unqualified, will henceforth mean "straight line."

48. A *Sect* is the part of a line between two definite points.

49. A *Plane Surface*, or a *Plane*, is a surface which divides space evenly, so that a piece of space from along one side of it will fit either side of any other portion.

50. A plane is generated by the motion of a line always passing through a fixed point and leaning on a fixed line.

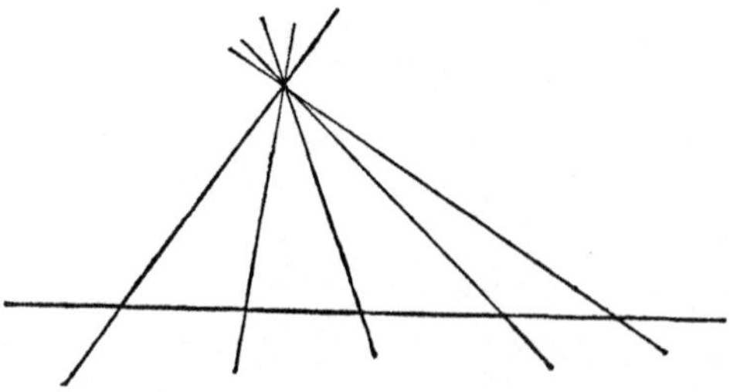

51. A *Figure* is any definite combination of points, lines, curves, or surfaces.

52. A *Plane Figure* is in one plane.

53. If a sect turns about one of its end points, the other end point describes a curve which is called a *Circle*. The fixed end point is called the *Center of the Circle*.

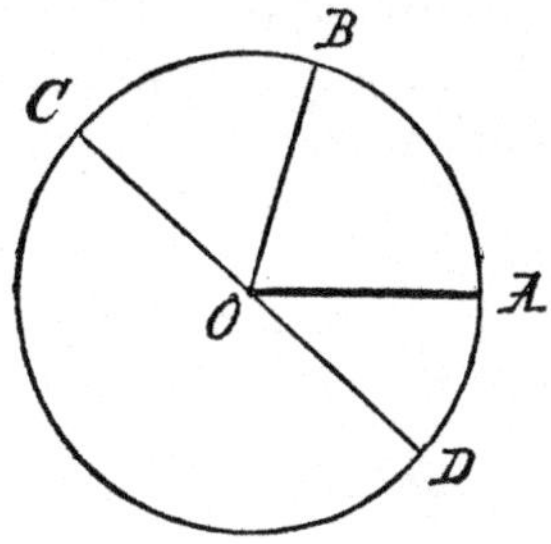

54. The *Radius of a Circle* is a sect drawn from the center to the circle.

55. A *Diameter of a Circle* is a sect drawn through the center, and terminated both ways by the circle.

56. An *Arc* is a part of a circle.

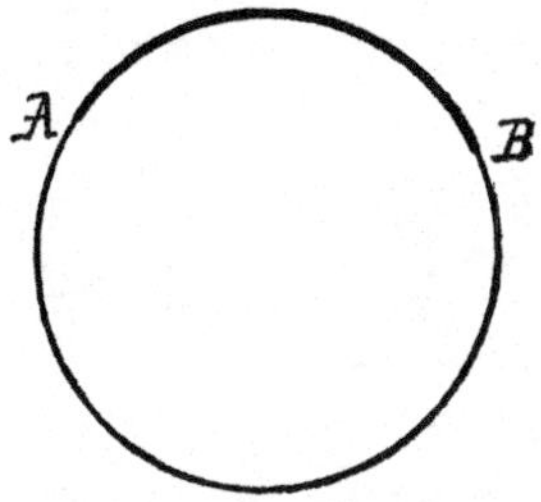

57. *Parallel Lines* are such as are in the same plane, and which, being produced ever so far both ways, do not meet.

58. When two lines are drawn from the same point, they are said to contain, or to make with each other, an *Angle*.

The point is called the *Vertex*, and the lines are called the *Arms*, of the angle. A line drawn from the vertex, and turning about the vertex in the plane of the angle from the position of coincidence with one arm to that of coincidence with the other, is said to turn through the angle; and the angle is greater as the quantity of turning is greater.

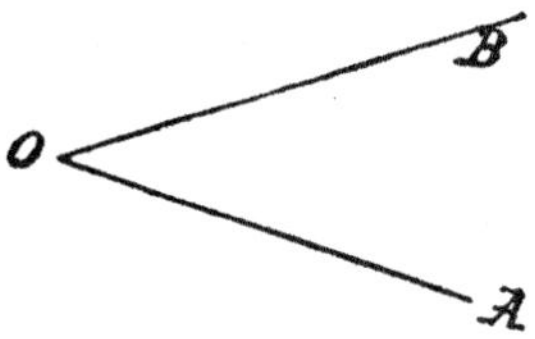

59. Since the line can turn from the one position to the other in either of two ways, two angles are formed by two lines drawn from a point.

Each of these angles is called the *Explement* of the other. If we say two lines going out from a point form an angle, we

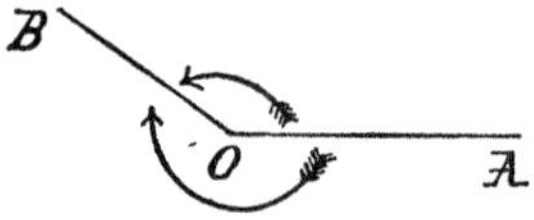

are fixing the attention upon one of the two explemental angles which they really form; and usually we mean the smaller angle.

60. Two angles are called *Equal* if they can be placed so that their arms coincide, and that both are described simultaneously by the turning of the same line about their common vertex.

61. If two angles have the vertex and an arm in common, and do not overlap, they are called *Adjacent Angles;* and the angle made by the other two arms on the side toward the common arm is called the *Sum of the Adjacent Angles.* Thus,

using the sign $\measuredangle$ for the word "angle," the sign of equality (=), and the sign of addition (+, plus),

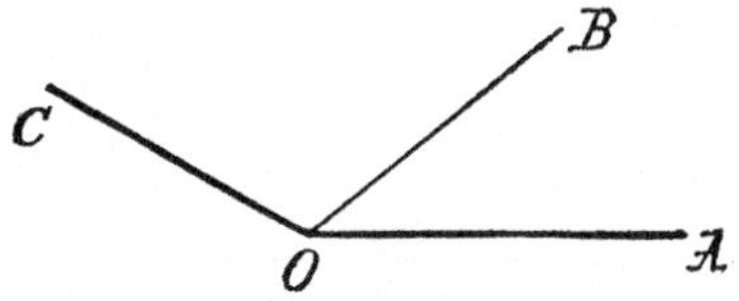

$$\measuredangle AOC = \measuredangle AOB + \measuredangle BOC.$$

62. A *Straight Angle* has its arms in the same line, and on different sides of the vertex.

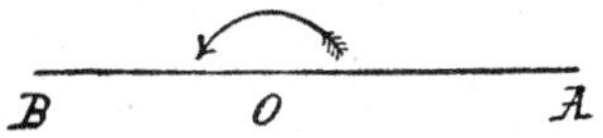

63. The sum of two adjacent angles which have their exterior arms in the same line on different sides of the vertex is a straight angle.

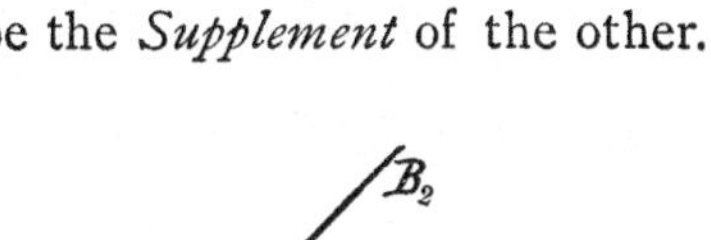

64. When the sum of any two angles is a straight angle, each is said to be the *Supplement* of the other.

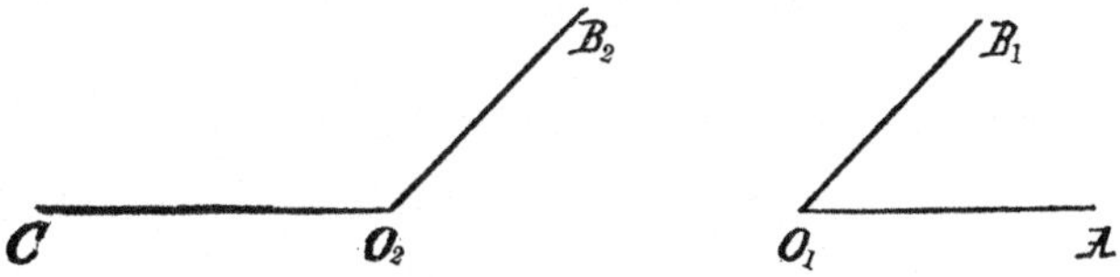

65. If two supplemental angles be added, their exterior

arms will form one line; and then the two angles are called *Supplemental Adjacent Angles*.

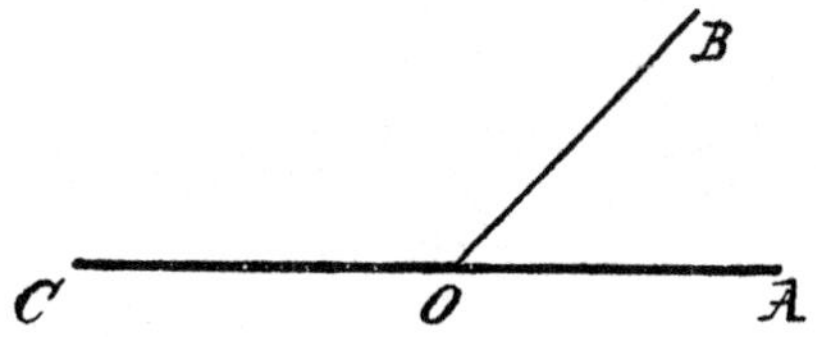

66. A *Right Angle* is half a straight angle.

67. A *Perpendicular* to a line is a line that makes a right angle with it.

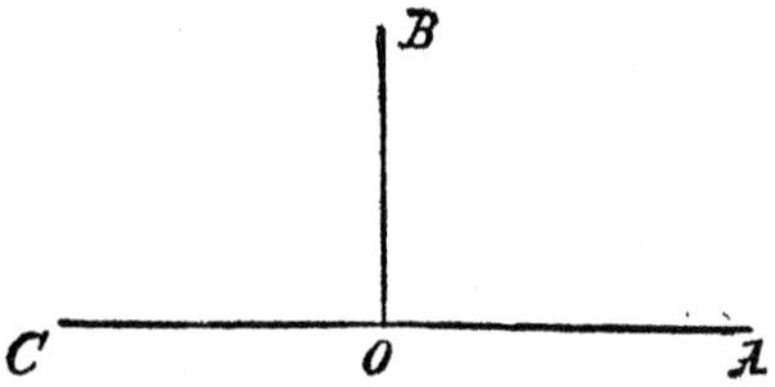

68. When the sum of two angles is a right angle, each is said to be the *Complement* of the other.

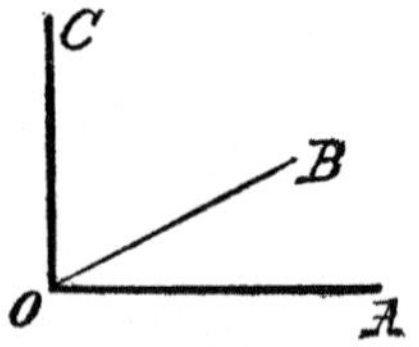

69. An *Acute Angle* is one which is less than a right angle.

70. An *Obtuse Angle* is one which is greater than a right angle, but less than a straight angle.

71. The whole angle which a sect must turn through, about one of its end points, to take it all around into its first position,

or, in one plane, the whole amount of angle round a point, is called a *Perigon.*

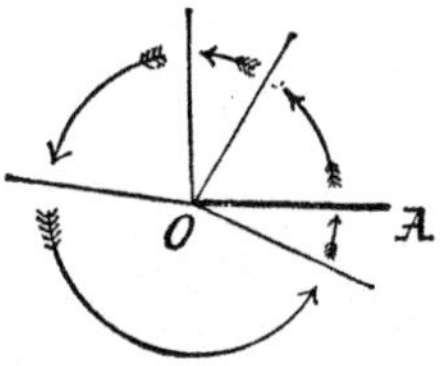

72. Since the angular magnitude about a point is neither increased nor diminished by the number of lines which radiate from the point, the sum of all the angles about a point in a plane is a perigon.

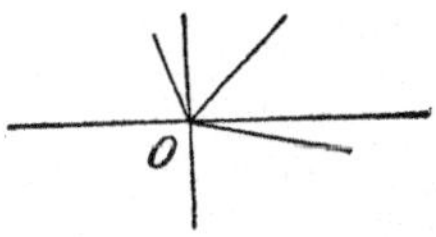

73. A *Reflex Angle* is one which is greater than a straight angle, but less than a perigon.

74. Acute, obtuse, and reflex angles, in distinction from right angles, straight angles, and perigons, are called *Oblique Angles;* and intersecting lines which are not perpendicular to each other are called *Oblique Lines.*

75. When two lines intersect, a pair of angles contained by the same two lines on different sides of the vertex, having no arm in common, are called *Vertical Angles.*

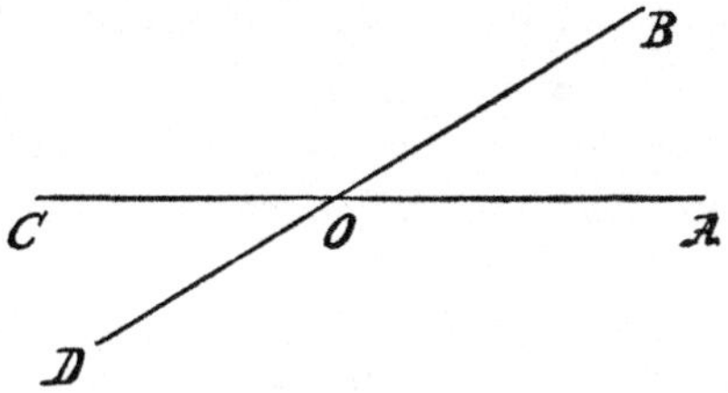

76. That which divides a magnitude into two equal parts is said to halve or bisect the magnitude, and is called a *Bisector.*

77. If we imagine a figure moved, we may also suppose it to leave its outline, or *Trace,* in the first position.

78. A *Triangle* is a figure formed by three lines, each intersecting the other two.

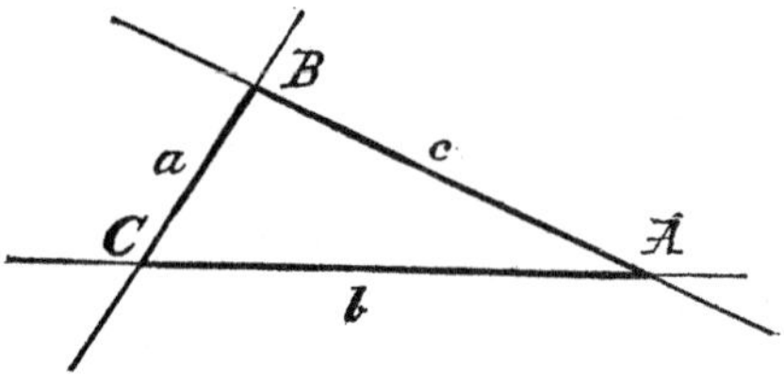

79. The three points of intersection are the three *Vertices* of the triangle.

80. The three sects joining the vertices are the *Sides* of the triangle. The side opposite A is named a; the side opposite B is b.

81. An *Interior Angle* of a triangle is one between two of the sects.

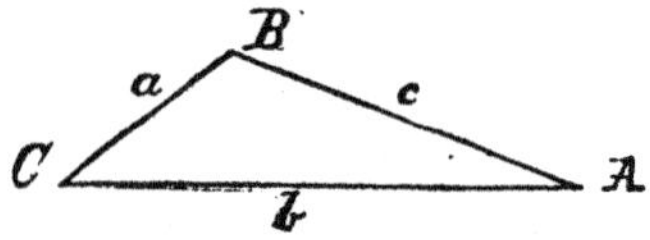

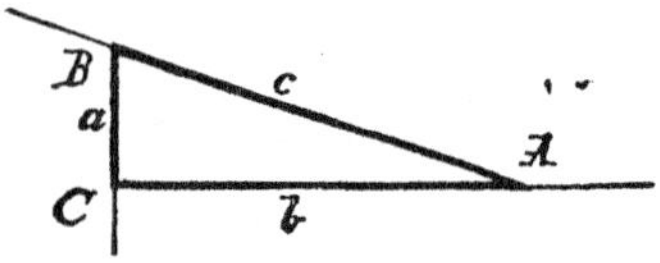

82. An *Exterior Angle* of a triangle is one between either sect and a line which is a continuation of another side.

83. Magnitudes which are identical in every respect except the place in space where they may be, are called *Congruent.*

84. Two magnitudes are *Equivalent* which can be cut into parts congruent in pairs.

II. Properties of Distinct Things.

85. The whole is greater than its part.

86. The whole is equal to the sum of all its parts.

87. Things which are equal to the same thing are equal to one another.

88. If equals be added to equals, the sums are equal.

89. If equals be taken from equals, the remainders are equal.

90. If to equals unequals are added, the sums are unequal, and the greater sum comes from adding the greater magnitude.

91. If equals are taken from unequals, the remainders are unequal, and the greater remainder is obtained from the greater magnitude.

92. Things that are double of the same thing, or of equal things, are equal to one another.

93. Things which are halves of the same thing, or of equal things, are equal to one another.

III. Some Geometrical Assumptions about Euclid's Space.

94. A solid or a figure may be imagined to move about in space without any other change. Magnitudes which will coincide with one another after any motion in space, are congruent; and congruent magnitudes can, after proper turning, be made to coincide, point for point, by superposition.

95. Two lines cannot meet twice; that is, if two lines have two points in common, the two sects between those points coincide.

96. If two lines have a common sect, they coincide throughout. Therefore through two points, only one distinct line can pass.

97. If two points of a line are in a plane, the line lies wholly in the plane.

98. All straight angles are equal to one another.

99. Two lines which intersect one another cannot both be parallel to the same line.

IV. The Assumed Constructions.

100. Let it be granted that a line may be drawn from any one point to any other point.

101. Let it be granted that a sect or a terminated line may be produced indefinitely in a line.

102. Let it be granted that a circle may be described around any point as a center, with a radius equal to a given sect.

103. REMARK. — Here we are allowed the use of a straight edge, not marked with divisions, and the use of a pair of compasses; the edge being used for drawing and producing lines, the compasses for describing circles and for the transferrence of sects.

But it is more important to note the implied restriction, namely, that no construction is allowable in elementary geometry which cannot be effected by combinations of these primary constructions.

SYMBOLS USED.

$\sim$	similar.	$<$	less than.
$=$	equivalent.	$>$	greater than.
$\cong$	congruent.	$\triangle$	triangle.
$\parallel$	parallel to.	▱	parallelogram.
$\perp$	perpendicular to.	$\odot$	circle.
$+$	plus.	rt.	right.
$\therefore$	therefore.	st.	straight.
$\measuredangle$	angle.		

CHAPTER III.

PRIMARY RELATIONS OF LINES, ANGLES, AND TRIANGLES.

I. Angles about a Point.

THEOREM I.

104. *All right angles are equal.*

Such a proposition is proved in geometry by showing it to be true of any two right angles we choose to take.

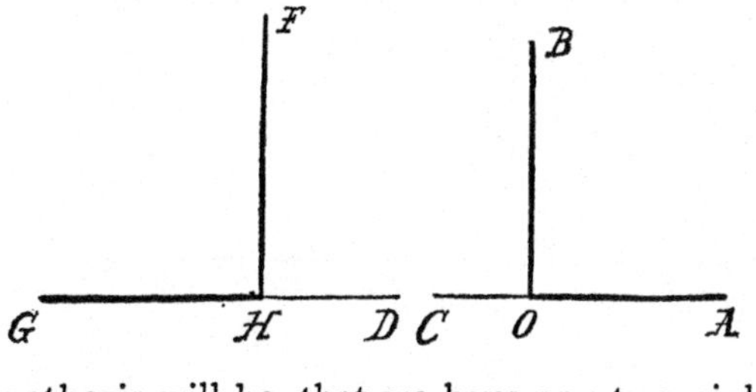

The hypothesis will be, that we have any two right angles; as, for instance, right ∡ *AOB* and right ∡ *FHG.*

By the definition of a right angle (66), the hypothesis will mean, ∡ *AOB* is half a straight angle at *O*, and ∡ *FHG* is half a straight angle at *H.*

The conclusion is, that ∡ *AOB* and ∡ *FHG* are equal.

The proof consists in stating the equality of the two straight angles of which the angles *AOB* and *FHG* are halves, referring to the assumption (in 98) that all straight angles are equal, and then stating that therefore the right ∡ *AOB* equals the right ∡ *FHG*, because, by 93, things which are halves of equal things are equal.

Now, we may restate and condense this as follows, using the abbreviations rt. for "right," and st. for "straight," and the symbol ∴ for the word "therefore":

HYPOTHESIS. $\measuredangle AOB$ *is rt.* $\measuredangle$; *also* $\measuredangle FHG$ *is rt.* $\measuredangle$.
CONCLUSION. $\measuredangle AOB = \measuredangle FHG$.
PROOF. $\measuredangle AOB$ is half a st. $\measuredangle$; also $\measuredangle FHG$ is half a st. $\measuredangle$.
(66. A right angle is half a straight angle.)

By 98, all straight angles are equal,

∴ $\measuredangle AOB = \measuredangle FHG$.

(93. Halves of equals are equal.)

105. COROLLARY. From a point on a line, there cannot be more than one perpendicular to that line.

THEOREM II.

106. *All perigons are equal.*

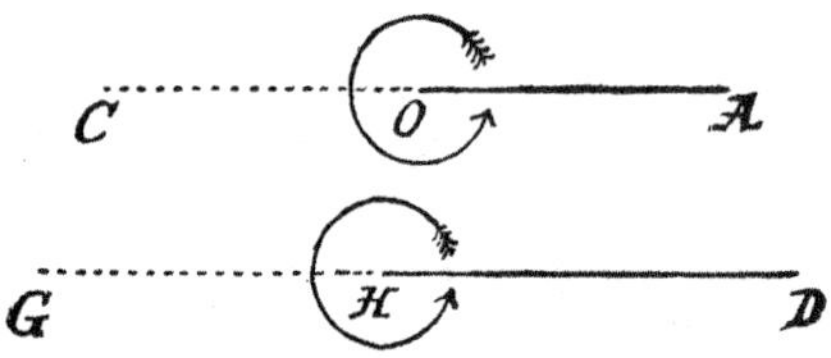

HYPOTHESIS. $\measuredangle AOA$ *is a perigon; also* $\measuredangle DHD$ *is a perigon.*
CONCLUSION. $\measuredangle AOA = \measuredangle DHD$.
PROOF. Any line through the vertex of a perigon divides it into two straight angles.
By 98, all straight angles are equal,

∴ Perigon at O equals perigon at H.

(92. Doubles of equals are equal.)

107. COROLLARY. From the preceding demonstration, it follows that half a perigon is a straight angle.

THEOREM III.

108. *If two lines intersect each other, the vertical angles are equal.*

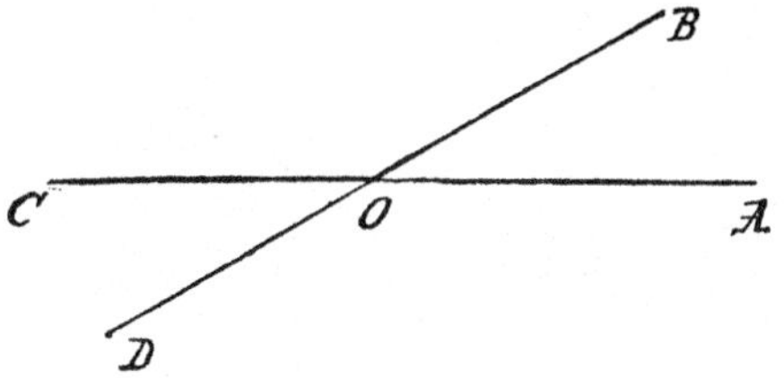

HYPOTHESIS. *AC and BD are two lines intersecting at O.*

CONCLUSIONS. $\measuredangle\, AOB = \measuredangle\, COD.$
$\measuredangle\, BOC = \measuredangle\, DOA.$

PROOF. Because the sum of the angles AOB and BOC is the angle AOC, and by hypothesis AO and OC are in one line,

$$\therefore\quad \measuredangle\, AOB + \measuredangle\, BOC = \text{st. } \measuredangle.$$

In the same way,

$$\measuredangle\, BOC + \measuredangle\, COD = \text{st. } \measuredangle,$$

$$\therefore\quad \measuredangle\, AOB + \measuredangle\, BOC = \measuredangle\, BOC + \measuredangle\, COD.$$

Take away from these equal sums the common angle BOC, and we have

$$\measuredangle\, AOB = \measuredangle\, COD.$$

(89. If equals be taken from equals, the remainders are equal.)

In the same way we may prove

$$\measuredangle\, BOC = \measuredangle\, DOA.$$

EXERCISES. 1. Two angles are formed at a point on one side of a line. Show that the lines which bisect these angles contain a right angle.

THEOREM IV.

109. *If four lines go out from a point so as to make each angle equal to the one not adjacent to it, the four lines will form only two intersecting lines.*

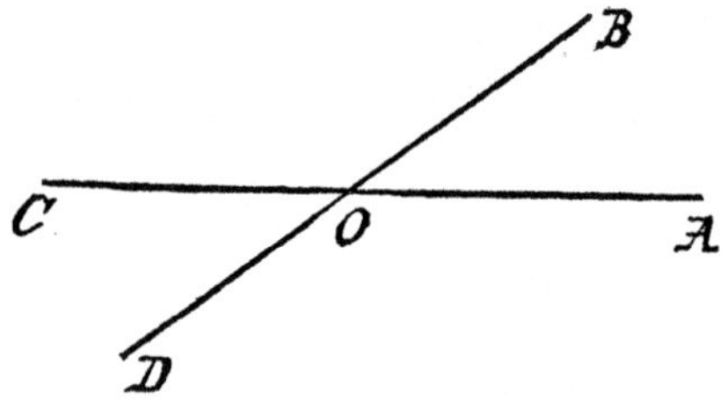

HYPOTHESIS. ***Let*** OA, OB, OC, OD, ***be four lines, with the common point*** O; ***and let*** $\measuredangle AOB = \measuredangle COD$, ***and*** $\measuredangle BOC = \measuredangle DOA$.

CONCLUSIONS. AO and OC are in one line.
BO and OD are in one line.

PROOF. $\measuredangle AOB + \measuredangle BOC + \measuredangle COD + \measuredangle DOA =$ a perigon,

(71. The sum of all the angles about a point in a plane is a perigon.)

By hypothesis, $\measuredangle AOB = \measuredangle COD$, and $\measuredangle BOC = \measuredangle DOA$,

$\therefore$ twice $\measuredangle AOB +$ twice $\measuredangle BOC =$ a perigon;

$\therefore$ $\measuredangle AOB + \measuredangle BOC =$ a st. $\measuredangle$.

(107. Half a perigon is a straight angle.)

$\therefore$ AO and OC form one line.

(65. If two supplemental angles be added, their exterior arms will form one line.)

In the same way we may prove that $\measuredangle AOB$ and $\measuredangle DOA$ are supplemental adjacent angles, and so that BO and OD form one line.

110. COROLLARY. The four bisectors of the four angles formed by two lines intersecting, form a pair of lines perpendicular to each other.

II. Angles about Two Points.

111. A line cutting across other lines is called a *Transversal.*

112. If in a plane two lines are cut in two distinct points by a transversal, at each of the points four angles are determined.

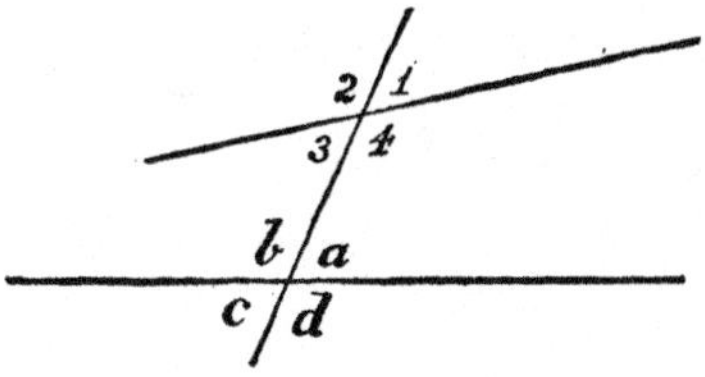

Of these eight angles, four are between the two given lines (namely, *4*, *3*, *a*, *b*), and are called *Interior Angles;* the other four lie outside the two lines, and are called *Exterior Angles.*

Angles, one at each point, which lie on the same side of the transversal, the one exterior, the other interior, like *1* and *a*, are called *Corresponding Angles.*

Two angles on opposite sides of the transversal, and both interior or both exterior, like *3* and *a*, are called *Alternate Angles.*

Theorem V.

113. *If two corresponding or two alternate angles are equal, or if two interior or two exterior angles on the same side of the transversal are supplemental, then every angle is equal to its corresponding and to its alternate angle, and is supplemental to the angle on the same side of the transversal which is interior or exterior according as the first is interior or exterior.*

FIRST CASE. — HYPOTHESIS. $\measuredangle a = \measuredangle 1$.

CONCLUSIONS. $\measuredangle a = \measuredangle 1 = \measuredangle 3 = \measuredangle c$.

$\measuredangle b = \measuredangle 2 = \measuredangle 4 = \measuredangle d$.

$\measuredangle a + \measuredangle 4 = \measuredangle 1 + \measuredangle d = \measuredangle 2 + \measuredangle c = \measuredangle b + \measuredangle 3 = \text{st.} \measuredangle$.

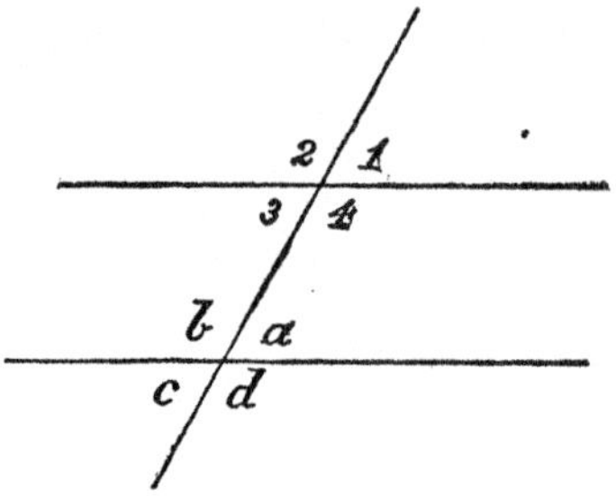

PROOF. $\measuredangle a = \measuredangle c$, $\measuredangle b = \measuredangle d$, $\measuredangle 1 = \measuredangle 3$, $\measuredangle 2 = \measuredangle 4$.

(108. If two lines intersect, the vertical angles are equal.)

Moreover, since, by hypothesis, $\measuredangle a = \measuredangle 1$, their supplements are equal, or $\measuredangle b = \measuredangle 2$.

(98. All straight angles are equal.)

(89. If equals be taken from equals, the remainders are equal.)

$\therefore \measuredangle a = \measuredangle 1 = \measuredangle c = \measuredangle 3, \quad \measuredangle b = \measuredangle 2 = \measuredangle d = \measuredangle 4,$

and st. $\measuredangle = \measuredangle a + \measuredangle b = \measuredangle a + \measuredangle 4 = \measuredangle 1 + \measuredangle d = \measuredangle 2 + \measuredangle c$
$= \measuredangle b + \measuredangle 3$.

(88. If equals be added to equals, the sums are equal.)

SECOND CASE. If, instead of two corresponding, we have given two alternate angles equal, we substitute for one its vertical, which gives the First Case again.

THIRD CASE. — HYPOTHESIS. $\measuredangle a + \measuredangle 4 = \text{st.} \measuredangle$.

But $\measuredangle 4 + \measuredangle 1 = \text{st.} \measuredangle$,

$\therefore \measuredangle a = \measuredangle 1$,

which gives again the First Case.

FOURTH CASE. — HYPOTHESIS. $\measuredangle 1 + \measuredangle d = \text{st.} \measuredangle$.

But $\measuredangle a + \measuredangle d = \text{st.} \measuredangle$,

$\therefore \measuredangle a = \measuredangle 1$,

which gives again the First Case.

III. Triangles.

114. An *Equilateral Triangle* is one in which the three sides are equal.

115. An *Isosceles Triangle* is one which has two sides equal.

116. A *Scalene Triangle* has no two sides equal.

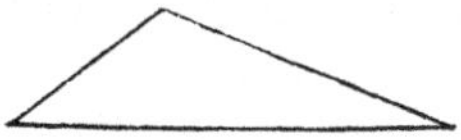

117. When one side of a triangle has to be distinguished from the other two, it may be called the *Base;* then that one of the vertices opposite the base is called *the Vertex.*

118. When we speak of the angles of a triangle, we mean the three interior angles.

119. A *Right-angled Triangle* has one of its angles a right angle. The side opposite the right angle is called the *Hypothenuse.*

120. An *Obtuse-angled Triangle* has one of its angles obtuse.

121. An *Acute-angled Triangle* has all three angles acute.

122. An *Equiangular Triangle* is one which has all three angles equal.

123. When two triangles have three angles of the one equal respectively to the three angles of the other, a pair of equal angles are called *Homologous Angles.* The pair of sides opposite homologous angles are called *Homologous Sides.*

Theorem VI.

124. *Two triangles are congruent if two sides and the included angle in the one are equal respectively to two sides and the included angle in the other.*

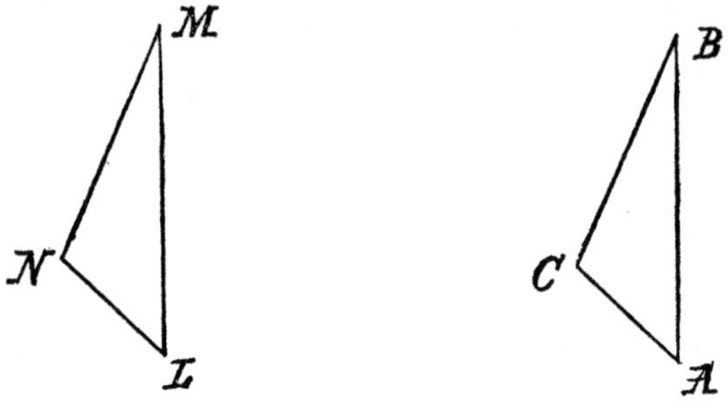

Hypothesis. *ABC* **and** *LMN* **two triangles, with** $AB = LM$, $BC = MN$, $\measuredangle B = \measuredangle M$.

Conclusion. The two triangles are congruent; or, using $\triangle$ for "triangle," and $\cong$ for "congruent,"

$$\triangle ABC \cong \triangle LMN.$$

Proof. Apply the triangle *LMN* to $\triangle ABC$ in such a manner that the vertex *M* shall rest on *B*, and the side *ML* on *BA*, and the point *N* on the same side of *BA* as *C*.

Then, because the side *ML* equals *BA*, the point *L* will rest upon *A*; because $\measuredangle M = \measuredangle B$, the side *MN* will rest upon the line *BC*; because the side *MN* equals *BC*, the point *N* will rest upon the point *C*. Now, since the point *L* rests upon *A*, and the point *N* rests upon *C*, therefore the side *LN* coincides with the side *AC*.

(95. If two lines have two points in common, the two sects between those points coincide.)

Therefore every part of one triangle will coincide with the corresponding part of the other, and the two are congruent.

(94. Magnitudes which will coincide are congruent.)

125. In any pair of congruent triangles, the homologous sides are equal.

Theorem VII.

126. *In an isosceles triangle the angles opposite the equal sides are equal.*

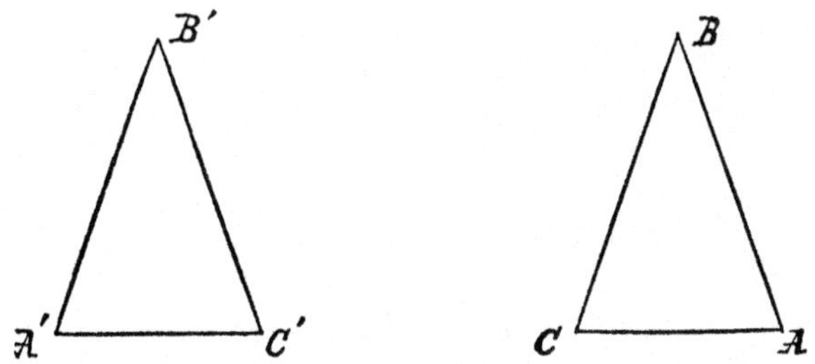

Hypothesis. ABC *a triangle, with* $AB = BC$.

Conclusion. $\measuredangle A = \measuredangle C$.

Proof. Imagine the triangle ABC to be taken up, turned over, and put down in a reversed position; and now designate the angular points A' (read A prime), B', C', to distinguish the triangle from its trace ABC left behind.

Then, in the triangles ABC, $C'B'A'$,

$$AB = C'B' \text{ and } BC = B'A',$$

since, by hypothesis, $AB = CB$.

Also $$\measuredangle B = \measuredangle B',$$

$$\therefore \measuredangle A = \measuredangle C'.$$

(124. Triangles are congruent if two sides and the included angle are equal in each.)

But $\measuredangle C' = \measuredangle C$,

$$\therefore \measuredangle A = \measuredangle C.$$

(87. Things equal to the same thing are equal to one another.)

127. Corollary. Every equilateral triangle is also equiangular.

Exercises. 2. The bisectors of vertical angles are in the same line.

Theorem VIII.

128. *Two triangles are congruent if two angles and the included side in the one are equal respectively to two angles and the included side in the other.*

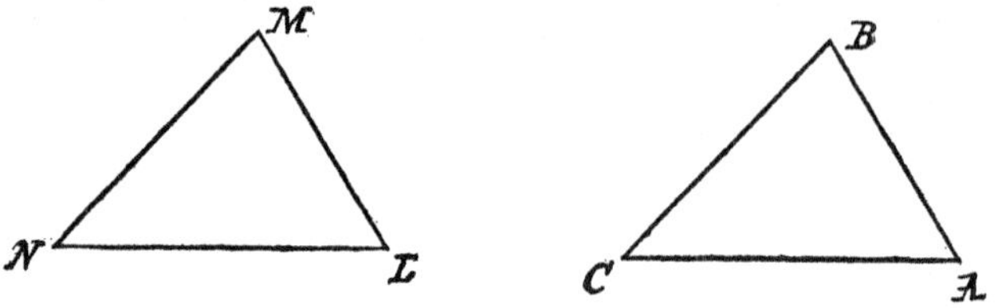

Hypothesis. *ABC and LMN two triangles, with* $\measuredangle A = \measuredangle L$, $\measuredangle C = \measuredangle N$, $AC = LN$.

Conclusion. The two triangles are congruent.

$$\triangle ABC \cong \triangle LMN.$$

Proof. Apply the triangle LMN to the triangle ABC so that the point L shall rest upon A, and the side LN lie along the side AC, and the point M lie on the same side of AC as B.

Then, because the side $AC = LN$,

$\therefore$ point N will rest upon point C.

Because $\measuredangle L = \measuredangle A$,

$\therefore$ side LM will rest upon the line AB.

Because $\measuredangle N = \measuredangle C$,

$\therefore$ side NM will rest upon the line CB.

Because the sides LM and NM rest respectively upon the lines AB and CB,

$\therefore$ the vertex M, resting upon both the lines AB and CB, must rest upon the vertex B, the only point common to the two lines.

Therefore the triangles coincide in all their parts, and are congruent.

THEOREM IX.

129. *Two triangles are congruent if the three sides of the one are equal respectively to the three sides of the other.*

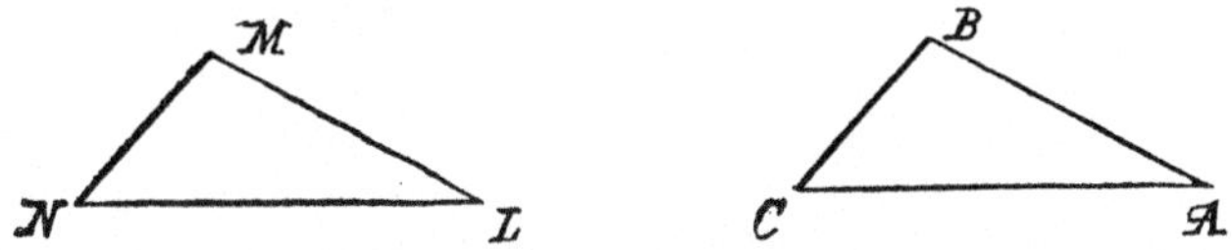

HYPOTHESIS. **Triangles ABC and LMN, having $AB = LM$, $BC = MN$, and $CA = NL$.**

CONCLUSION. $\triangle ABC \cong \triangle LMN$.

PROOF. Imagine $\triangle ABC$ to be applied to $\triangle LMN$ in such a way that LN coincides with AC, and the vertex M falls on the side of AC opposite to the side on which B falls, and join BM.

CASE I. When BM cuts the sect AC.

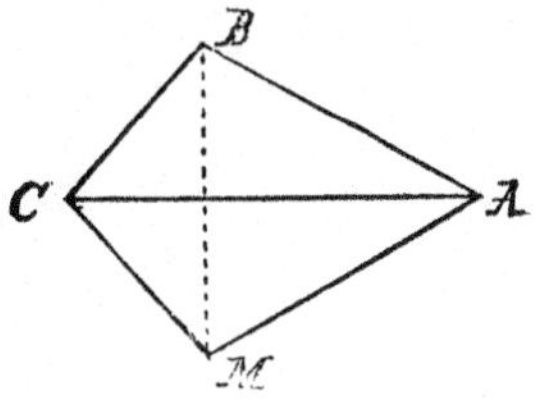

Then in $\triangle ABM$, because, by hypothesis, $AB = AM$,

$$\therefore \quad \measuredangle ABM = \measuredangle AMB.$$

(126. In an isosceles triangle the angles opposite the equal sides are equal.)

And for the same reason, in $\triangle BCM$, because $BC = CM$,

$$\therefore \quad \measuredangle CBM = \measuredangle CMB.$$

Therefore $\measuredangle ABM + \measuredangle CBM = \measuredangle AMB + \measuredangle CMB$;

(88. If equals be added to equals, the sums are equal.)

that is, $\measuredangle ABC = \measuredangle AMC$,

$$\therefore \quad \triangle ABC \cong \triangle LMN.$$

(124. Triangles are congruent if two sides and the included angle are equal in each.)

CASE II. When BM passes through one extremity of the sect AC, as C.

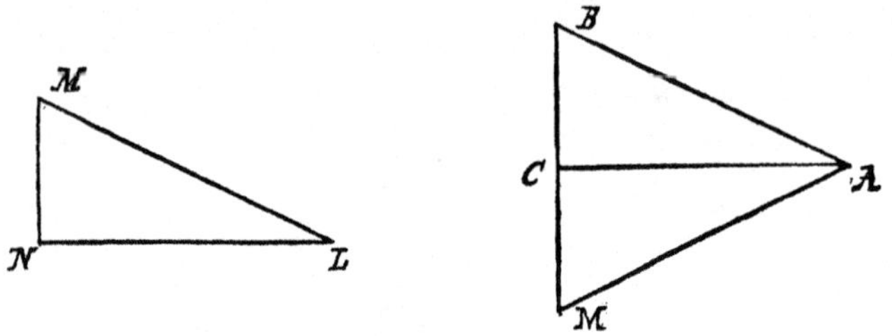

Then, by the first step in Case I.,

$$\measuredangle B = \measuredangle M;$$

and for the same reason as before,

$$\triangle ABC \cong \triangle LMN.$$

CASE III. When BM falls beyond the extremity of the sect AC.

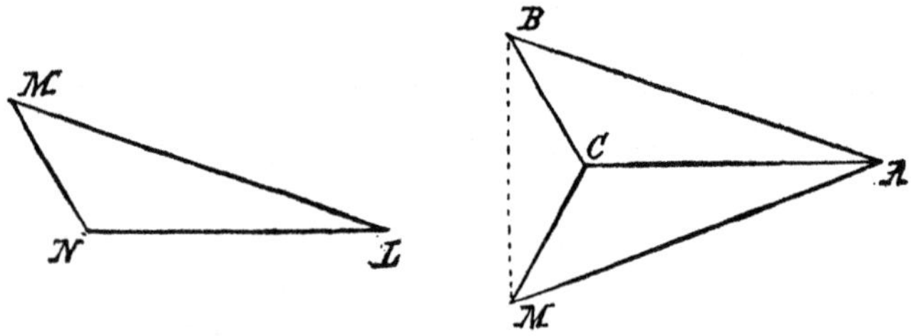

Then in $\triangle ABM$, because, by hypothesis, $AB = AM$,

$$\therefore \quad \measuredangle ABM = \measuredangle AMB.$$

And in $\triangle BCM$, because $BC = CM$,

$$\therefore \quad \measuredangle CMB = \measuredangle CBM.$$

Therefore the remainders, $\measuredangle ABC = \measuredangle AMC$;

(89. If equals be taken from equals, the remainders are equal.)

and so, as in Case I., $\triangle ABC \cong \triangle LMN$.

CHAPTER IV.

PROBLEMS.

130. A *Problem* is a proposition in which something is required to be done by a process of construction, which is termed the *Solution.*

131. The solution of a problem in elementary geometry consists, —

(1) In indicating how the ruler and compasses are to be used in effecting the construction required.

(2) In proving that the construction so given is correct.

(3) In discussing the limitations, which sometimes exist, within which alone the solution is possible.

PROBLEM I.

132. *To describe an equilateral triangle upon a given sect.*

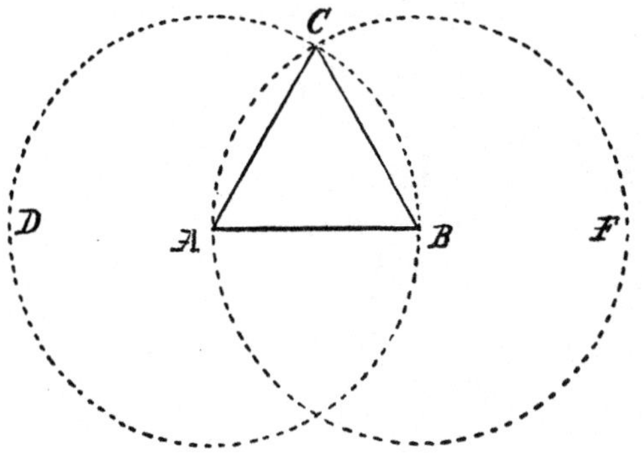

GIVEN, *the sect AB.*

REQUIRED, *to describe an equilateral triangle on* AB.

CONSTRUCTION. With center A and radius AB, describe the circle BCD.

With center B and radius BA, describe ⊙ ACF (using ⊙ for "circle").

(102. A circle may be described with any center and radius.)

Join a point C, at which the circles cut one another, to the points A and B.

(100. A line may be drawn from any one point to any other point.)

Then will ABC be an equilateral triangle.

PROOF. Because A is the center of ⊙ BCD,

∴ $AB = AC$, being radii of the same circle;

and because B is the center of the ⊙ ACF,

∴ $BA = BC$, being radii of the same circle.

Therefore $AC = AB = BC$,

(87. Things equal to the same thing are equal to one another.)

and an equilateral triangle has been described on AB.

PROBLEM II.

133. *On a given line, to mark off a sect equal to a given sect.*

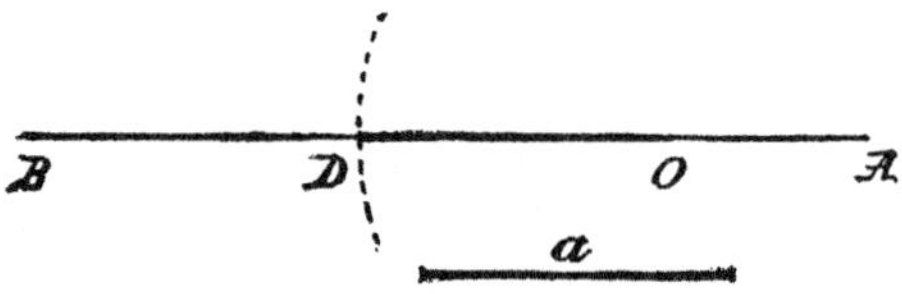

GIVEN, AB, *a line*; a, *a sect.*

REQUIRED, *to mark off a sect on* AB *equal to* a.

CONSTRUCTION. From any point O as a center, on AB, describe the arc of a circle with a radius equal to a.

If D be the point in which the arc intersects AB, then OD will be the required sect.

PROOF. All the radii of the circle around O are, by construction, equal to a.

OD is one of these radii, therefore it is equal to a.

PROBLEM III.

134. *To bisect a given angle.*

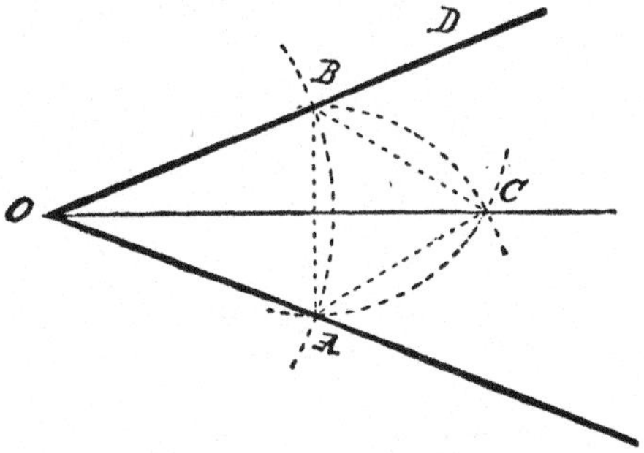

GIVEN, *the angle AOD.*

REQUIRED, *to bisect it.*

CONSTRUCTION. From O as a center, with any radius, OA, describe the arc of a circle, cutting the arms of the angle in the points A and B.

Join AB. On AB, on the side remote from O, describe, by 132, an equilateral triangle, ABC.

Join OC. The line OC will bisect the given angle AOD.

PROOF. In the triangles CAO and CBO we have

$$OA = OB,$$

$$CA = CB \text{ by construction,}$$

and the side CO common.

$$\therefore \quad \triangle CAO \cong \triangle CBO,$$

(129. Triangles with the three sides respectively equal are congruent.)

$$\therefore \quad \measuredangle COA = \measuredangle COB,$$

$$\therefore \quad CO \text{ is the bisector of } \measuredangle AOB.$$

EXERCISES. 1. Describe an isosceles triangle having each of the equal sides double the base.

2. Having given the hypothenuse and one of the sides of a right-angled triangle, construct the triangle.

PROBLEM IV.

135. *Through a given point on a given line, to draw a perpendicular to this line.*

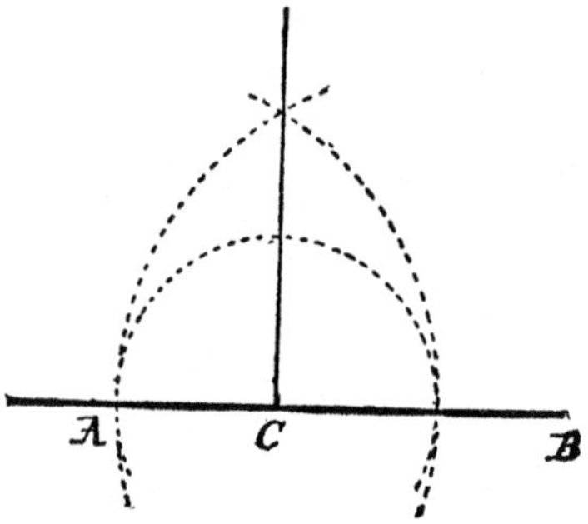

GIVEN, *the line AB, and the point C in it.*
REQUIRED, *to draw from* C *a perpendicular to* AB.
SOLUTION. By 134, bisect the st. ∡ *ACB.*

EXERCISES. 5. Solve 134 without an equilateral triangle.

6. Divide a given angle into four equal parts.

3. Find a point in a line at a given distance from a given point. When is the problem impossible?

4. Find a point in a line equally distant from two given points without the line.

5. From a given point without a given line, draw a line making with the given line an angle equal to one-half a right angle.

6. *A* is a point without a given line, *BC*. Construct an isosceles triangle, having *A* as vertex, whose base shall lie along *BC*, and be equal to a given sect.

11. If a right-angle triangle have one acute angle double the other, the hypothenuse is double one side.

PROBLEM V.

136. *To bisect a given sect.*

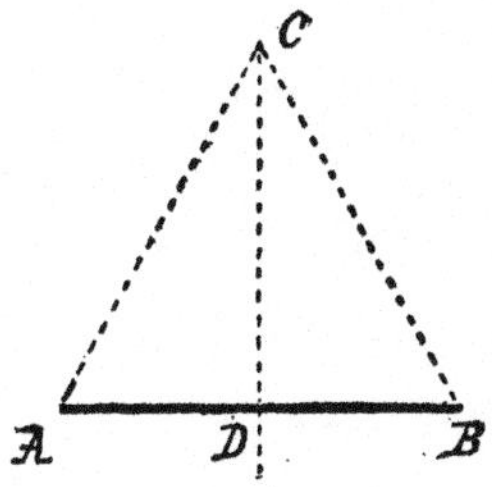

GIVEN, **the sect** *AB*.

REQUIRED, *to bisect it.*

CONSTRUCTION. On *AB* describe an equilateral triangle *ABC*, by 132.

By 134, bisect the angle *ACB* by the line meeting *AB* in *D*. Then *AB* shall be bisected in *D*.

PROOF. In the triangles *ACD* and *BCD*

$$\measuredangle ACD = \measuredangle BCD,$$

and $AC = BC$ by construction, and *CD* is common.

$$\therefore \quad AD = BD,$$

(124. Triangles are congruent if two sides and the included angle are equal in each.)

$$\therefore \quad AB \text{ is bisected at } D.$$

137. COROLLARY I. The line drawn to bisect the angle at the vertex of an isosceles triangle, also bisects the base, and is perpendicular to it.

138. COROLLARY II. The line drawn from the vertex of an isosceles triangle to bisect the base, is perpendicular to it, and also bisects the vertical angle.

EXERCISES. 12. Construct a right-angled isosceles triangle on a given sect as hypothenuse.

PROBLEM VI.

139. *From a point without a given line, to drop a perpendicular upon the line.*

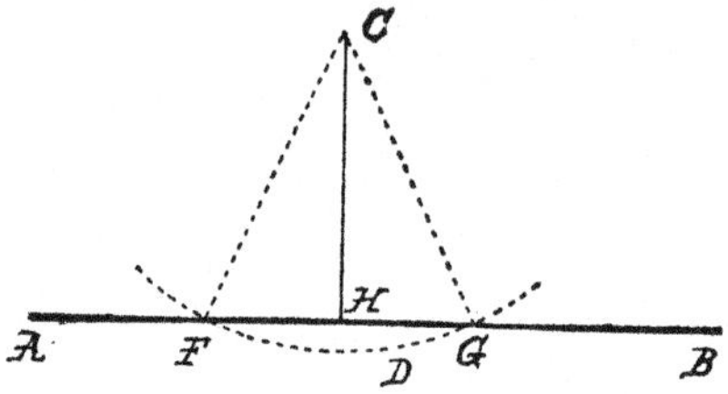

GIVEN, *the line AB, and the point C without it.*

REQUIRED, *to drop from* C *a perpendicular upon* AB.

CONSTRUCTION. Take any point, *D*, on the other side of *AB* from *C*, and, by 102, from the center *C*, with radius *CD*, describe the arc *FDG*, meeting *AB* at *F* and *G*.

By 136, bisect *FG* at *H*. Join *CH*.

$$CH \text{ shall be } \perp AB$$

(using $\perp$ for the words "perpendicular to").

PROOF. Because in the triangles *CHF* and *CHG*, by construction, $CF = CG$, and $HF = HG$, and *CH* is common,

$$\therefore \measuredangle CHF = \measuredangle CHG.$$

(129. Triangles with three sides respectively equal are congruent.)

$\therefore$ $\measuredangle$ *FHC*, being half the st. $\measuredangle$ *FHG*, is a rt. $\measuredangle$; and *CH* is perpendicular to *AB*.

140. COROLLARY. The line drawn from the vertex of an isosceles triangle perpendicular to the base, bisects it, and also bisects the vertical angle.

EXERCISES. 13. Instead of bisecting *FG*, would it do to bisect the angle *FCG*?

CHAPTER V.

INEQUALITIES.

141. The symbol $>$ is called the *Sign of Inequality*, and $a > b$ means that a is greater than b; so $a < b$ means that a is less than b.

THEOREM X.

142. *An exterior angle of a triangle is greater than either of the two opposite interior angles.*

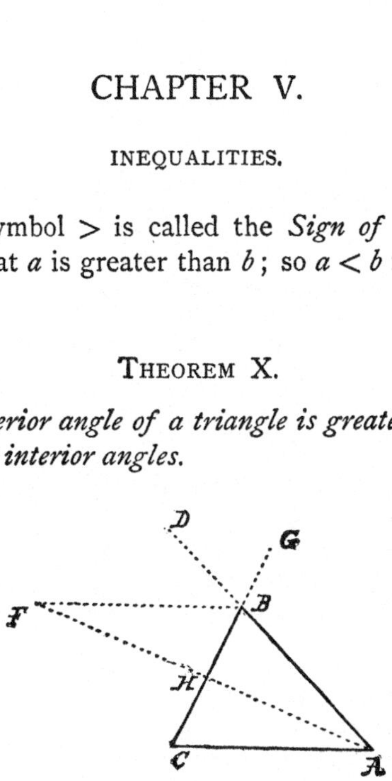

HYPOTHESIS. ***Let the side AB of the triangle ABC be produced to D.***

CONCLUSIONS. $\measuredangle CBD > \measuredangle BCA$.

$\measuredangle CBD > \measuredangle BAC$.

PROOF. By 136, bisect BC in H.

By 100 and 101, join AH, and produce it to F; making, by 133, $HF = AH$.

By 100, join BF.

Then, in the triangles AHC and FHB, by construction,

$$CH = HB, \quad AH = HF, \quad \text{and} \quad \measuredangle AHC = \measuredangle FHB;$$

(108. Vertical angles are equal.)

$$\therefore \quad \triangle AHC \cong \triangle FHB,$$

(124. Triangles are congruent if two sides and the included angle are equal in each.)

$$\therefore \quad \measuredangle HCA = \measuredangle HBF.$$

Now, $\measuredangle CBD > \measuredangle HBF$,

(85. The whole is greater than its part.)

$$\therefore \quad \measuredangle CBD > \measuredangle HCA.$$

Similarly, if CB be produced to G, it may be shown that

$$\measuredangle ABG > \measuredangle BAC,$$

$\therefore$ the vertical $\measuredangle CBD > \measuredangle BAC$.

THEOREM XI.

143. *Any two angles of a triangle are together less than a straight angle.*

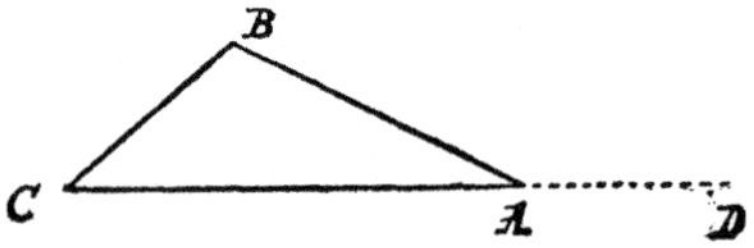

HYPOTHESIS. Let ABC be any $\triangle$.

CONCLUSION. $\measuredangle A + \measuredangle B <$ st. $\measuredangle$.

PROOF. Produce CA to D.

Then $\measuredangle BAD > \measuredangle B$,

(142. An exterior angle of a triangle is greater than either opposite interior angle.)

$$\therefore \quad \measuredangle BAD + \measuredangle BAC > \measuredangle B + \measuredangle BAC,$$

$$\therefore \quad \text{st.} \measuredangle > \measuredangle B + \measuredangle A.$$

144. COROLLARY I. No triangle can have more than one right angle or obtuse angle.

145. COROLLARY II. There can be only one perpendicular from a point to a line.

THEOREM XII.

146. *If one side of a triangle be greater than a second, the angle opposite the first must be greater than the angle opposite the second.*

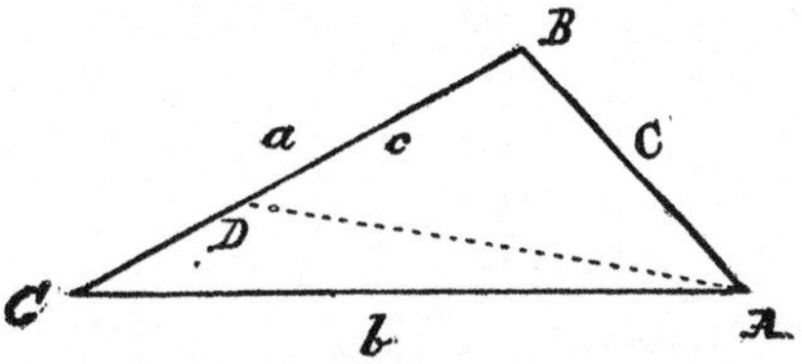

HYPOTHESIS. $\triangle ABC$, with side $a >$ side c.

CONCLUSION. $\measuredangle BAC > \measuredangle C$.

PROOF. By 133, from a cut off $BD = c$.

By 100, join AD.

Then, because $BD = c$, $\therefore \measuredangle BDA = \measuredangle BAD$.

(126. In an isosceles triangle the angles opposite the equal sides are equal.)

And, by 142, the exterior $\measuredangle BDA$ of $\triangle CDA >$ the opposite interior $\measuredangle C$, $\therefore$ also $\measuredangle BAD > \measuredangle C$.

Still more is $\measuredangle BAC > \measuredangle C$.

147. COROLLARY. If one side of a triangle be less than the second, the angle opposite the first will be less than the angle opposite the second.

For, if $a < b$, $\therefore b > a$; $\therefore$ by 146, $\measuredangle B > \measuredangle A$, $\therefore \measuredangle A < \measuredangle B$.

148. INVERSES. We have now proved,

By 146, if $a > b$, $\therefore \measuredangle A > \measuredangle B$;
By 126, if $a = b$, $\therefore \measuredangle A = \measuredangle B$;
By 147, if $a < b$, $\therefore \measuredangle A < \measuredangle B$.

Therefore, by 33, the inverses are necessarily true, namely,

If $\measuredangle A > \measuredangle B$, $\therefore a > b$;
If $\measuredangle A = \measuredangle B$, $\therefore a = b$;
If $\measuredangle A < \measuredangle B$, $\therefore a < b$.

149. COROLLARY. An equiangular triangle is also equilateral.

THEOREM XIII.

150. *The perpendicular is the least sect between a given point and a given line.*

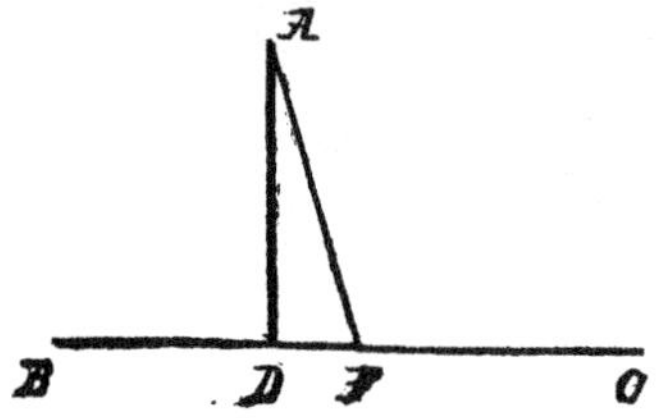

HYPOTHESIS. ***Let A be a given point, and BC a given line.***

CONSTRUCTION. By 139, from A drop $AD \perp BC$.

By 100, join A to F, any point of BC except D.

CONCLUSION. $AD < AF$.

PROOF. Since, by construction, ∡ ADF is rt. ∡,

$$\therefore \measuredangle ADF > \measuredangle AFD,$$

(143. Any two angles of a triangle are together less than a straight angle.)

$$\therefore AF > AD.$$

(148. If angle ADF is greater than angle AFD, therefore side AF is greater than side AD.)

151. Except the perpendicular, any sect from a point to a line is called an *Oblique*.

EXERCISES. 14. Determine the least path from one point to another, subject to the condition that it shall meet a given straight line.

15. The four sides of a quadrilateral figure are together greater than its two diagonals.

THEOREM XIV.

152. *Two obliques from a point, making equal sects from the foot of the perpendicular, are equal, and make equal angles with the line.*

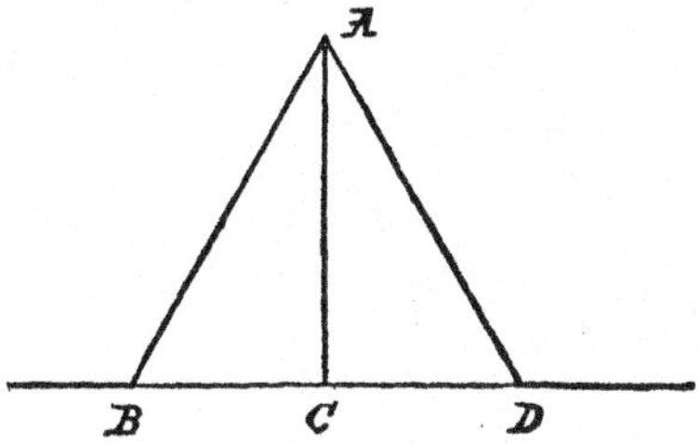

HYPOTHESIS. $AC \perp BD$.
$BC = DC$.

CONCLUSIONS. $AB = AD$.
$\measuredangle ABC = \measuredangle ADC$.

PROOF. $\triangle ACB \cong \triangle ACD$.

(124. Triangles are congruent if two sides and the included angle are equal in each.)

153. COROLLARY. For every oblique, there can be drawn one equal, and on the other side of the perpendicular.

EXERCISES. 16. *ABC* is a triangle whose angle *A* is bisected by a line meeting *BC* at *D*. Prove *AB* greater than *BD*, and *AC* greater than *CD*.

17. Prove that the sum of the sects from any point to the three angles of a triangle is greater than one-half the perimeter of the triangle.

18. The two sides of a triangle are together greater than twice the line drawn from the vertical angle to the middle point of the base.

THEOREM XV.

154. *Of any two obliques between a given point and line, that which makes the greater sect from the foot of the perpendicular is the greater.*

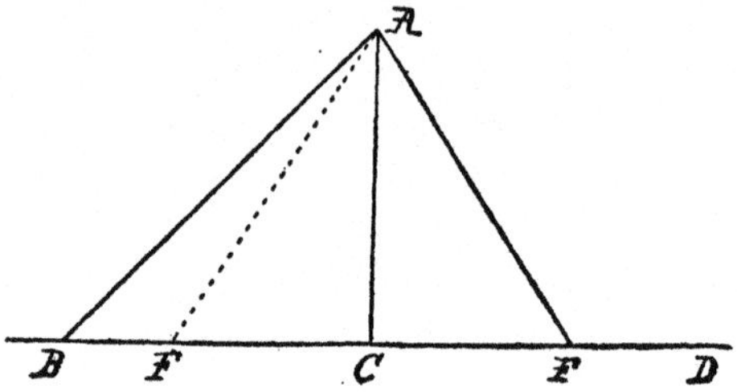

HYPOTHESIS. $AC \perp BD$.

$CB > CF$.

CONCLUSION. $AF < AB$.

PROOF. By 153, take F on the same side of the perpendicular as B; then, in $\triangle\, AFC$, because, by hypothesis, $\angle\, ACF$ is rt.,

$$\therefore \quad \angle\, AFC < \text{rt. } \angle,$$

(143. Any two angles of a triangle are together less than a straight angle.)

$$\therefore \quad \angle\, AFB > \text{rt. } \angle,$$

since $\angle\, AFC + \angle\, AFB = \text{st. } \angle$.

$$\therefore \quad \angle\, ABF < \angle\, AFB,$$

(143. Any two angles of a triangle are together less than a straight angle.)

$$\therefore \quad AF < AB.$$

(148. If angle A is less than angle B, therefore a is less than b.)

155. COROLLARY. No more than two of all the sects can be equal.

PROOF. For no two sects on the same side of the perpendicular can be equal.

THEOREM XVI.

156. *Any two sides of a triangle are together greater than the third side.*

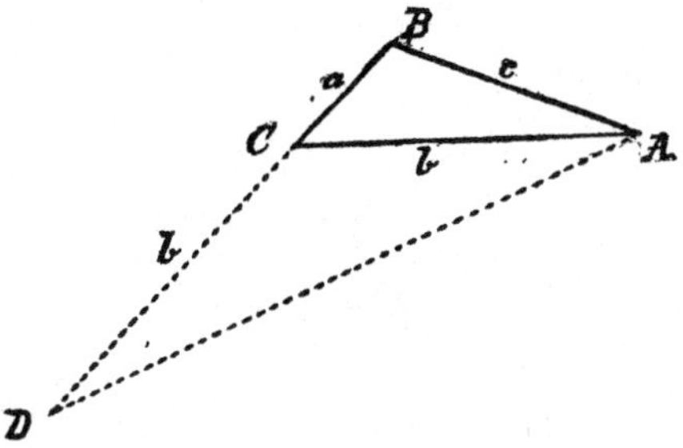

HYPOTHESIS. ***Let*** ABC ***be a triangle.***

CONCLUSIONS. $a + b > c$.

$a + c > b$.

$b + c > a$.

PROOF. By 101, produce BC to D; making, by 133, $CD = b$.
By 100, join AD.
Then, because $AC = CD$,

$$\therefore \quad \measuredangle CDA = \measuredangle CAD.$$

Now, $\measuredangle BAD > \measuredangle CAD$,

(85. The whole is greater than its part.)

$$\therefore \text{ also } \measuredangle BAD > \measuredangle BDA.$$

By 148, $\quad \therefore \quad BD > BA$.

But $BD = a + b$, and BA is c,

$$\therefore \quad a + b > c.$$

Similarly it may be shown that $a + c > b$, and that $b + c > a$.

157. COROLLARY. Any side of a triangle is greater than the difference between the other two sides.

Theorem XVII.

158. *If from the ends of one of the sides of a triangle two sects be drawn to a point within the triangle, these will be together less than the other two sides of the triangle, but will contain a greater angle.*

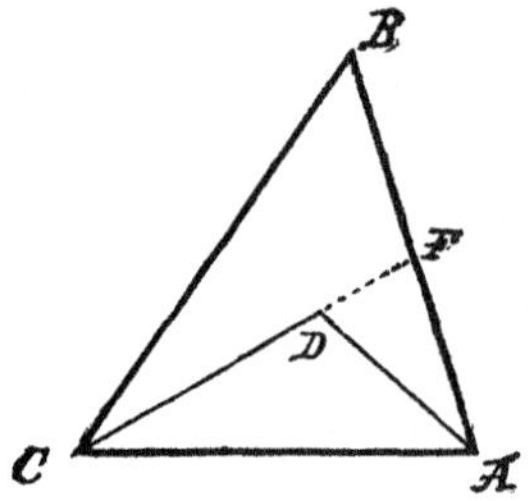

Hypothesis. D *is a point within* $\triangle\ ABC$.

Conclusions. (I.) $AB + BC > AD + DC$.

(II.) $\measuredangle\ ADC \quad > B$.

Proof. (I.) Join AD and CD.

Produce CD to meet AB in F.

Then $(CB + BF) > (CF)$, and $[DF + FA] > [DA]$.

(156. In a triangle any two sides are together greater than the third.)

$$\therefore \quad CB + BA = (CB + BF) + FA > (CF) + FA$$
$$= CD + [DF + FA] > CD + [DA].$$

(II.) Next, in $\triangle\ AFD$

$$\measuredangle\ ADC > \measuredangle\ AFD.$$

(142. Exterior angle of a triangle is greater than opposite interior angle.)

And for the same reason, in $\triangle\ CFB$

$$\measuredangle\ AFD > \measuredangle\ B, \qquad \therefore \quad \measuredangle\ ADC > \measuredangle\ B.$$

Exercises. 19. The perimeter of a triangle is greater than the sum of sects from a point within to the vertices.

Theorem XVIII.

159. *If two triangles have two sides of the one equal respectively to two sides of the other, but the included angle of the first greater than the included angle of the second, then the third side of the first will be greater than the third side of the second.*

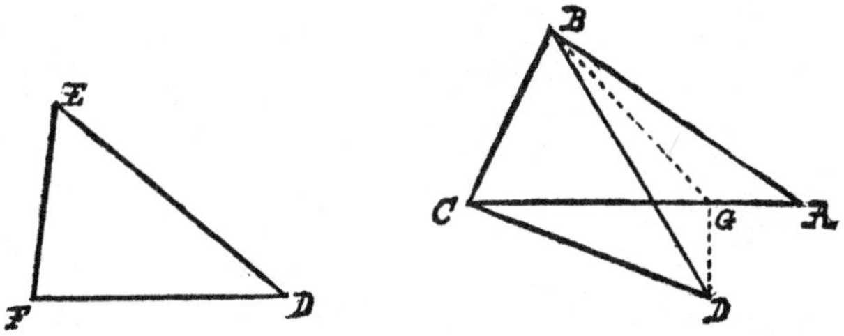

Hypothesis. ABC and DEF are two triangles, in which

$$AB = DE,$$
$$BC = EF,$$
$$\measuredangle ABC > \measuredangle DEF.$$

Conclusion. $AC > DF$.

Proof. Place the triangles so that EF shall coincide with BC, and the point D fall on the same side of BC as A.

By 134, bisect $\measuredangle ABD$, and let G be the point in which the bisector meets AC.

By 100, join DG.

Then, in the triangles ABG and BDG, by construction, $\measuredangle ABG = \measuredangle DBG$; by hypothesis, $AB = BD$; and BG is common.

$$\therefore \quad \triangle ABG \cong \triangle DBG,$$

(124. Two triangles are congruent if two sides and the included angle are equal in each.)

$$\therefore \quad AG = DG,$$

$$\therefore \quad CA = CG + GA = CG + GD.$$

But $CG + GD > CD$,

(156. Any two sides of a triangle are together greater than the third.)

$$\therefore \quad CA > FD.$$

160. COROLLARY. If $c = f$, $a = d$, and $\measuredangle B < \measuredangle E$, then $\measuredangle E > \measuredangle B$; and so, by 159,

$$e > b, \quad \therefore \quad b < e.$$

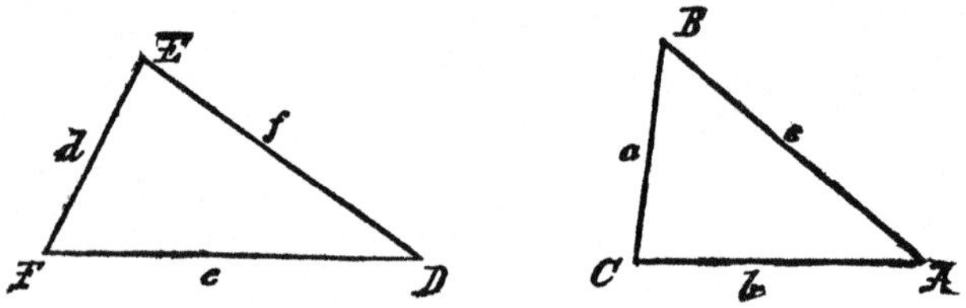

161. INVERSES. We have now proved, when $a = d$, and $c = f$,

By 159, if $\measuredangle B > \measuredangle E$, $\quad \therefore \quad b > e$;
By 124, if $\measuredangle B = \measuredangle E$, $\quad \therefore \quad b = e$;
By 160, if $\measuredangle B < \measuredangle E$, $\quad \therefore \quad b < e$.

Therefore, by 33, Rule of Inversion,

If $b > e$, $\quad \therefore \quad \measuredangle B > \measuredangle E$;
If $b = e$, $\quad \therefore \quad \measuredangle B = \measuredangle E$;
If $b < e$, $\quad \therefore \quad \measuredangle B < \measuredangle E$.

EXERCISES. 20. In $\triangle ABC$, $AB < AC$. D is the middle point of BC. Prove the angle ADB acute.

PROBLEM VII.

162. *To construct a triangle of which the sides shall be equal to three given sects, but any two whatever of these sects must be greater than the third.*

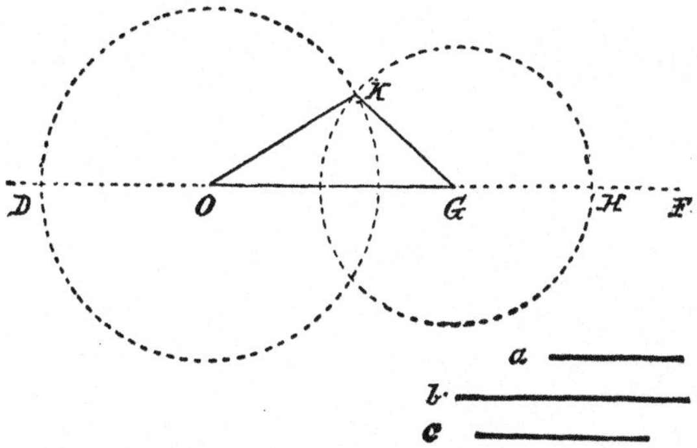

GIVEN, ***the three sects*** *a, b, c,* ***any two whatever greater than the third.***

REQUIRED, *to make a triangle having its sides equal respectively to* a, b, c.

CONSTRUCTION. On a line *DF*, by 133, take a sect *OG* equal to one of the given sects, as *b*. From *O* as a center, with a radius equal to one of the remaining sects, as *c*, describe, by 102, a circle.

From *G* as a center, with a radius equal to the remaining sect *a*, describe an arc intersecting the former circle at *K*. Join *KO* and *KG*. *KGO* will be the triangle required.

PROOF. By the construction and the equality of all radii of the same circle, the three sides *GK, KO,* and *OG* are equal respectively to *a, b, c.*

LIMITATION. It is necessary that any two of the sects should be greater than the third, or, what amounts to the same, the difference of any two sides less than the third. For, if *a* and *c* were together less than *b*, the circles in the figure would not meet; and, if they were together equal to *b*, the point *K* would be on OG, and the triangle would become a sect. But as we have proved, in 156, that any two sides of a triangle are together greater than the third side, our solution of Problem VII will apply to any three sects that could possibly form a triangle.

163. COROLLARY. If a triangle be made of hinged rods, though the hinges be entirely free, the rods cannot turn upon them; the triangle is rigid.

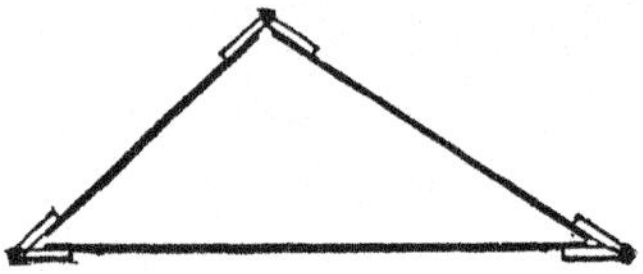

PROBLEM VIII.

164. *At a given point in a given line, to make an angle equal to a given angle.*

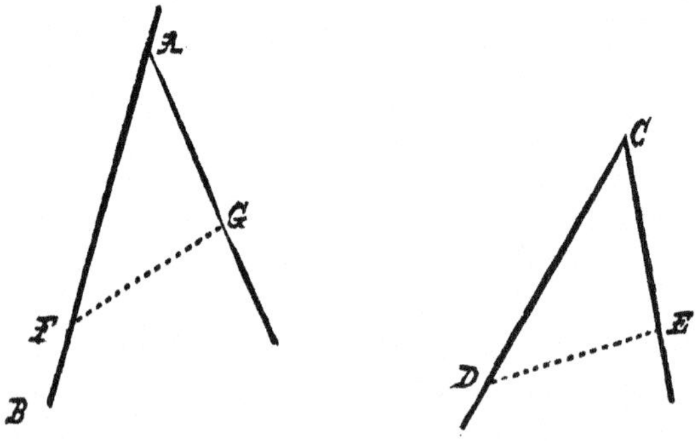

GIVEN, *the point A, and the line AB and* $\measuredangle C$.

REQUIRED, *at* A, *with arm* AB, *to make an angle* = $\measuredangle$ C.

CONSTRUCTION. By 100, join any two points in the arms of $\measuredangle C$, thus forming a $\triangle DCE$. By 162, make a triangle whose three sides shall be equal to the three sides of DCE; making the sides equal to CD and CE intersect at A.

$$\therefore \measuredangle FAG = \measuredangle DCE.$$

(129. Triangles with three sides respectively equal are congruent.)

CHAPTER VI.

PARALLELS.

THEOREM XIX.

165. *If two lines cut by a transversal make alternate angles equal, the lines are parallel.*

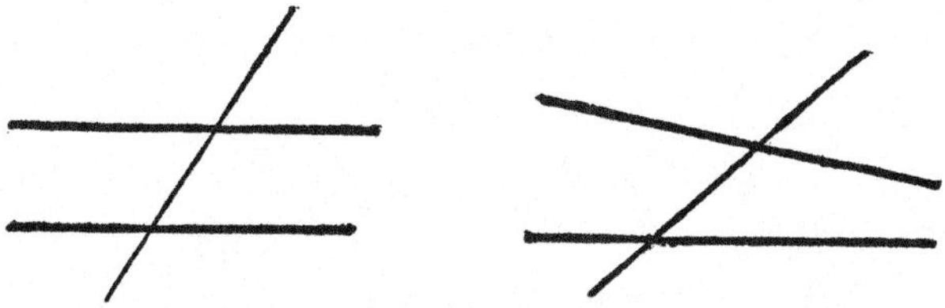

This is the contranominal of 142, part of which may be stated thus: If two lines which meet are cut by a transversal, their alternate angles are unequal.

166. COROLLARY. If two lines cut by a transversal make corresponding angles equal, or interior or exterior angles on the same side of the transversal supplemental, the lines are parallel.

For we know, by 113, that either of these suppositions makes also the alternate angles equal.

EXERCISES. 21. From a given point without a given line, draw a line making an angle with the given line equal to a given angle.

PROBLEM IX.

167. *Through a given point, to draw a line parallel to a given line.*

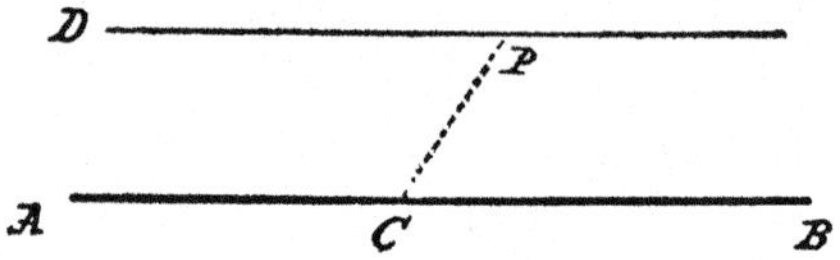

GIVEN, *a line AB and a point P.*
REQUIRED, *to draw a line through* P || AB

(using || for the words "parallel to").

CONSTRUCTION. In *AB* take any point, as *C*. By 100, join *PC*.
At *P* in the line *CP*, by 164, make ∡ *CPD* = ∡ *PCB*.

PD is || *AB*.

PROOF. By construction, the transversal *PC* makes alternate angles equal; ∡ *CPD* = ∡ *PCB*,

∴ by 165, *PD* is || *AB*.

EXERCISES. 22. Draw, through a given point between two lines which are not parallel, a sect, which shall be terminated by the given lines, and bisected at the given point.

23. If a line bisecting the exterior angle of a triangle be parallel to the base, show that the triangle is isosceles.

24. Sects from the middle point of the hypothenuse of a right-angled triangle to the three vertices are equal.

25. Through two given points draw two lines, forming, with a given line, a triangle equiangular with a given triangle.

26. *AC, BD,* are equal sects, drawn from the extremities of the sect *AB* on opposite sides of it, such that the angles *BAC, ABD,* are together equal to a straight angle. Prove that *AB* bisects *CD*.

THEOREM XX.

168. *If a transversal cuts two parallels, the alternate angles are equal.*

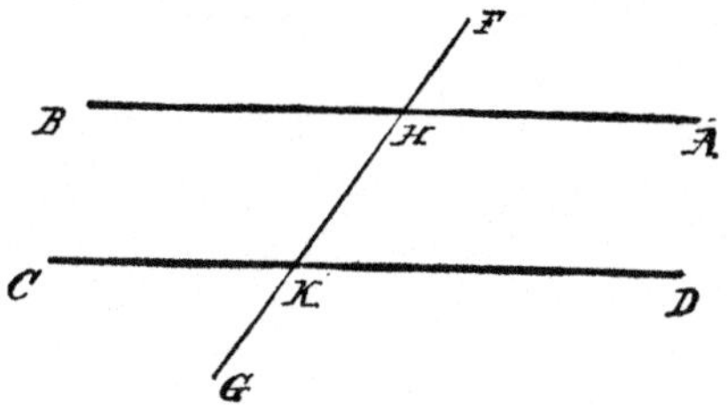

HYPOTHESIS. $AB \parallel CD$ cut at H and K by transversal FG.

CONCLUSION. $\angle HKD = \angle KHB$.

PROOF. A line through H, making alternate angles equal, is parallel to CD, by 165.

By our assumption, 99, two different lines through H cannot both be parallel to CD;

$\therefore$ by 31, the line through $H \parallel CD$ is identical with the line which makes alternate angles equal.

But, by hypothesis, AB is parallel to CD through H,

$$\therefore \angle KHB = \angle HKD.$$

169. COROLLARY I. If a transversal cuts two parallels, it makes the alternate angles equal; and therefore, by 113, the corresponding angles are equal, and the two interior or two exterior angles on the same side of the transversal are supplemental.

170. COROLLARY II. If a line be perpendicular to one of two parallels, it will be perpendicular to the other also.

171. CONTRANOMINAL OF 168. If alternate angles are unequal, the two lines meet.

So, if the interior angles on the same side of the transversal are not supplemental, the two lines meet; and as, by 143, they cannot meet on that side of the transversal where the two interior angles are greater than a straight angle, therefore they must meet on the side where the two interior angles are together less than a straight angle.

172. CONTRANOMINAL OF 99. Lines in the same plane parallel to the same line cannot intersect, and so are parallel to one another.

THEOREM XXI.

173. *Each exterior angle of a triangle is equal to the sum of the two interior opposite angles.*

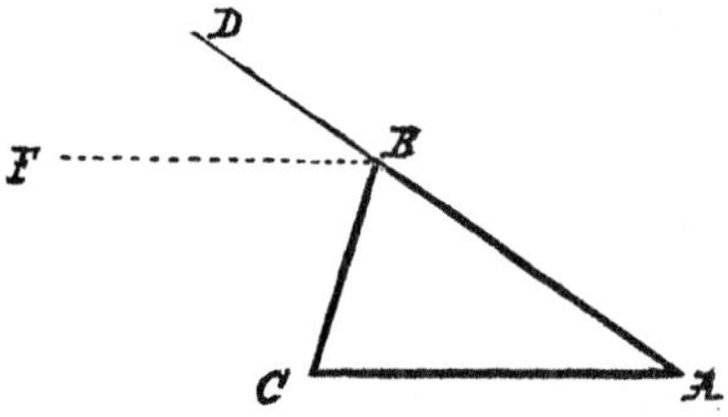

HYPOTHESIS. *ABC* **any** $\triangle$, **with side** *AB* **produced to** *D*.

CONCLUSION. $\measuredangle CBD = \measuredangle A + \measuredangle C$.

PROOF. From *B*, by 167, draw $BF \parallel AC$. Then, by 168, $\measuredangle C = \measuredangle CBF$, and by 169, $\measuredangle A = \measuredangle DBF$;

$\therefore$ by adding, $\measuredangle A + \measuredangle C = \measuredangle FBD + \measuredangle CBF = \measuredangle CBD$.

EXERCISES. 27. Each angle of an equilateral triangle is two-thirds of a right angle. Hence show how to trisect a right angle.

28. If any of the angles of an isosceles triangle be two-thirds of a right angle, the triangle must be equilateral.

Theorem XXII.

174. *The sum of the three interior angles of a triangle is equal to a straight angle.*

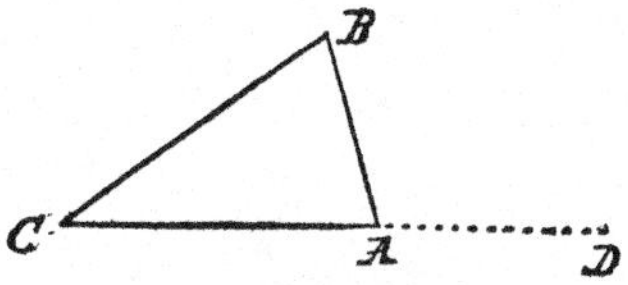

Hypothesis. *ABC* any △.
Conclusion. ∡ *CAB* + ∡ *B* + ∡ *C* = st. ∡.
Proof. Produce a side, as *CA*, to *D*.
By 173, ∡ *B* + ∡ *C* = ∡ *DAB*.
Add to both sides ∡ *BAC*.

∴ ∡ *CAB* + ∡ *B* + ∡ *C* = ∡ *DAB* + ∡ *BAC* = st. ∡.

175. Corollary. In any right-angled triangle the two acute angles are complemental.

CHAPTER VII.

TRIANGLES.

THEOREM XXIII.

176. *Two triangles are congruent if two angles and an opposite side in the one are equal respectively to two angles and the corresponding side in the other.*

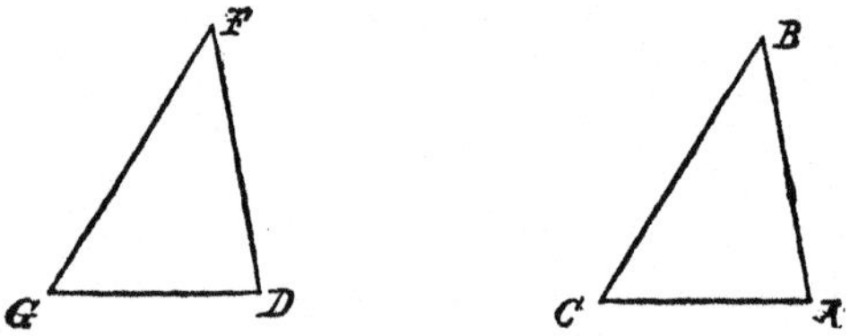

HYPOTHESIS. ABC *and* DFG $\triangle$*s, with*

$$\measuredangle A = \measuredangle D,$$
$$\measuredangle C = \measuredangle G,$$
$$AB = DF.$$

CONCLUSION. $\triangle ABC \cong \triangle DFG.$

PROOF. By 174, $\measuredangle A + \measuredangle B + \measuredangle C = \text{st.} \measuredangle = \measuredangle D + \measuredangle F + \measuredangle G,$

By hypothesis, $\measuredangle A + \measuredangle C = \measuredangle D + \measuredangle G,$

$$\therefore \quad \measuredangle B = \measuredangle F,$$

(89. If equals be taken from equals, the remainders are equal.)

$$\therefore \quad \triangle ABC \cong \triangle DFG.$$

(128. Triangles are congruent if two angles and the included side are equal in each.)

THEOREM XXIV.

177. *If two triangles have two sides of the one equal respectively to two sides of the other, and the angles opposite to one pair of equal sides equal, then the angles opposite to the other pair of equal sides are either equal or supplemental.*

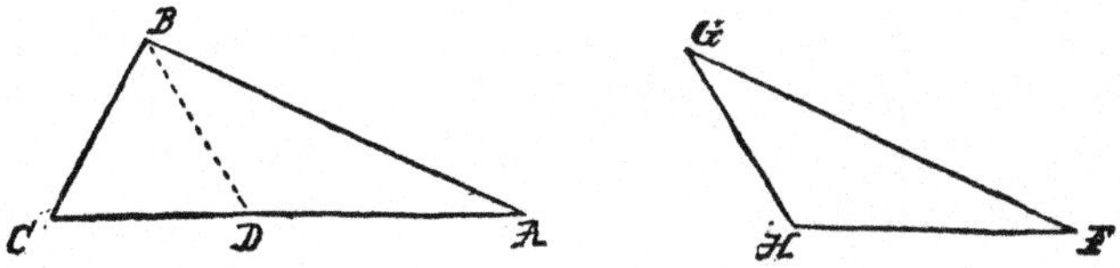

The angles included by the equal sides must be either equal or unequal.

CASE I. If they are equal, the third angles are equal.

(174. The sum of the angles of a triangle is a straight angle.)

CASE II. If the angles included by the equal sides are unequal, one must be the greater.

HYPOTHESIS. *ABC and FGH △s, with*

$$\measuredangle A = \measuredangle F,$$
$$AB = FG,$$
$$BC = GH,$$
$$\measuredangle ABC > \measuredangle G.$$

CONCLUSION. $\measuredangle C + \measuredangle H =$ st. $\measuredangle$.

PROOF. By 164, make the $\measuredangle ABD = \measuredangle G$,

$$\therefore \triangle ABD \cong \triangle FGH,$$

(128. Triangles are congruent if two angles and the included side are equal in each.)

$$\therefore \measuredangle BDA = \measuredangle H,$$

and

$$BD = GH.$$

But, by hypothesis, $BC = GH$,

$$\therefore \quad BC = BD,$$

$$\therefore \quad \measuredangle BDC = \measuredangle C.$$

(126. In an isosceles triangle the angles opposite equal sides are equal.)

But $\measuredangle BDA + \measuredangle BDC =$ st. $\measuredangle$,

$$\therefore \quad \measuredangle H + \measuredangle C = \text{st. } \measuredangle.$$

178. COROLLARY I. If two triangles have two sides of the one respectively equal to two sides of the other, and the angles opposite to one pair of equal sides equal, then, if one of the angles opposite the other pair of equal sides is a right angle, or if they are oblique but not supplemental, or if the side opposite the given angle is not less than the other given side, the triangles are congruent.

179. COROLLARY II. Two right-angled triangles are congruent if the hypothenuse and one side of the one are equal respectively to the hypothenuse and one side of the other.

On the Conditions of Congruence of Two Triangles.

180. A triangle has three sides and three angles.

The three angles are not all independent, since, whenever two of them are given, the third may be determined by taking their sum from a straight angle.

In four cases we have proved, that, if three independent parts of a triangle are given, the other parts are determined; in other words, that there is only one triangle having those parts:—

(124) Two sides and the angle between them.

(128) Two angles and the side between them.

(129) The three sides.

(176) Two angles and the side opposite one of them.

In the only other case, two sides and the angle opposite one of them, if the side opposite the given angle is shorter than the other given side, two different triangles may be formed, each of which will have the given parts. This is called the ambiguous case.

Suppose that the side a and side c, and $\measuredangle C$, are given. If $a > c$, then, making the $\measuredangle C$, and cutting off $CB = a$, taking B

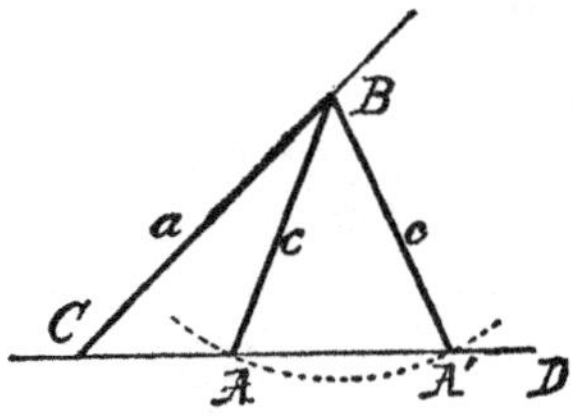

as center, and describing an arc with radius equal to c, it may cut CD in two points; and the two unequal triangles ABC and $A'BC$ will satisfy the required conditions.

In these the angles opposite the side BC are supplemental, by 177.

Loci.

181. The aggregate of all points and only those points which satisfy a given condition, is called the *Locus* of points satisfying that condition.

182. Hence, in order that an aggregate of points, L, may be properly termed the locus of a point satisfying an assigned condition, C, it is necessary and sufficient to demonstrate the following pair of inverses:—

(1) If a point satisfies C, it is upon L.

(2) If a point is upon L, it satisfies C.

We know from 18, that, instead of (1), we may prove its contranominal:—

(3) If a point is not upon L, it does not satisfy C.

Also, that, instead of (2), we may prove the obverse of (1):—

(4) If a point does not satisfy *C*, it is not upon *L*.

THEOREM XXV.

183. *The locus of the point to which sects from two given points are equal is the perpendicular bisector of the sect joining them.*

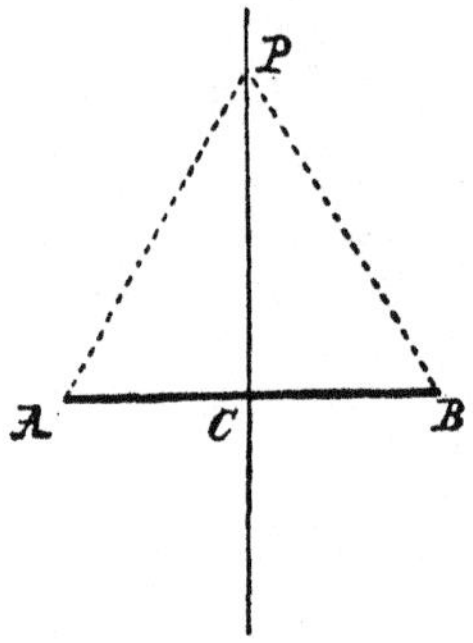

HYPOTHESIS. *AB a sect, C its mid-point.*

$PA = PB$.

CONCLUSION. $PC \perp AB$.

PROOF. Join *PC*.

Then $\triangle ACP$ is $\cong \triangle BCP$,

(129. Triangles with the three sides respectively equal are congruent.)

$\therefore \measuredangle ACP = \measuredangle BCP$,

$\therefore$ *CP* is the perpendicular bisector of *AB*.

184. INVERSE.

HYPOTHESIS. *P, any point on* $CP \perp AB$ *at its mid-point C.*

CONCLUSION. $PA = PB$.

PROOF. $\triangle ACP \cong \triangle BCP$.

(124. Two triangles are congruent if two sides and the included angle are equal in each.)

THEOREM XXVI.

185. *The locus of points from which perpendiculars on two given intersecting lines are equal, consists of the two bisectors of the angles between the given lines.*

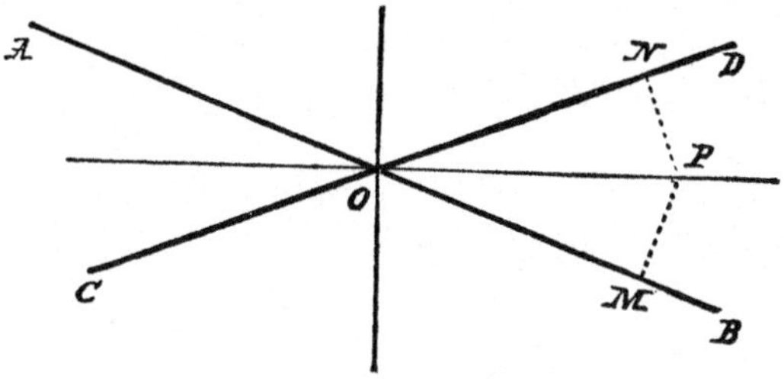

HYPOTHESIS. **Given AB and CD intersecting at O, and P any point such that $PM \perp AB = PN \perp CD$.**

CONCLUSION. $\measuredangle POM = \measuredangle PON$.

PROOF. $\triangle POM \cong \triangle PON$.

(179. Right triangles having hypothenuse and one side equal are congruent.)

Therefore all points from which perpendiculars on two intersecting lines are equal, lie on the bisectors of the angles between them.

186. INVERSE.

HYPOTHESIS. **P, any point on bisector.**

CONCLUSION. $PM = PN$.

PROOF. $\triangle POM \cong PON$.

(176. Triangles having two angles and a corresponding side equal in each are congruent.)

Hence, by 110, the two bisectors of the four angles between AB and CD are the locus of a point from which perpendiculars on AB and CD are equal.

Intersection of Loci.

187. Where it is required to find points satisfying two conditions, if we leave out one condition, we may find a locus of points satisfying the other condition.

Thus, for each condition we may construct the corresponding locus; and, if these two loci have points in common, these points, and these only, satisfy both conditions.

THEOREM XXVII.

188. *There is one, and only one, point from which sects to three given points not on a line are equal.*

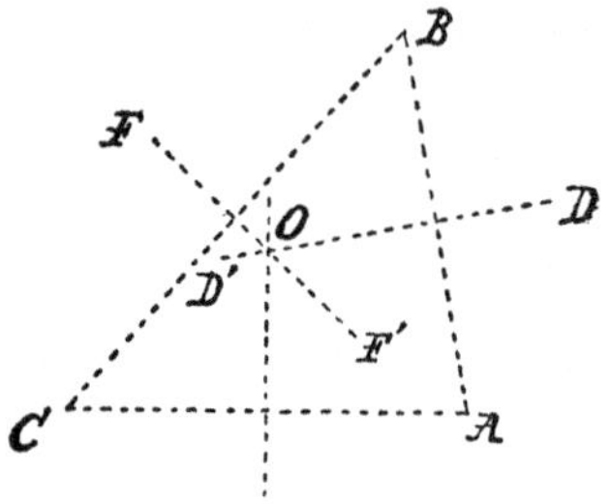

HYPOTHESIS. ***Given A, B, and C, not in a line.***

CONSTRUCTION AND PROOF. By 183, every point to which sects from A and B are equal, lies on DD', the perpendicular bisector of AB. Again: the locus of points to which sects from B and C are equal is the perpendicular bisector FF'.

Now, DD' and FF' intersect, since, if they were parallel, AB, which is perpendicular to one of them, would be perpendicular to the other also, by 170, and ABC would be one line. Let them intersect in O. By 95, they cannot intersect again.

189. Thus, $OB = OA$,
and

$$OC = OA,$$

$$\therefore \quad OB = OC.$$

Therefore, by 183, O is on the perpendicular bisector of AC. Therefore

COROLLARY. The three perpendicular bisectors of the sides of a triangle meet in a point from which sects to the three vertices of the triangle are equal.

PROBLEM X.

190. *To find points from which perpendiculars on three given lines which form a triangle are equal.*

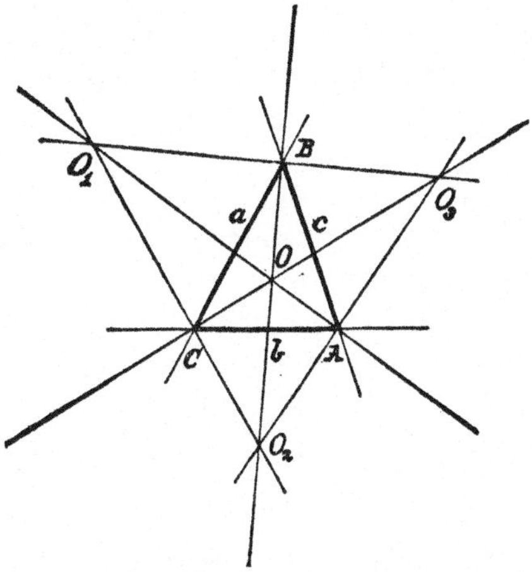

GIVEN, *a, b, c, three lines intersecting in the three distinct points A, B, C.*

SOLUTION. The locus of points from which perpendiculars on the two lines *a* and *b* are equal consists of the two bisectors of the angles between the lines.

The locus of points from which perpendiculars on the two lines *b* and *c* are equal, similarly consists of the two lines bisecting the angles between *b* and *c*. Our two loci consist thus of two pairs of lines. Each line of one pair cuts each line of the second pair in one point; so that we get four points, O, O_1, O_2, O_3, common to the two loci.

191. By 185, the four intersection points of the bisectors of the angles between *a* and *b*, and between *b* and *c*, lie on the bisectors of the angles between *a* and *c*. Therefore

COROLLARY. The six bisectors of the interior and exterior angles of a triangle meet four times, by threes, in a point.

CHAPTER VIII.

POLYGONS.

I. Definitions.

192. A *Polygon* is a figure formed by a number of lines of which each is cut by the following one, and the last by the first.

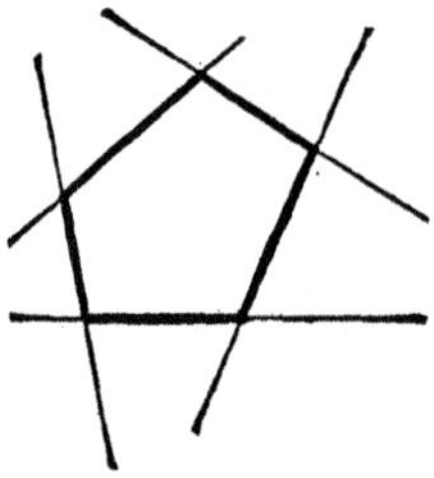

193. The common points of the consecutive lines are called the *Vertices* of the polygon.

194. The sects between the consecutive vertices are called the *Sides* of the polygon.

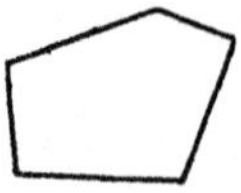

195. The sum of the sides, a broken line, makes the *Perimeter* of the polygon.

196. The angles between the consecutive sides and towards the enclosed surface are called the *Interior Angles* of the polygon. Every polygon has as many interior angles as sides.

197. A polygon is said to be *Convex* when no one of its interior angles is reflex.

198. The sects joining the vertices not consecutive are called *Diagonals* of the polygon.

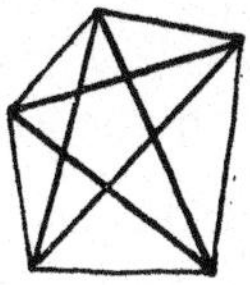

199. When the sides of a polygon are all equal to one another, it is called *Equilateral.*

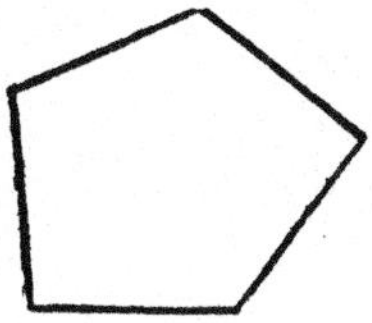

200. When the angles of a polygon are all equal to one another, it is called *Equiangular.*

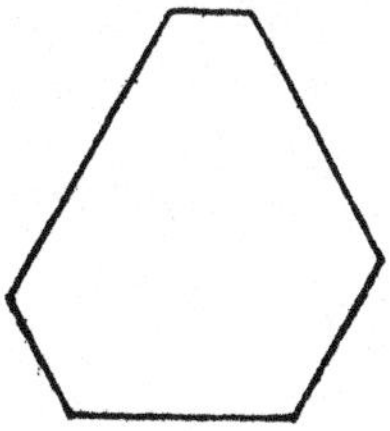

201. A polygon which is both equilateral and equiangular is called *Regular.*

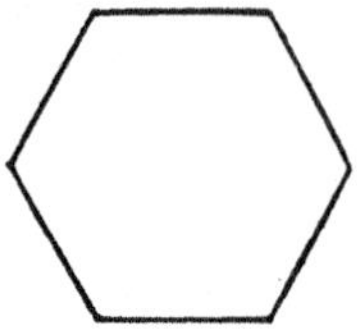

202. Two polygons are *Mutually Equilateral* if the sides of the one are equal respectively to the sides of the other taken in the same order.

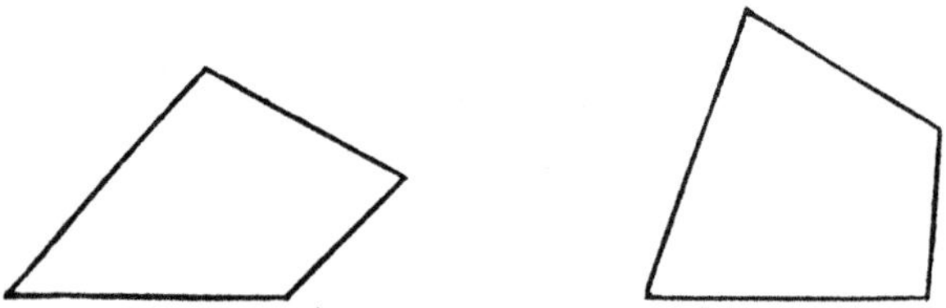

203. Two polygons are *Mutually Equiangular* if the angles of the one are equal respectively to the angles of the other taken in the same order.

204. Two polygons may be mutually equiangular without being mutually equilateral.

205. Except in the case of triangles, two polygons may be mutually equilateral without being mutually equiangular.

206. A polygon of three sides is a *Trigon* or *Triangle;* one of four sides is a *Tetragon* or *Quadrilateral;* one of five sides is a *Pentagon;* one of six sides is a *Hexagon;* one of seven sides is a *Heptagon;* one of eight, an *Octagon;* of nine, a *Nonagon;* of ten, a *Decagon;* of twelve, a *Dodecagon;* of fifteen, a *Quindecagon.*

207. The *Surface* of a polygon is that part of the plane enclosed by its perimeter.

208. A *Parallelogram* is a quadrilateral whose opposite sides are parallel.

209. A *Trapezoid* is a quadrilateral with two sides parallel.

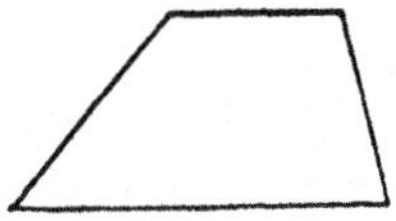

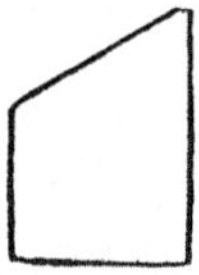

II. General Properties.

Theorem XXVIII.

210. *If two polygons be mutually equilateral and mutually equiangular, they are congruent.*

Proof. Superposition: they may be applied, the one to the other, so as to coincide.

Exercises. 29. Is a parallelogram a trapezoid? How could a triangle be considered a trapezoid?

THEOREM XXIX.

211. *The sum of the interior angles of a polygon is two less straight angles than it has sides.*

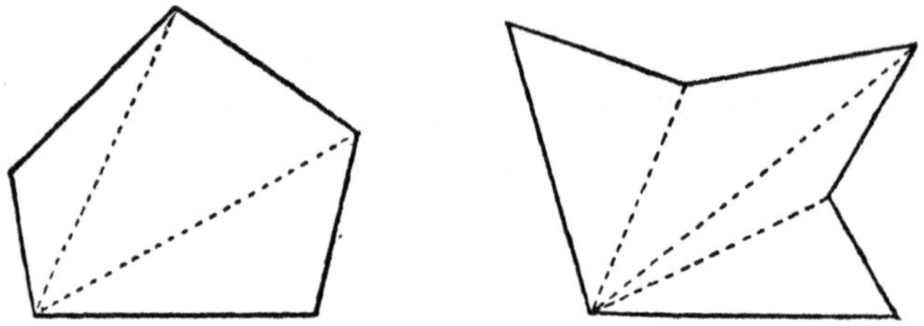

HYPOTHESIS. ***A polygon of n sides.***

CONCLUSION. Sum of ∡'s = $(n - 2)$ st. ∡'s.

PROOF. If we can draw all the diagonals from any one vertex with out cutting the perimeter, then we have a triangle for every side of the polygon, except the two which make our chosen vertex. Thus, we have $(n - 2)$ triangles, whose angles make the interior angles of the polygon.

But, by 174, the sum of the angles in each triangle is a straight angle,

∴ Sum of ∡'s in polygon = $(n - 2)$ st. ∡'s.

212. COROLLARY I. From each vertex of a polygon of n sides are $(n - 3)$ diagonals.

213. COROLLARY II. The sum of the angles in a quadrilateral is a perigon.

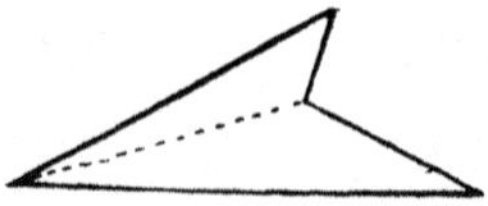

214. COROLLARY III. Each angle of an equiangular polygon of n sides is $\frac{(n - 2) \text{ st. ∡'s}}{n}$.

THEOREM XXX.

215. *In a convex polygon, the sum of the exterior angles, one at each vertex, made by producing each side in order, is a perigon.*

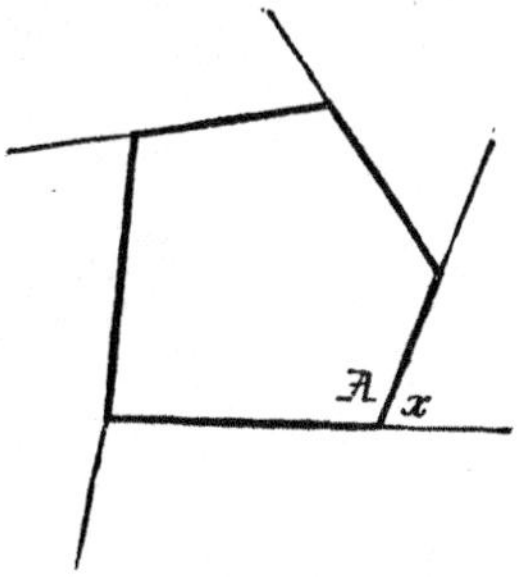

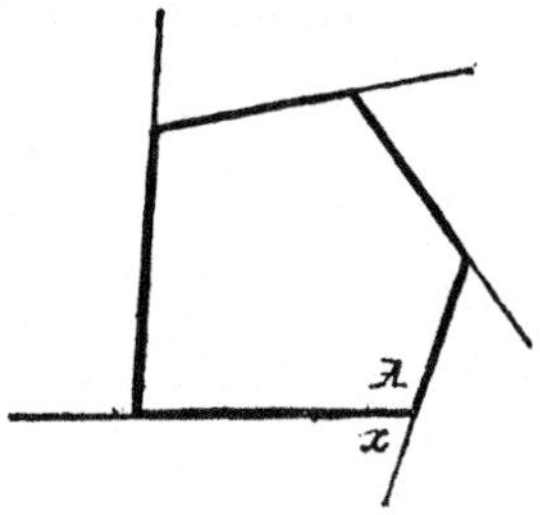

HYPOTHESIS. ***A convex polygon of n sides, each in order produced at one end.***

CONCLUSION. Sum of exterior ∡'s = perigon.

PROOF. Every interior angle, as ∡ A, and its adjacent exterior angle, as ∡ x, together = st. ∡.

∴ all interior ∡'s + all exterior ∡'s = n st. ∡'s.

But, by 211, all interior ∡'s = $(n - 2)$ st. ∡'s.

∴ sum of all exterior ∡'s = a perigon.

EXERCISES. 30. How many diagonals can be drawn in a polygon of n sides?

31. The exterior angle of a regular polygon is one-third of a right angle: find the number of sides in the polygon.

32. The four bisectors of the angles of a quadrilateral form a second quadrilateral whose opposite angles are supplemental.

33. Divide a right-angled triangle into two isosceles triangles.

34. In a right-angled triangle, the sect from the mid-point of the hypothenuse to the right angle is half the hypothenuse.

III. Parallelograms.

THEOREM XXXI.

216. *If two opposite sides of a quadrilateral are equal and parallel, it is a parallelogram.*

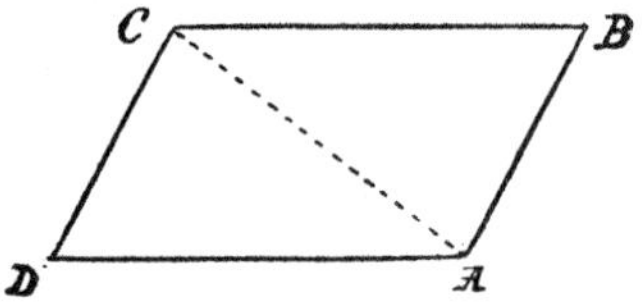

HYPOTHESIS. *ABCD* ***a quadrilateral, with*** $AB =$ ***and*** $\parallel$ CD.
CONCLUSION. $AD \parallel BC$.
PROOF. Join AC. Then $\angle BAC = \angle ACD$.

(168. If a transversal cuts two parallels, the alternate angles are equal.)

By hypothesis, $AB = CD$, and AC is common,

$$\therefore \quad \angle BCA = \angle CAD,$$

(124. Triangles are congruent if two sides and the included angle are equal in each.)

$$\therefore \quad BC \parallel AD.$$

(165. Lines making alternate angles equal are parallel.)

EXERCISES. 35. Find the number of elements required to determine a parallelogram.

36. The four sects which connect the mid-points of the consecutive sides of any quadrilateral, form a parallelogram.

37. The perpendiculars let fall from the extremities of the base of an isosceles triangle on the opposite sides will include an angle supplemental to the vertical angle of the triangle.

38. If BE bisects the angle B of a triangle ABC, and CE bisects the exterior angle ACD, the angle E is equal to one-half the angle A.

THEOREM XXXII.

217. *The opposite sides and angles of a parallelogram are equal to one another, and each diagonal bisects it.*

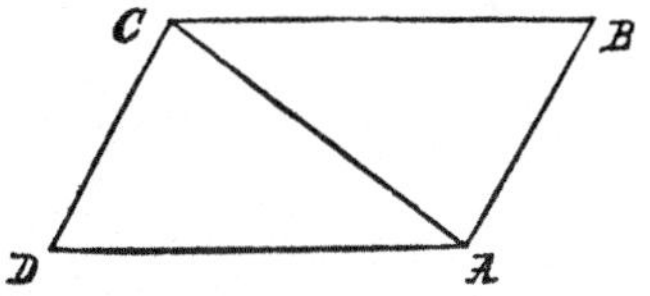

HYPOTHESIS. **Let** AC **be a diagonal of** ▱ $ABCD$ (using the sign ▱ for the word "parallelogram").

CONCLUSIONS. $AB = CD.$

$BC = DA.$

$\measuredangle DAB = \measuredangle BCD.$

$\measuredangle B = \measuredangle D.$

$\triangle ABC \cong \triangle CDA.$

PROOF. $\measuredangle BAC = \measuredangle ACD$, and

$\measuredangle CAD = \measuredangle BCA$;

(168. If a transversal cuts parallels, the alternate angles are equal.)

therefore, adding,

$$\measuredangle BAC + \measuredangle CAD = \measuredangle BAD = \measuredangle ACD + \measuredangle ACB = \measuredangle BCD.$$

Again, the side AC included between the equal angles is common,

$$\therefore \quad \triangle ABC \cong \triangle CDA,$$

(128. Triangles are congruent if two angles and the included side are equal in each.)

$$\therefore \quad AB = CD, \qquad BC = DA, \qquad \text{and} \qquad \measuredangle B = \measuredangle D.$$

EXERCISES. 39. If one angle of a parallelogram be right, all the angles are right.

40. If two parallelograms have one angle of the one equal to one angle of the other, the parallelograms are mutually equiangular.

218. Corollary I. Any pair of parallels intercept equal sects of parallel transversals.

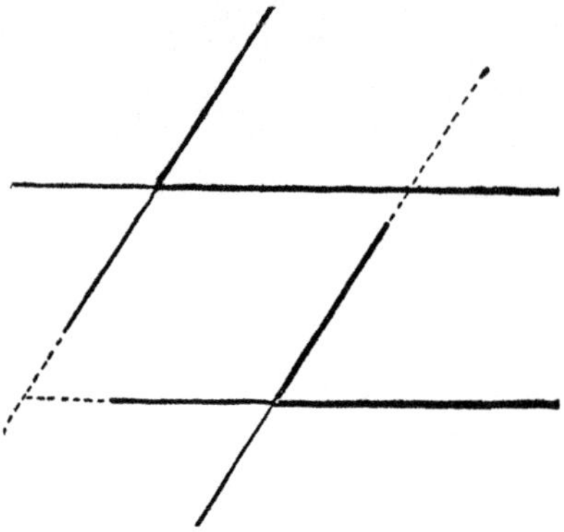

219. Corollary II. If two lines be respectively parallel to two other lines, any angle made by the first pair is equal or supplemental to any angle made by the second pair.

220. Corollary III. If two angles have their arms respectively perpendicular, they are equal or supplemental.

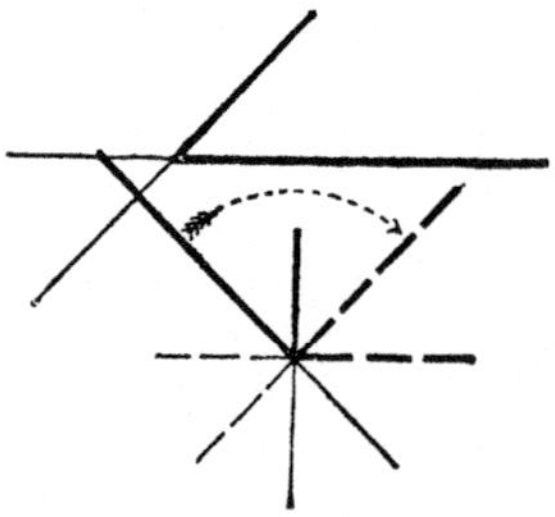

For, revolving one of the angles through a right angle around its vertex, its arms become perpendicular to their traces, and therefore parallel to the arms of the other given angle.

221. Corollary IV. If one of the angles of a parallelogram is a right angle, all its angles are right angles, and it is called a *Rectangle.*

222. Corollary V. If two consecutive sides of a parallelogram are equal, all its sides are equal, and it is called a *Rhombus.*

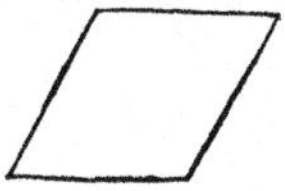

223. A *Square* is a rectangle having consecutive sides equal.

224. Both diagonals, AC and BD, being drawn, it may with a few exceptions be proved that a quadrilateral which has any two of the following properties will also have the others:

1. $AB \parallel CD$.
2. $BC \parallel DA$.
3. $AB = CD$.
4. $BC = DA$.
5. $\measuredangle DAB = \measuredangle BCD$.
6. $\measuredangle ABC = \measuredangle CDA$.
7. The bisection of AC by BD.
8. The bisection of BD by AC.
9. The bisection of the $\square$ by AC.
10. The bisection of the $\square$ by BD.

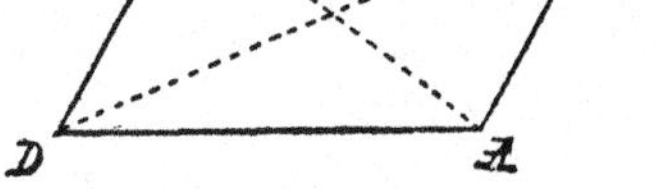

These ten combined in pairs will give forty-five pairs; with each of these pairs it may be required to establish any of the eight other properties, and thus three hundred and sixty questions respecting such quadrilaterals may be raised.

For example, from 1 and 2, 217 proves 3,

$$\therefore \quad \triangle ABE \cong \triangle CDE,$$

(128. Two triangles are congruent if two angles and the included side are equal in each.)

∴ *the diagonals of a parallelogram bisect each other.*

The inverse is to prove 1 and 2 from 7 and 8:

If the diagonals of a quadrilateral bisect each other, the figure is a parallelogram.

225. Since by 170 two lines perpendicular to one of two parallels are perpendicular to the other, and by 166 are parallel, and so by 218 the sects intercepted on them are equal, therefore all perpendicular sects between two parallels are equal.

EXERCISES. 41. If the mid-points of the three sides of a triangle be joined, the four resulting triangles are equal.

42. If the diagonals of a parallelogram be equal, the parallelogram is a rectangle.

43. If the diagonals of a parallelogram cut at right angles, it is a rhombus.

44. If the diagonals of a parallelogram bisect the angles, it is a rhombus.

45. If the diagonals of a rectangle cut at right angles, it is a square.

THEOREM XXXIII.

226. *The three perpendiculars from the vertices of a triangle to the opposite sides meet in a point.*

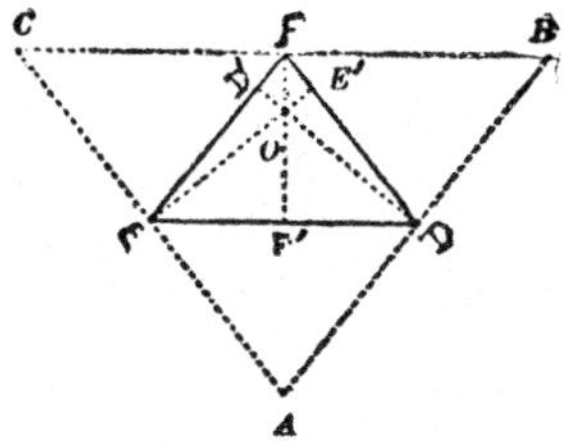

HYPOTHESIS. ***Let*** DD', EE', FF', ***be the three perpendiculars from the vertices*** D, E, F, ***to the opposite sides of*** $\triangle DEF$.

CONCLUSION. DD', EE', FF', meet in a point O.

PROOF. By 167, through D, E, F, draw $AB, BC, CA \parallel FE, DF, DE$.

Then the figures $ADFE$, $DBFE$, are parallelograms.

By 217,

$$\therefore \quad AD = FE = DB,$$

$$\therefore \quad D \text{ is mid-point of } AB.$$

In same way, E and F are mid-points of AC and BC.

But, since DD', EE', FF', are respectively $\perp$ EF, FD, DE,

$$\therefore \quad \text{they must be } \perp AB, AC, \text{ and } BC,$$

(170. A line perpendicular to one of two parallels is perpendicular to the other.)

$\therefore$ by 189, they meet in a point.

EXERCISES. 46. The intersection of the sects joining the mid-points of opposite sides of any quadrilateral is the mid-point of the sect joining the mid-points of the diagonals.

THEOREM XXXIV.

227. *If three or more parallels intercept equal sects on one transversal, they intercept equal sects on every transversal.*

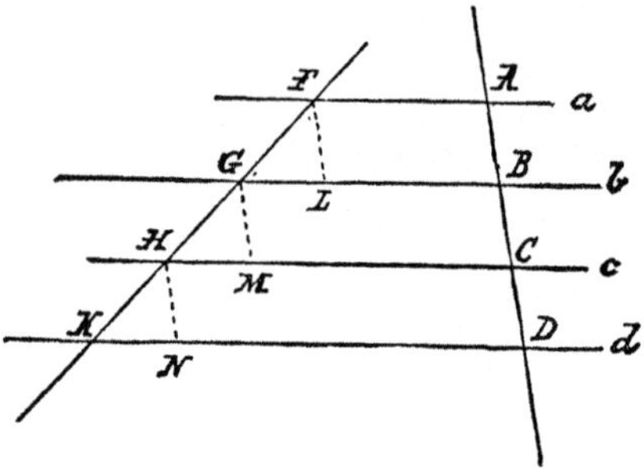

HYPOTHESIS. $AB = BC = CD =$, ***etc., are sects on the transversal*** AC, ***intercepted by the parallels*** a, b, c, d, ***etc.***

FG, GH, HK, ***etc., are corresponding sects on any other transversal.***

CONCLUSION. $FG = GH = HK =$, etc.

PROOF. From F, G, and H, draw, by 167, FL, GM, and HN, all parallel to AD;

$\therefore$ they are all equal, because $AB = BC = CD =$, etc.

(217. Opposite sides of a parallelogram are equal.)

But $\measuredangle GFL = \measuredangle HGM = \measuredangle KHN$, and

$\measuredangle FGL = \measuredangle GHM = \measuredangle HKN$,

(169. If a transversal cuts two parallels, the corresponding angles are equal.)

$$\therefore \quad \triangle FGL \cong \triangle GHM \cong \triangle HKN,$$

(176. Triangles are congruent if they have two angles and a corresponding side equal in each.)

$$\therefore \quad FG = GH = HK.$$

228. COROLLARY. The intercepted part of each parallel will differ from the neighboring intercepts by equal sects.

THEOREM XXXV.

229. *The line drawn through the mid-point of one of the non-parallel sides of a trapezoid parallel to the parallel sides, bisects the remaining side.*

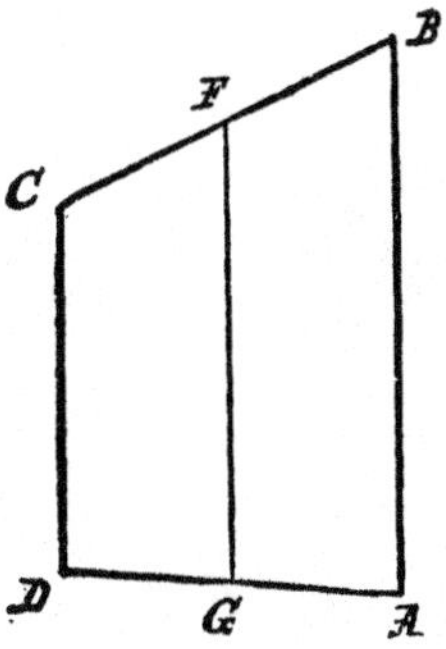

HYPOTHESIS. *In the trapezoid* $ABCD$, $BF = FC$, *and* $AB \parallel CD \parallel FG$.

CONCLUSION. $AG = GD$.

PROOF. By 227, if parallels intercept equal sects on any transversal, they intercept equal sects on every transversal.

230. COROLLARY I. The line joining the mid-points of the non-parallel sides of a trapezoid is parallel to the parallel sides, for we have just proved it identical with a line drawn parallel to them through one mid-point.

231. COROLLARY II. If through the mid-point of one side of a triangle a line be drawn parallel to a second side, it will bisect the third side; and, inversely, the sect joining the mid-points of any two sides of a triangle is parallel to the third side and equal to half of it.

232. A sect from any vertex of a triangle to the mid-point of the opposite side is called a *Medial* of the triangle.

233. Three or more lines which intersect in the same point are said to be *Concurrent.*

234. Three or more points which lie on the same line are said to be *Collinear.*

THEOREM XXXVI.

235. *The three medials of a triangle are concurrent in a trisection point of each.*

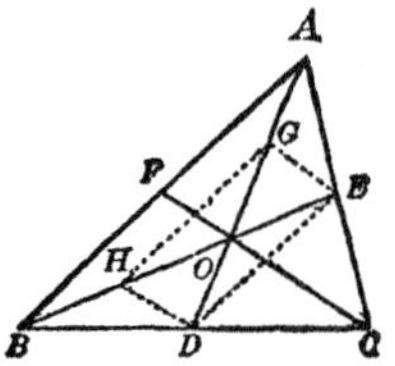

PROOF. Let two medials, AD and BE, meet in O. Take G the mid-point of OA, and H of OB. Join GH, HD, DE, EG.

By 231, in $\triangle ABO$, GH is equal and parallel to $\frac{1}{2}AB$; so is DE in $\triangle ABC$;

$\therefore$ by 216, $GHDE$ is a parallelogram.

But, by 224, the diagonals of a parallelogram bisect each other,

$$\therefore \quad AG = GO = OD, \quad \text{and} \quad BH = HO = OE.$$

So any medial cuts any other at its point of trisection remote from its vertex,

$\therefore$ the three are concurrent.

236. The intersection point of medials is called the *Centroid* of the triangle. The intersection point of perpendiculars from the vertices to the opposite sides is called the *Orthocenter* of the triangle. The intersection point of perpendiculars erected at the mid-points of the sides is called the *Circumcenter* of the triangle.

IV. Equivalence.

237. *Two plane figures are called equivalent if we can prove that they must contain the same extent of surface, even if we do not show how to cut them into parts congruent in pairs.*

238. The base of a figure is that one of its sides on which we imagine it to rest.

Any side of a figure may be taken as its base.

239. The altitude of a figure is the perpendicular from its highest point to the line of its base.

So the altitude of a parallelogram is the perpendicular dropped from any point of one side to the line of the opposite side.

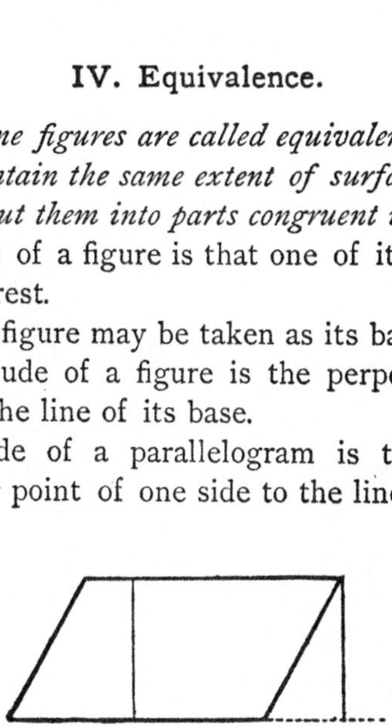

240. The words "altitude" and "base" are correlative; thus, a triangle may have three different altitudes.

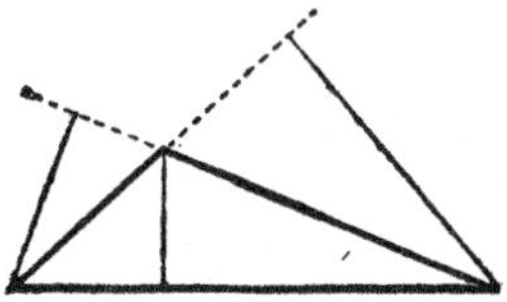

PROBLEM XI.

241. *To describe a square upon a given sect.*

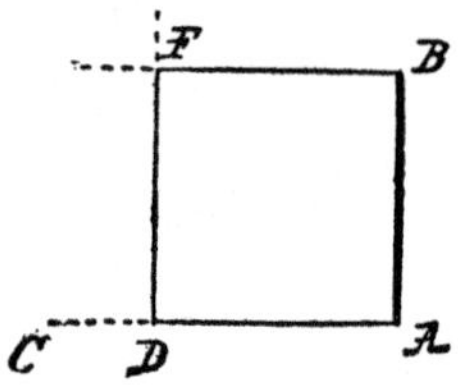

GIVEN, *the sect AB.*

REQUIRED, *to describe a square on* AB.

CONSTRUCTION. By 135, from A draw $AC \perp AB$.
By 133, in AC make $AD = AB$.
By 167, through D draw $DF \parallel AB$.
By 167, through B draw $BF \parallel AD$.

AF will be the required square.

PROOF. By construction, AF is a parallelogram,

$$\therefore \quad AB = FD, \text{ and } AD = BF,$$

$$\measuredangle F = \measuredangle A, \text{ and } \measuredangle B = \measuredangle D.$$

(217. In a parallelogram the opposite sides and angles are equal.)

But, by construction, $AB = AD$,

$\therefore$ AF is equilateral.

Again, $\measuredangle A + \measuredangle B =$ st. $\measuredangle$.

(169. If two parallel lines are cut by a transversal, the two interior angles are supplemental.)

But, by construction, $\measuredangle A$ is rt.,

$$\therefore \quad \measuredangle A = \measuredangle B,$$

and all four angles are rt.

242. COROLLARY. If a parallelogram have one angle right, it is a rectangle.

Theorem XXXVII.

243. *In any right-angled triangle, the square on the hypothenuse is equivalent to the sum of the squares of the other two sides.*

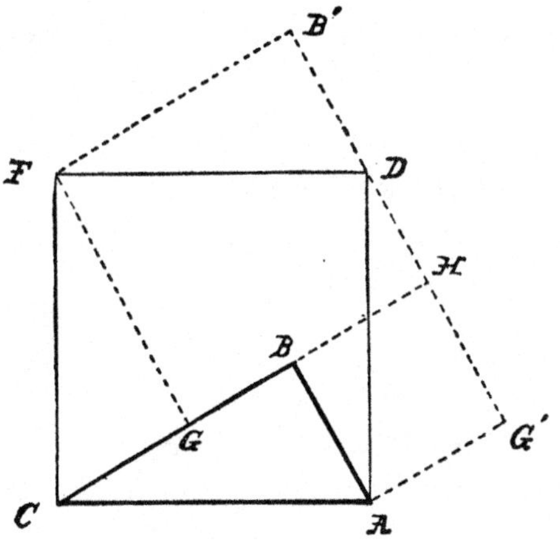

Hypothesis. $\triangle ABC$, *right-angled at* B.

Conclusion. Square on AB + square on BC = square on AC.

Proof. By 241, on hypothenuse AC, on the side toward the $\triangle ABC$, describe the square $ADFC$.

On the greater of the other two sides, as BC, by 133, lay off $CG = AB$. Join FG.

Then, by construction, $CA = FC$, and $AB = CG$, and $\measuredangle CAB = \measuredangle FCG$, since each is the complement of ACB;

$$\therefore \quad \triangle ABC \cong \triangle CGF.$$

Translate the $\triangle ABC$ upward, keeping point A on sect AD, and point C on sect CF, until A is on D, and C on F; then call B' the position of B.

Likewise translate CGF to the right until C is on A, and F on D; then call G' the position of G.

The resulting figure, $AG'DB'FGBA$, will be the squares on the other two sides, AB and BC.

For, since the sum of the angles at $D =$ st. $\measuredangle$,

$\therefore$ $G'D$ and DB' arc in one line.

Produce GB to meet this line at H.

Then $GF = FB' = BC$, and rt. $\measuredangle\, B' = \measuredangle\, GFB' = \measuredangle\, FGH$,

$\therefore$ $GFB'H$ equals square on BC.

Again, $G'A = AB$, and rt. $\measuredangle\, G' = \measuredangle\, G'AB = \measuredangle\, ABH$,

$\therefore$ $AG'HB$ is the square on AB,

$\therefore$ sq. of $AC =$ sq. of $AB +$ sq. of BC.

244. Corollary. Given any two squares placed side by side, with bases AB and BC in line; to cut this figure into three pieces, two of which being translated without rotation, the figure shall be transformed into one square.

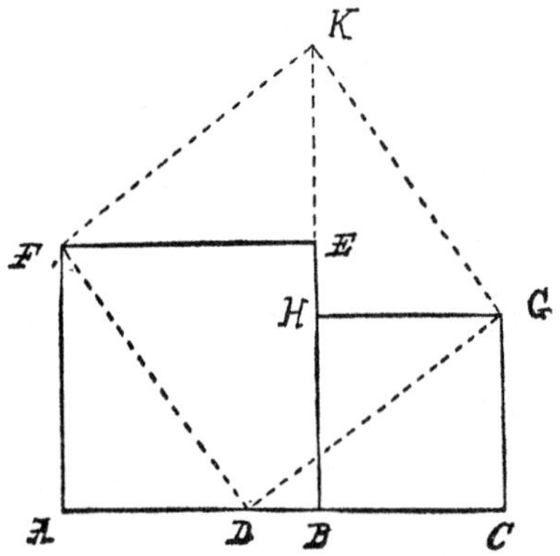

In AC take $AD = BC$, and join D to the corners of the squares opposite B. Two right-angled triangles are thus produced, with hypothenuses perpendicular to one another. Translate each triangle along the hypothenuse of the other.

THEOREM XXXVIII.

245. *Two triangles are congruent if they are mutually equiangular, and have corresponding altitudes equal.*

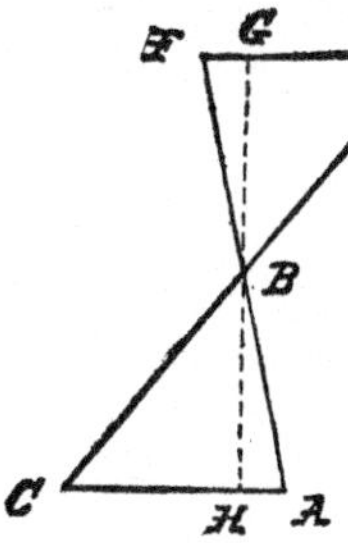

HYPOTHESIS. *△s ABC and BDF mutually equiangular, and altitude BH = altitude GB.*

CONCLUSION. △ *ABC* ≅ △ *BDF.*

PROOF. Rt. △ *BDG* ≅ rt. △ *BCH,*

(176. Triangles are congruent when two angles and a side in one are equal to two angles and a corresponding side in the other.)

$$\therefore \quad BD = BC,$$

$$\therefore \quad \triangle ABC \cong \triangle BDF.$$

(124. Triangles are congruent when two angles and the included side are equal in each.)

EXERCISES. 47. If from the vertex of any triangle a perpendicular be drawn to the base, the difference of the squares on the two sides of the triangle is equal to the difference of the squares on the parts of the base.

48. Show how to find a square triple a given square.

49. Five times the square on the hypothenuse of a right-angled triangle is equivalent to four times the sum of the squares on the medials to the other two sides.

THEOREM XXXIX.

246. *If two consecutive sides of one rectangle be respectively equal to two consecutive sides of another, the rectangles are congruent.*

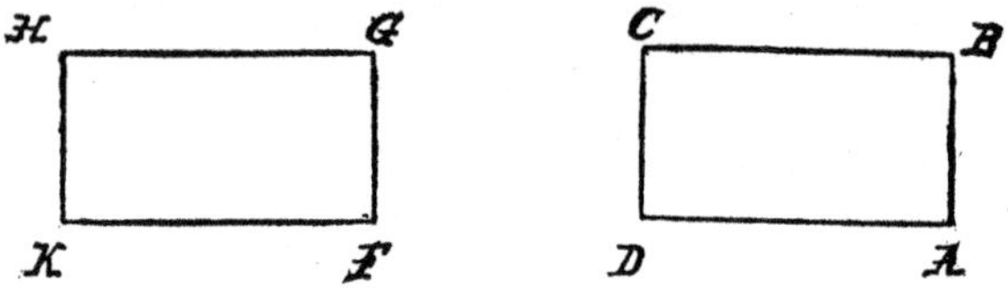

HYPOTHESIS. ***Two quadrilaterals*** *ABCD* ***and*** *FGHK*, ***with all*** ***∡s rt.***

$$AB = FG, BC = GH.$$

CONCLUSION. $ABCD \cong FGHK.$

PROOF. $AB \parallel CD$, and $AD \parallel BC$,

(166. If interior angles on same side of transversal are supplemental, lines are parallel.)

∴ a rectangle is a parallelogram,

∴ $AB = CD$, and $AD = BC$.

(217. Opposite sides of a parallelogram are equal.)

In the same way, $FG = HK$, and $GH = FK$,

∴ $ABCD \cong FGHK$.

(210. If two polygons be mutually equilateral and mutually equiangular, they are congruent.)

247. COROLLARY. A rectangle is completely determined by two consecutive sides; so if two sects, *a* and *b*, are given, we

may speak of the rectangle of *a* and *b*, or we may call it the rectangle *ab*. It can be constructed by the method given in 241 to describe a square.

Theorem XL.

248. *A parallelogram is equivalent to the rectangle of its base and altitude.*

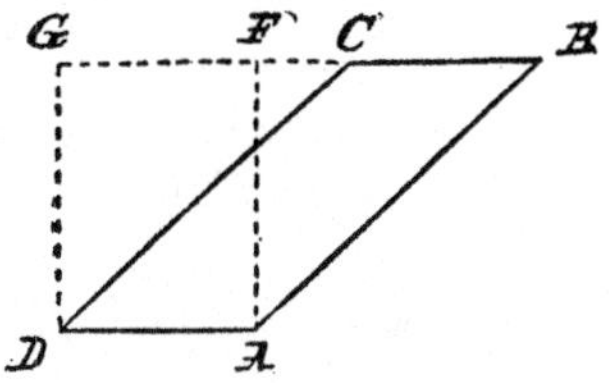

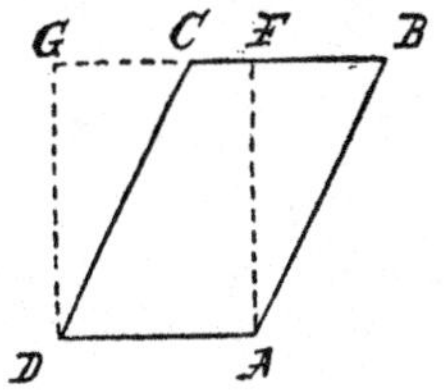

Hypothesis. $ABCD$ ***any ▱, of which side*** AD ***is taken as base;*** $AF =$ ***altitude of ▱.***

Conclusion. ▱ $ABCD =$ rt. ▱ $AFGD$.

Proof. $AF = DG$, and $AB = DC$.

(217. Opposite sides of a parallelogram are equal.)

By construction, ∡ G and ∡ F are rt.,

$$\therefore \quad \triangle ABF \cong \triangle DCG.$$

(179. Right triangles are congruent if hypothenuse and one side are respectively equal in each.)

From the trapezoid $ABGD$ take away △ DCG, and then is left ▱ $ABCD$. From the same trapezoid take the equal △ ABF, and there is left the rt. ▱ $AFGD$.

$$\therefore \quad ▱ ABCD = \text{rt. } ▱ AFGD.$$

(89. If equals be taken from equals, the remainders are equal.)

249. Corollary. All parallelograms having equal bases and equal altitudes are equivalent, because they are all equivalent to the same rectangle.

Exercises. 50. Equivalent parallelograms on the same base and on the same side of it are between the same parallels.

EXERCISES. 51. Prove 248 for the case when C and F coincide.

52. If through the vertices of a triangle lines be drawn parallel to the opposite sides, and produced until they meet, the resulting figure will contain three equivalent parallelograms.

53. On the same base and between the same parallels as a given parallelogram, construct a rhombus equivalent to the parallelogram.

54. Divide a given parallelogram into two equivalent parallelograms.

55. Of two parallelograms between the same parallels, that is the greater which stands on the greater base. Prove also an inverse of this.

56. Equivalent parallelograms situated between the same parallels have equal bases.

57. Of parallelograms on equal bases, that is the greater which has the greater altitude.

58. A trapezoid is equivalent to a rectangle whose base is half the sum of the two parallel sides, and whose altitude is the perpendicular between them.

59. The sect joining the mid-points of the non-parallel sides of a trapezoid is half their sum.

60. If E and F are the mid-points of the opposite sides, AD, BC, of a parallelogram $ABCD$, the lines BE, DF, trisect the diagonal AC.

61. Any line drawn through the intersection of the diagonals of a parallelogram to meet the sides, bisects the surface.

62. The squares described on the two diagonals of a rhombus are together equivalent to the squares on the four sides.

63. Bisect a given parallelogram by a line passing through any given point.

64. In 244, what two rotations might be substituted for the two translations?

THEOREM XLI.

ALTERNATIVE PROOF.

250. *All parallelograms having equal bases and equal altitudes are equivalent.*

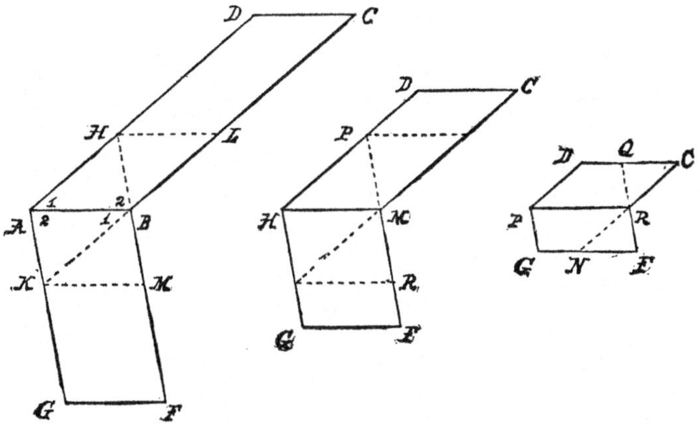

HYPOTHESIS. ***Two ▱s with equal bases and equal altitudes.***

CONCLUSION. They can be cut into parts congruent in pairs.

CONSTRUCTION. Place the parallelograms on opposite sides of their coincident equal bases, *AB*. Produce a side, as *FB*, which when continued will enter the other parallelogram. If it cuts out of the parallelogram at *H* before reaching the side *CD* opposite *AB*, then will the other cutting side, as *CB*, when produced, also leave the ▱ *ABFG* before reaching the side *FG* opposite *AB*; that is, the point, *K*, where *CB* cuts the line *AG* will be on the sect *AG*.

For, through *H* and *K* draw *HL* and *KM* ∥ *AB*.

Then, in △s *ABH* and *ABK*

$$\measuredangle 1 = \measuredangle 1, \quad \measuredangle 2 = \measuredangle 2,$$

(168. If a transversal cuts two parallels, the alternate angles are equal.)

and side *AB* is common;

$$\therefore \quad \triangle ABH \cong \triangle ABK.$$

But altitude of $\triangle AHB <$ altitude of $\square ABCD$,

$\therefore$ altitude of $\triangle ABK <$ altitude of $\square ABFG$, and K is on sect AG.

Now, $\triangle ABH \cong \triangle BHL$, and $\triangle ABK \cong \triangle BKM$.

(217. The diagonal of a parallelogram makes two congruent triangles.)

Taking away these four congruent triangles, we have left two parallelograms, *HLCD* and *KMFG*, with equal bases, and equal but diminished altitudes. Treat these in the same way as the parallelograms first given; and so continue until a produced side, as *FRQ*, and so the other, *CRN*, also, reaches the side opposite the base before leaving the parallelogram.

Then, as before, the $\triangle$s *FRN* and *QRC* are mutually equiangular; but now we know their corresponding altitudes are equal.

$\therefore$ by 241, they are congruent,

$\therefore FN = CQ$,

$\therefore GN = DQ$.

Also, $QR = RF = PG$, and $NR = RC = PD$,

$$\measuredangle G = \measuredangle FRP = \measuredangle RQD,$$
$$\measuredangle D = \measuredangle CRP = \measuredangle RNG,$$
$$\measuredangle DPR = \measuredangle PRN,$$
$$\measuredangle GPR = \measuredangle PRQ;$$

and therefore the remaining trapezoids, $PRQD \cong PRNG$.

(210. If two polygons be mutually equilateral and mutually equiangular, they are congruent.)

251. Corollary. Since a rectangle is a parallelogram, therefore a parallelogram is equivalent to the rectangle of its base and altitude.

Exercises. 65. How do you know that *HLCD* and *KMFG* have equal altitudes?

66. How do you know, that, if Q is on the sect CD, N is on the sect FG?

THEOREM XLII.

252. *A triangle is equivalent to half the rectangle of its base and altitude.*

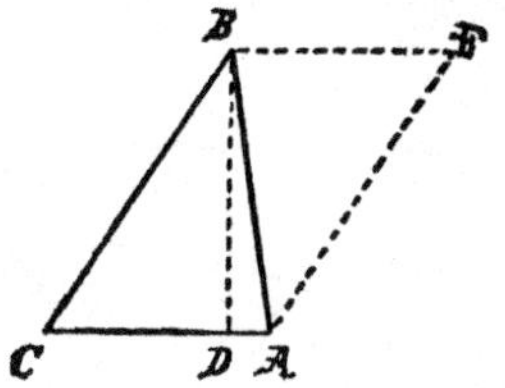

HYPOTHESIS. $\triangle ABC$, *with base* AC *and altitude* BD.

CONCLUSION. $\triangle ABC$ = half the rectangle of AC and BD.

PROOF. Through A draw $AF \parallel CB$, and through B draw $BF \parallel CA$, meeting AF in F;

$$\therefore \quad \triangle ACB \cong \triangle BFA.$$

(217. The diagonal of a parallelogram bisects it.)

But $\square AFBC$ = rectangle of AC and BD,

(248. A parallelogram is equivalent to the rectangle of its base and altitude.)

$$\therefore \quad \triangle ABC = \tfrac{1}{2} \square AFBC = \tfrac{1}{2} \text{ rectangle of } AC \text{ and } BD.$$

253. COROLLARY I. All triangles on the same base having their vertices in the same line parallel to the base, are equivalent.

254. COROLLARY II. Triangles having their vertices in the same point, and for their bases equal sects of the same line, are equivalent.

255. COROLLARY III. If a parallelogram and a triangle be upon the same base and between the same parallels, the parallelogram is double the triangle.

EXERCISES. 67. Equivalent triangles on equal bases have equal altitudes.

THEOREM XLIII.

256. *If through any point on the diagonal of a parallelogram two lines be drawn parallel to the sides, the two parallelograms, one on each side of the diagonal, will be equivalent.*

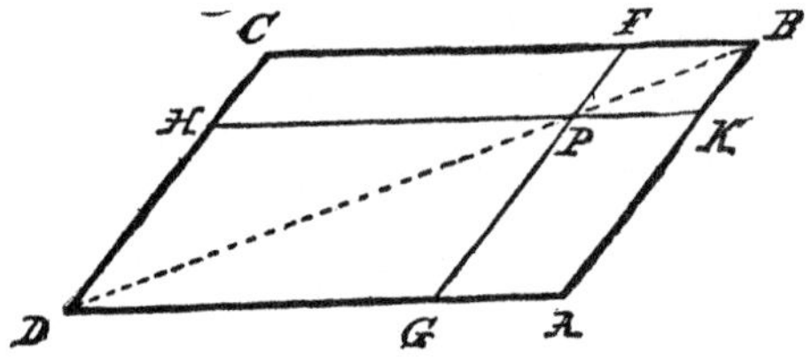

HYPOTHESIS. *P any point on diagonal BD of ▱ ABCD; FG and HK lines through P ∥ AB and BC respectively, and meeting the four sides in the points F, G, H, K.*

CONCLUSION. ▱ *AKPG* = ▱ *PFCH.*

PROOF. △ *ABD* = △ *BCD.*

△ *KBP* = △ *BFP.*

△ *GPD* = △ *PHD.*

(217. The diagonal of a parallelogram bisects it.)

From △ *ABD* take away △ *KBP* and △ *GPD*, and we have left ▱ *AKPG.* From the equal △ *BCD* take away the equal △s *BFP* and *PHD*, and we have left the ▱ *PFCH.*

∴ ▱ *AKPG* = ▱ *PFCH.*

(89. If equals be taken from equals, the remainders will be equal.)

257. In figures like the preceding, parallelograms like *KBFP* and *GPHD* are called parallelograms about the diagonal *BD*; while ▱s *AKPG* and *PFCH* are called complements of parallelograms about the diagonal *BD.*

EXERCISES. 68. When are the complements of the parallelograms about a diagonal of a parallelogram congruent?

V. Problems.

Problem XII.

258. *To describe a parallelogram equivalent to a given triangle, and having an angle equal to a given angle.*

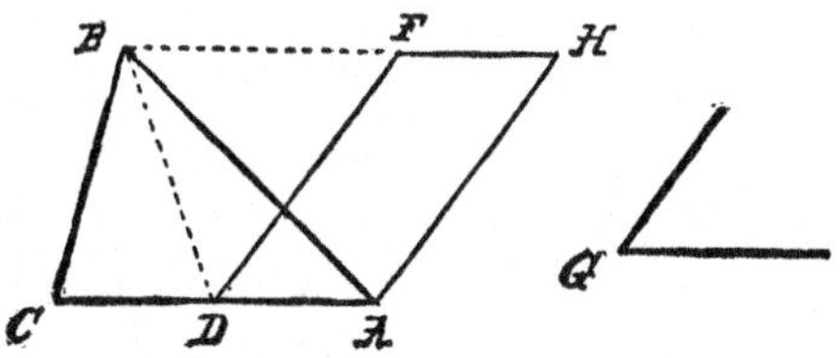

Given, △ *ABC and* ∡ *G.*

Required, *to describe a parallelogram* = △ ABC, *while having an angle* = ∡ G.

Construction. Bisect *AC* in *D*.

At *D*, by 164, make ∡ *ADF* = ∡ *G.*

By 167, through *A* draw *AH* ∥ *DF.*

By 167, through *B* draw *BFH* ∥ *CA.*

AHFD will be the parallelogram required.

Proof. Join *DB*. Then △ *ABD* = △ *BCD,*

(254. Triangles having their vertices in same point, and for bases equal sects of the same line, are equivalent.)

∴ △ *ABC* is double △ *ABD.*

But ▱ *AHFD* is also double △ *ABD,*

(255. If a parallelogram and a triangle be upon the same base and between the same parallels, the parallelogram is double the triangle.)

∴ ▱ *AHFD* = △ *ABC,*

and, by construction, ∡ *ADF* = ∡ *G.*

PROBLEM XIII.

259. *On a given sect as base, to describe a parallelogram equivalent to a given triangle, and having an angle equal to a given angle.*

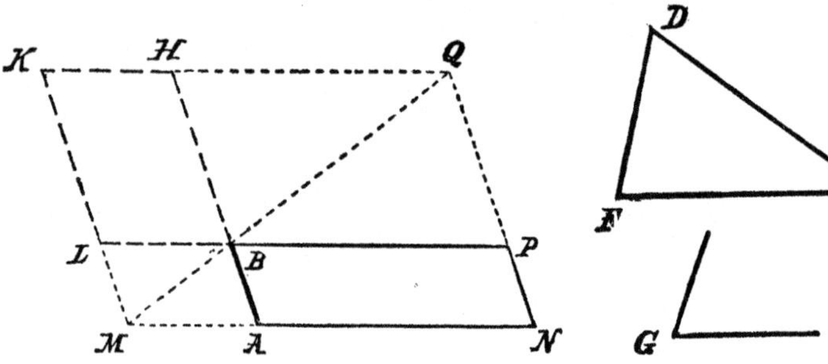

GIVEN, **the sect** AB, $\triangle\ CDF$, $\measuredangle\ G$.

REQUIRED, *to describe on* AB *a parallelogram* $=\triangle$ CDF, *while having an angle* $=\measuredangle$ G.

CONSTRUCTION. By 258, make $\square\ BHKL = \triangle\ CDF$, and having $\measuredangle\ HBL = \measuredangle\ G$, and place it so that BH is on line AB produced.

Produce KL. Draw $AM \parallel BL$. Join MB.

$$\measuredangle\ HKM + \measuredangle\ KMA = \text{st.}\ \measuredangle,$$

(169. If a transversal cuts two parallels, it makes the two interior angles supplemental.)

$$\therefore\ \measuredangle\ HKM + \measuredangle\ KMB < \text{st.}\ \measuredangle.$$

Therefore KH and MB meet if produced through H and B.

(171. If a transversal cuts two lines, and the interior angles are not supplemental, the lines meet.)

Let them meet in Q. Through Q draw $QN \parallel HA$, and produce LB and MA to meet QN in P and N.

$$\therefore\ \square\ NPBA = \square\ BHLK,$$

(256. Complements of parallelograms about the diagonal are equivalent.)

$$\therefore\ \square\ NPBA = \triangle\ CFD, \quad \text{and} \quad \measuredangle\ ABP = \measuredangle\ G.$$

260. COROLLARY. Thus we can describe on a given base a rectangle equivalent to a given triangle.

PROBLEM XIV.

261. *To describe a triangle equivalent to a given polygon.*

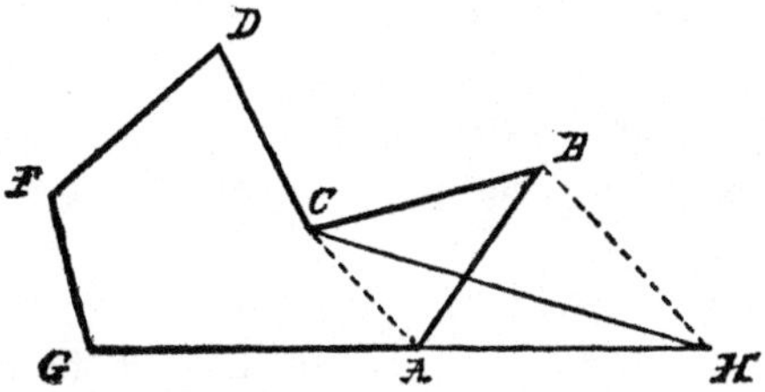

GIVEN, *a polygon ABCDFG.*

REQUIRED, *to construct an equivalent triangle.*

CONSTRUCTION. Join the ends of any pair of adjacent sides, as *AB* and *BC*, by the sect *CA*.

Through the intermediate vertex, *B*, draw a line ∥ *CA*, meeting *GA* produced in *H*. Join *HC*.

Polygon *ABCDFGA* = polygon *HCDFGH*,

and we have obtained an equivalent polygon having fewer sides.

PROOF. △ *ABC* = △ *AHC*.

(253. Triangles having the same base and equal altitudes are equivalent.)

Add to each the polygon *ACDFGA*,

∴ *ABCDFGA* = *HCDFGH*.

In the same way the number of sides may be still further diminished by one until reduced to three.

262. COROLLARY I. Hence we can describe on a given base a parallelogram equivalent to a given polygon, and having an angle equal to a given angle.

263. COROLLARY II. Thus we can describe on a given base a rectangle equivalent to a given polygon.

264. REMARK. To compare the surfaces of different polygons, we need only to construct rectangles equivalent to the given polygons, and all on the same base.

Then, by comparing the altitudes, we are enabled to judge of the surfaces.

VI. Axial and Central Symmetry.

265. If two figures coincide, every point A in the one coincides with a point A' in the other. These points are said to correspond.

Hence to every point in one of two congruent figures there corresponds one, and only one, point in the other; those points being called "corresponding" which coincide if one of the two figures is superimposed upon the other. Hence, calling those parts corresponding which coincide if the whole figures are made to coincide, it follows, that corresponding parts of congruent figures are themselves congruent.

Symmetry with Regard to an Axis.

266. If we start with two figures in the position of coincidence, and take in the common plane any line x, we may turn the plane of one figure about this line x until its plane, after half a revolution, coincides again with the plane of the other figure.

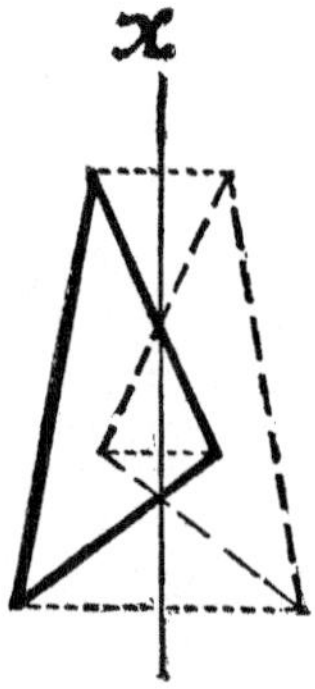

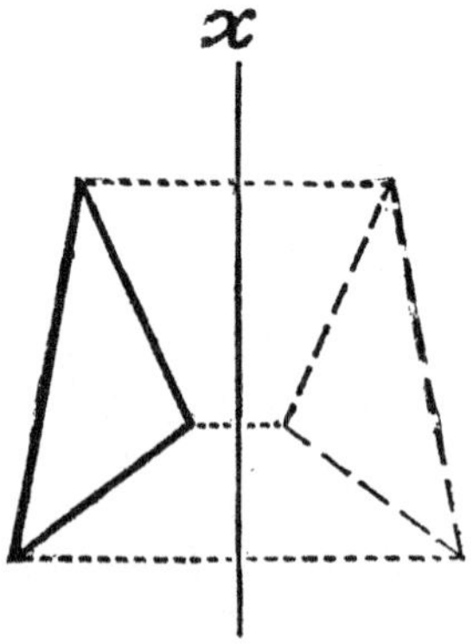

The two figures themselves will then have distinct positions in the same plane; but they will have this property, that they can be made to coincide by folding the plane over along the line x.

Two figures in the same plane which have this property are said to be symmetrical with regard to the line x as an axis of symmetry.

SYMMETRY WITH REGARD TO A CENTER.

267. If we take in the common plane of two coincident figures any point X, we may turn the one figure about this point so that its plane slides over the plane of the other figure without ever separating from it.

Let this turning be continued until one line to X, and therefore the whole moving figure, has been turned through a straight angle about X.

Then the two congruent figures still lie in the same plane, and have such positions that one can be made to coincide with the other by turning it in the plane through a straight angle about the fixed point X.

Two figures which have this property are said to be symmetrical with regard to the point X as center of symmetry.

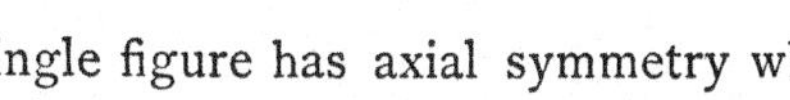

268. Any single figure has axial symmetry when it can be divided by an axis into two figures symmetrical with respect to that axis, or has central symmetry when it has a center such that every line drawn through it cuts the figure in two points symmetrical with respect to this center.

THEOREM XLIV.

269. *If a figure has two axes of symmetry perpendicular to each other, then their intersection is a center of symmetry.*

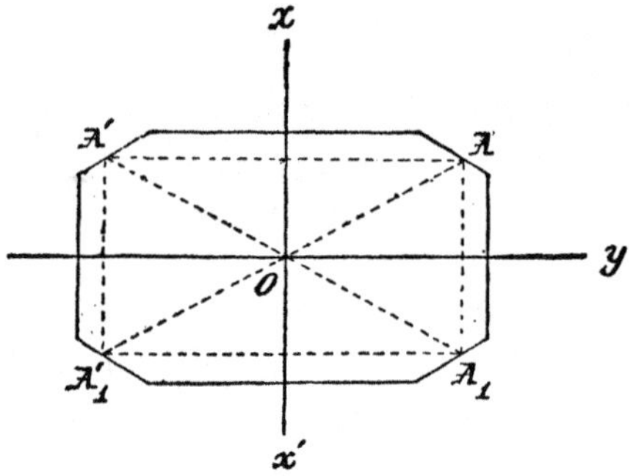

For, if x and y be two axes at right angles, then to a point A will correspond a point A' with regard to x as axis.

To these will correspond points A_1 and A_1' with regard to y as axis. These points A_1 and A_1', will correspond to each other with regard to x. To see this, let us first fold over along y; then A falls on A_1, and A' on A_1'.

If we now, without folding back, fold over along x, A, and with it A_1, will fall on A', which coincides with A_1'.

At the same time OA and OA_1' coincide, so that the angles AOx and $A_1'Ox'$ are equal, where x' denotes the continuation of x beyond O. It follows, that AOA_1' are in a line, and that the sect AA_1' is bisected at O, or O is a center of symmetry for AA_1', and similarly for A_1 and A'.

BOOK II.

RECTANGLES.

270. A *Continuous Aggregate* is an assemblage in which two adjacent parts have the same boundary.

271. A *Discrete* or *Discontinuous Aggregate* is one in which two adjacent parts have different boundaries.

A pile of cannon balls is a discrete aggregate. We know that any adjacent two could be painted different colors, and so they have direct independent boundaries.

Our fingers are a discontinuous aggregate.

272. All counting belongs first to the fingers.

273. There is implied and bound up in the word "number" the assumption that a group of things comes ultimately to the same finger, in whatever order they are counted.

This proposition is involved in the meaning of the phrase "distinct things."

Any one and any other of them make two. If they are attached to two of my fingers in a certain order, they can also be attached to the same fingers in the other order. Thus, one order of a group of three distinct things can be changed into any other order while using the same fingers, and so on with a group of four, etc.

274. By generalizing the use of the fingers in counting, man has made for himself a counting apparatus, which each one carries around in his mind. This counting apparatus, the natural series of numbers, was made from a discrete aggregate, and so will only correspond exactly to discrete aggregates.

275. In a row of shot, we can find between any two, only a finite number of others, and sometimes none at all.

Just so in regard to any two numbers. A row of six shot can be divided into two equal parts; but the half, which is three, we cannot divide into two equal parts: and so in a series of numbers.

276. But in 136 we have shown how any sect whatever may be bisected, and the bisection point is the boundary of both parts. So a line is not a discrete aggregate of points. It is something totally different in kind from the natural series of numbers.

277. The science of numbers is founded on the hypothesis of the distinctness of things The science of space is founded on the entirely different hypothesis of continuity.

278. Numbers are essentially discontinuous, and therefore unsuited to express the relations of continuous magnitudes.

279. In arithmetic we are taught to add and multiply numbers: we will now show how the laws for the addition and multiplication of these discrete aggregates are applicable to sects, which are continuous aggregates.

THE COMMUTATIVE LAW FOR ADDITION.

280. In a sum of numbers we may change the order in which the numbers are added.

If x and y represent numbers, this law is expressed by the equation

$$x + y = y + x.$$

It depends entirely on the interchangeability of any pair of the units of numeration.

281. The sum of two sects is the sect obtained by placing them on the same line so as not to overlap, with one end point in common.

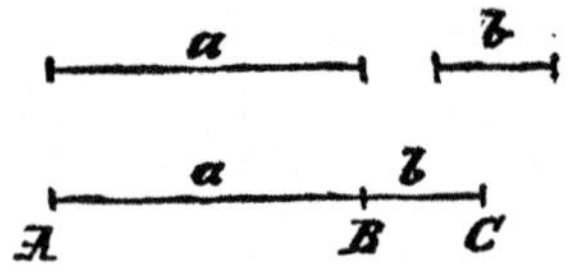

Thus, the sum of the sects a and b means the sect AC, which can be divided into two parts,

$$AB = a, \quad \text{and} \quad BC = b.$$

282. The commutative law holds for the summation of sects.

$$a + b = b + a.$$

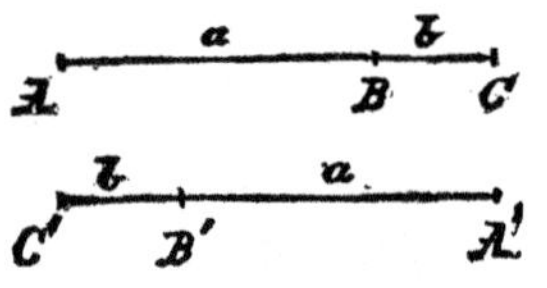

$$AC = C'A';$$

for AC revolved through a straight angle may be superimposed upon $C'A'$, and will coincide point for point.

The more general case, where three or more sects are added, follows from a repetition of the above.

Thus, the commutative law for addition in geometry depends entirely on the possibility of motion without deformation.

283. The sum of two rectangles is the hexagon formed by superimposing two sides, and bringing the bases into the same line.

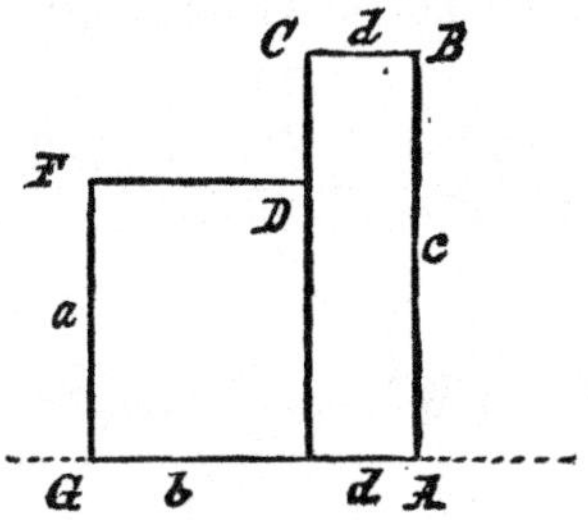

Thus, if two adjacent sides of one are *a* and *b*, and of the other *c* and *d*, the sum of the rectangles *ab* and *cd* is *ABCDFGA*.

284. The commutative law holds for the addition of rectangles; that is, the sum is independent of the order of summation.

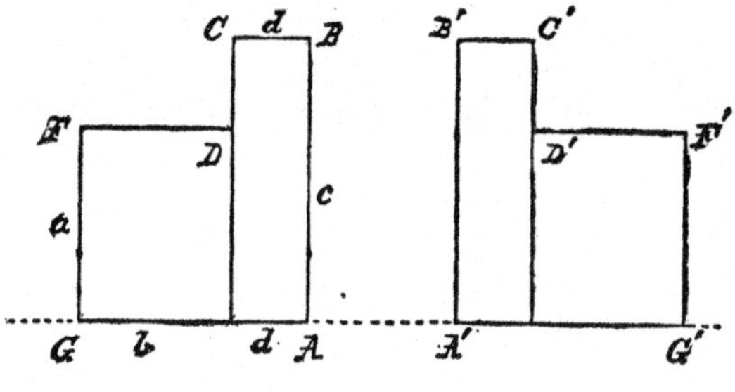

$$ab + cd = cd + ab;$$

for *ABCDFGA* turned over may be superimposed upon *A'B'C'D'F'G'A'*, and will coincide with it.

Since, by 263, we can describe on a given base a rectangle equivalent to a given polygon, the more general case, where three or more rectangles are added, follows from a repetition of the above.

THE ASSOCIATIVE LAW.

285. In getting a sum of numbers, we may add the numbers together in groups, and then add these groups.

If we use parentheses to mean that the terms enclosed have been added together before they are added to another term, this law may be expressed symbolically by the equation

$$x + (y + z) = x + y + z.$$

286. The associative law holds for the summation of sects.

$$a + (b + c) = a + b + c = AD.$$

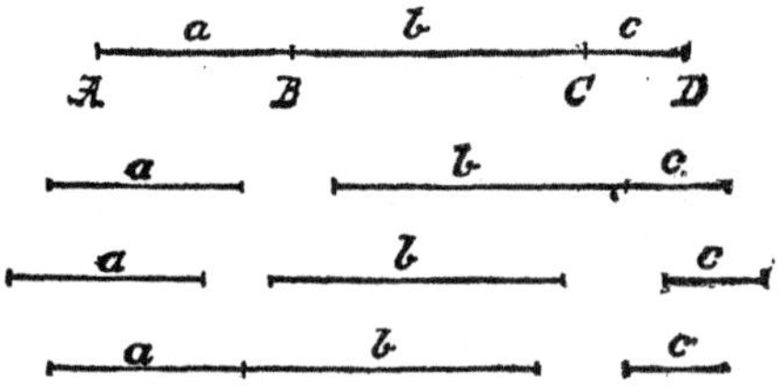

287. The associative law holds for the summation of rectangles.

$$ad + (bf + cg) = ad + bf + cg.$$

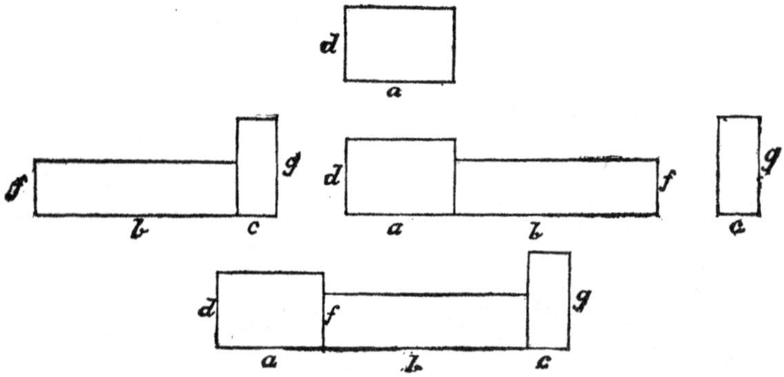

THE COMMUTATIVE LAW FOR MULTIPLICATION.

288. The product of numbers remains unaltered if the factors be interchanged.

$$xy = yx.$$

289. The commutative law holds for the rectangle of two sects.

If a and b are any two sects, rectangle ab = rectangle ba, or

$$ab = ba,$$

for rectangle ab may be so applied to rectangle ba as to coincide with it.

THE DISTRIBUTIVE LAW.

290. To multiply a sum of numbers by a number, we may multiply each term of the sum, and add the products thus obtained.

$$x(y + z) = xy + xz.$$

291. The distributive law holds when for numbers and products we substitute sects and rectangles.

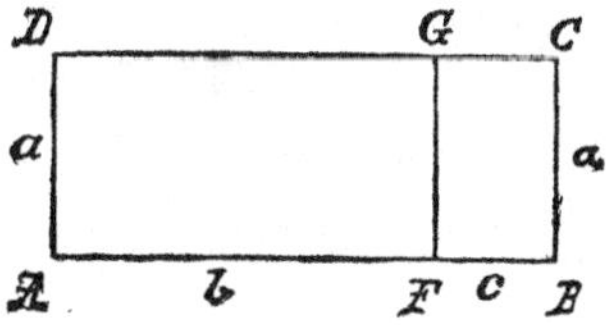

$$a(b + c) = ab + ac;$$

for if we add the rectangle ab to the rectangle ac, so that a side a in the one shall coincide with an equal side a in the other, the sum makes a rectangle whose base is $b + c$ and whose altitude is a; that is, the rectangle $a(b + c)$.

In the same way, by adding three rectangles of the same altitude, we get

$$a(b + c + a) = ab + ac + ad.$$

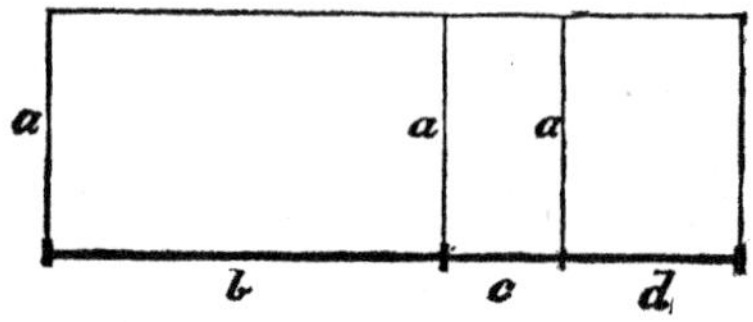

We may state this in words as follows:

If there be any two sects one of which is divided into any number of parts, the rectangle contained by the two sects is equivalent to the rectangles contained by the undivided sect and the several parts of the divided sect.

292. If $b + c = a$, then

$$ab + ac = a(b + c) = aa = a^2.$$

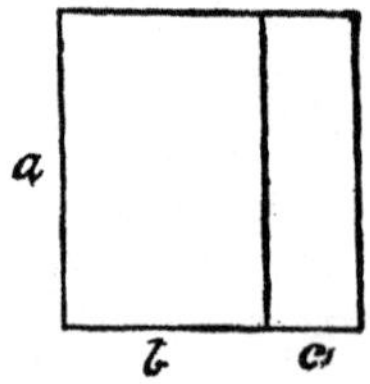

Therefore

If a sect be divided into any two parts, the rectangles contained by the whole and each of the parts are together equivalent to the square on the whole sect.

293. If $c = a$, then

$$a(b + c) = a(b + a) = ab + aa = ab + a^2.$$

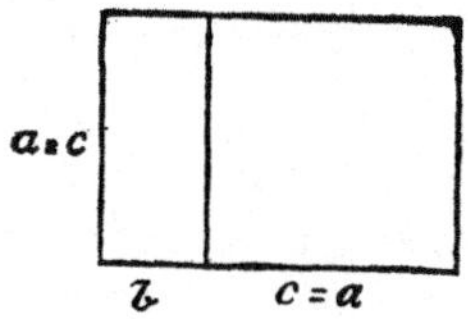

Therefore

If a sect be divided into any two parts, the rectangle contained by the whole and one of the parts is equivalent to the rectangle contained by the two parts, together with the square on the aforesaid part.

294. The rectangle of two equal sects is a square, and $(a + b)^2$ is only a condensed way of writing $(a + b)(a + b)$.

But, by the distributive law,

$$(a + b)(a + b) = (a + b)a + (a + b)b.$$

By the commutative law,

$$(a + b)a = a(a + b).$$

By the distributive law,

$$a(a + b) = aa + ab = a^2 + ab.$$

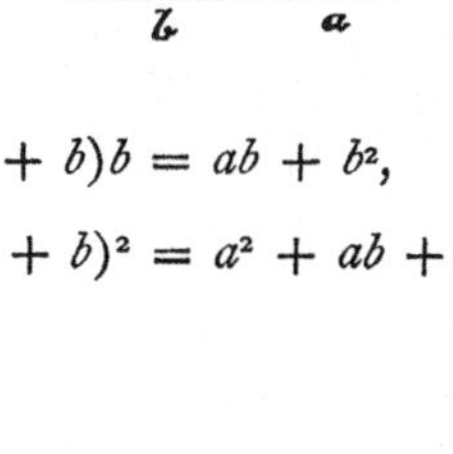

In the same way,

$$(a + b)b = ab + b^2,$$

$$\therefore\ (a + b)^2 = a^2 + ab + (ab + b^2).$$

By the associative law,

$$a^2 + ab + (ab + b^2) = a^2 + (ab + ab) + b^2 = a^2 + 2ab + b^2,$$

$$\therefore \quad (a + b)^2 = a^2 + 2ab + b^2.$$

Therefore

If a sect be divided into any two parts, the square on the whole sect is equivalent to the squares on the two parts, together with twice the rectangle contained by the two parts.

295. COROLLARY. The square on a sect is four times the square on half the sect.

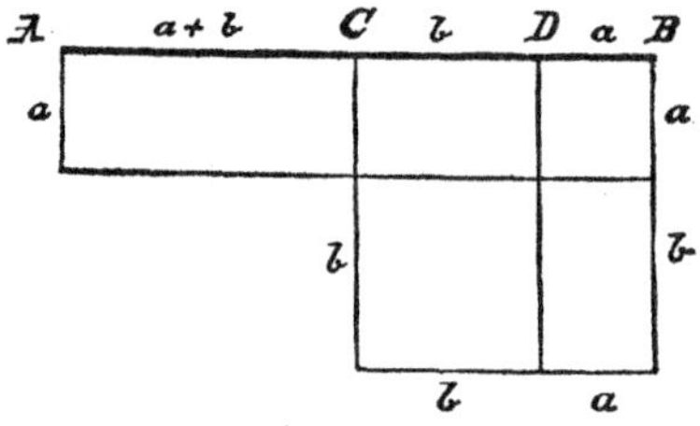

296. By the distributive law, and 294,

$$a(a + b + b) + b^2 = a^2 + ab + ab + b^2 = (a + b)^2.$$

From this, if we bisect the sect AB at C, and divide it into two unequal parts at D, and for BD put a, and for DC put b, we get the theorem:

If a sect is divided into two equal parts, and also into two unequal parts, the rectangle contained by the unequal parts, together with the square on the line between the points of section, is equivalent to the square on half the sect.

297. If $CD = b$, and $AC = a + b$, then their sum $AD = a + b + b$, and their difference is a; therefore the rectangle contained by their sum and difference equals $a(a + 2b)$, which, by 296, is the difference between $(a + b)^2$ and b^2.

Therefore

The rectangle contained by the sum and difference of any two sects is equivalent to the difference between the squares on those sects.

298. If we bisect the sect AB at C, and produce it to D, and for BD put a, and for BC put b, then the above equation, $a(a + b + b) + b^2 = (a + b)^2$, gives us the theorem:

If a sect be bisected, and produced to any point, the rectangle contained by the whole line thus produced and the part of it produced, together with the square on half the sect bisected, is equivalent to the square on the line which is made up of the half and the part produced.

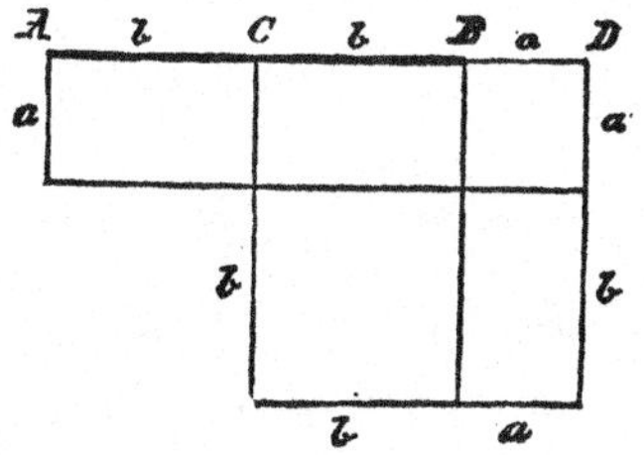

299. By 294,

$$(a + b)^2 + a^2 = a^2 + 2ab + b^2 + a^2.$$

By the associative law,

$$a^2 + 2ab + b^2 + a^2 = 2a^2 + 2ab + b^2.$$

By the distributive law,

$$2a^2 + 2ab + b^2 = 2a(a + b) + b^2.$$

By the commutative law,

$$2a(a + b) + b^2 = 2(a + b)a + b^2,$$

$$\therefore \quad (a + b)^2 + a^2 = 2(a + b)a + b^2.$$

Therefore

If a sect be divided into any two parts, the squares on the whole line and on one of the parts are equivalent to twice the rectangle contained by the whole and that part, together with the square on the other part.

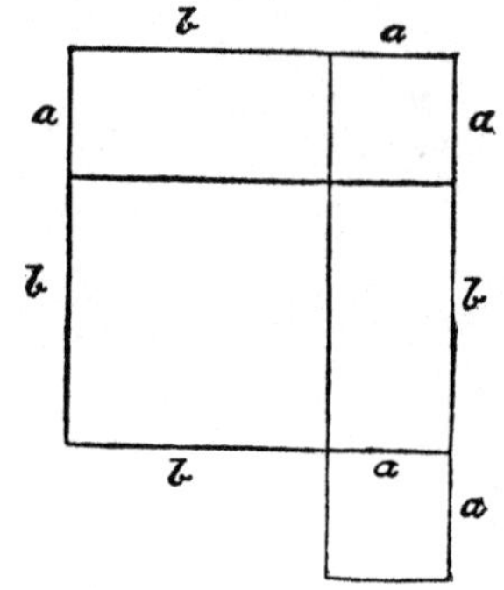

300. By the commutative and distributive laws, and 294,

$$4(a+b)a+b^2=4a^2+4ab+b^2=(2a+b)^2.$$

By the associative and commutative laws,

$$(2a+b)^2=(a+a+b)^2=([a+b]+a)^2,$$

$$\therefore\ 4(a+b)a+b^2=([a+b]+a)^2.$$

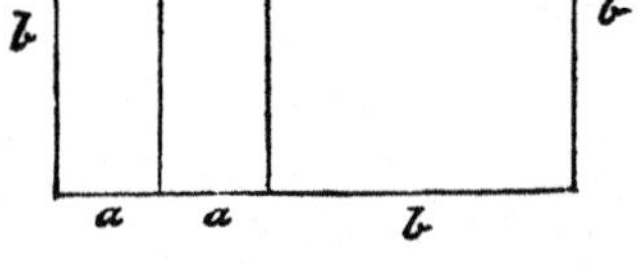

Therefore

If a sect be divided into any two parts, four times the rectangle contained by the whole sect and one of the parts, together with the square on the other part, is equivalent to the square on the line which is made up of the whole sect and the first part.

301. By the associative and commutative laws, and 294,

$$([a+b]+a)^2+b^2=(b+2a)^2+b^2$$
$$=b^2+4ab+4a^2+b^2=2a^2+4ab+2b^2+2a^2.$$

By the distributive law, and 294,

$$2a^2+4ab+2b^2+2a^2=2(a^2+2ab+b^2)+2a^2=2(a+b)^2+2a^2,$$

$$\therefore\quad ([a+b]+a)^2+b^2=2(a+b)^2+2a^2.$$

Therefore

If a sect be divided into two equal and also two unequal parts, the squares on the two unequal parts are together double the squares on half the sect and on the line between the points of section.

302. By 294, and the associative and distributive laws,

$$(a+b)^2+b^2=a^2+2ab+b^2+b^2$$
$$=\frac{a^2}{2}+\left(\frac{a^2}{2}+2ab+2b^2\right)=2\left(\frac{a}{2}\right)^2+2\left(\frac{a}{2}+b\right)^2.$$

Therefore

If a sect be bisected and produced to any point, the square on the whole line thus produced and the square on the part of it produced are together double the squares on half the sect and on the line made up of the half and the part produced.

303. The projection of a point on a line is the foot of the perpendicular from the point to the line.

304. The projection of a sect upon a line is the part between the perpendiculars dropped upon the line from the ends of the sect.

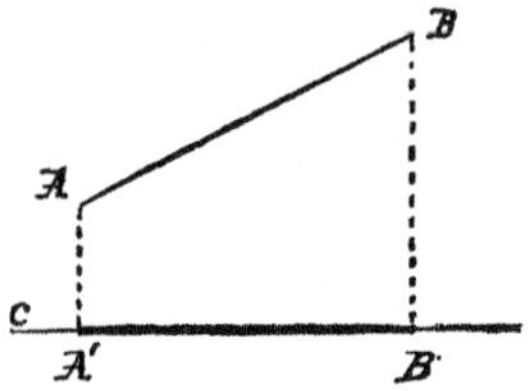

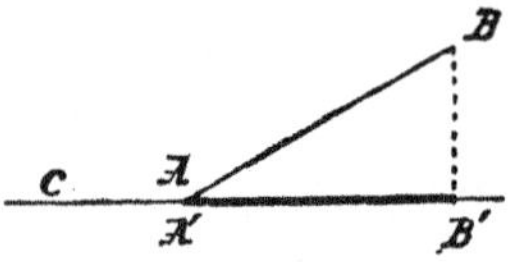

For example, $A'B'$ is the projection of the sect AB on the line c.

Theorem I.

305. *In an obtuse-angled triangle, the square on the side opposite the obtuse angle is greater than the sum of the squares on the other two sides by twice the rectangle contained by either of those sides and the projection of the other side upon it.*

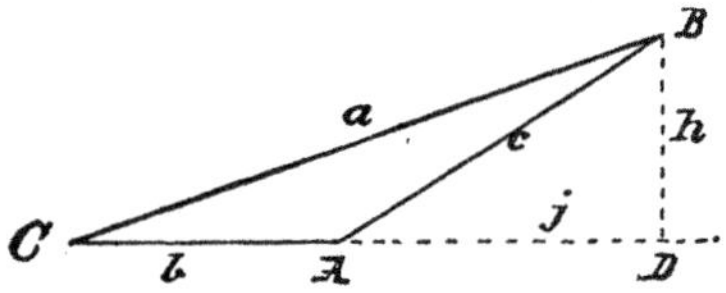

Hypothesis. $\triangle ABC$, with $\measuredangle CAB$ obtuse.

Conclusion. $a^2 = b^2 + c^2 + 2bj$.

Proof. By 294,

$$(b + j)^2 = b^2 + 2bj + j^2.$$

Adding h^2 to both sides,

$$(b + j)^2 + h^2 = b^2 + 2bj + j^2 + h^2.$$

But

$$(b + j)^2 + h^2 = a^2, \quad \text{and} \quad j^2 + h^2 = c^2,$$

(243. In a right triangle, the square of the hypothenuse equals the sum of the squares of the other sides.)

$$\therefore \quad a^2 = b^2 + 2bj + c^2.$$

THEOREM II.

306. *In any triangle, the square on a side opposite any acute angle is less than the sum of the squares on the other two sides by twice the rectangle contained by either of those sides and the projection of the other side upon it.*

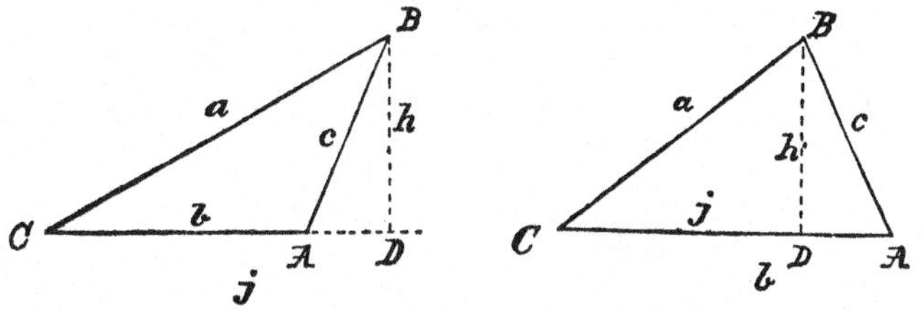

HYPOTHESIS. $\triangle ABC$, with $\measuredangle C$ acute.

CONCLUSION. $c^2 + 2bj = a^2 + b^2$.

PROOF. By 295,

$$b^2 + j^2 = 2bj + \overline{AD}^2.$$

Adding h^2 to both sides,

$$b^2 + j^2 + h^2 = 2bj + \overline{AD}^2 + h^2.$$

But

$$j^2 + h^2 = a^2, \quad \text{and} \quad \overline{AD}^2 + h^2 = c^2,$$

(243. In a right triangle, the square of the hypothenuse equals the sum of the squares of the other sides.)

$$\therefore \; b^2 + a^2 = 2bj + c^2.$$

307. Having now proved that in a triangle,

By 243, if $\measuredangle A = \text{rt.} \measuredangle$, $\therefore \; a^2 = b^2 + c^2$;
By 305, if $\measuredangle A > \text{rt.} \measuredangle$, $\therefore \; a^2 > b^2 + c^2$;
By 306, if $\measuredangle A < \text{rt.} \measuredangle$, $\therefore \; a^2 < b^2 + c^2$.

Therefore, by 33, Rule of Inversion,

If $a^2 = b^2 + c^2$, $\therefore \; \measuredangle A = \text{rt.} \measuredangle$;
If $a^2 > b^2 + c^2$, $\therefore \; \measuredangle A > \text{rt.} \measuredangle$;
If $a^2 < b^2 + c^2$, $\therefore \; \measuredangle A < \text{rt.} \measuredangle$.

THEOREM III.

308. *In any triangle, if a medial be drawn from the vertex to base, the sum of the squares on the two sides is equivalent to twice the square on half the base, together with twice the square on the medial.*

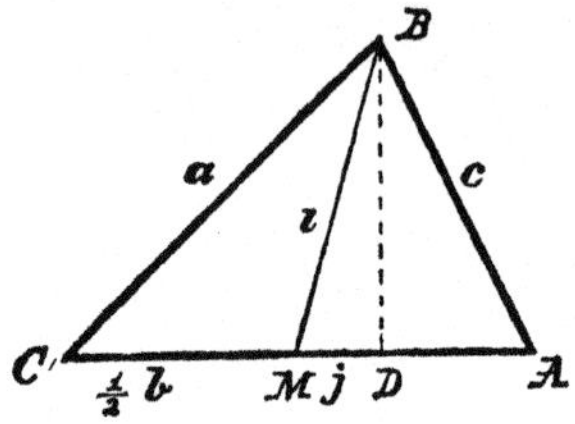

HYPOTHESIS. $\triangle ABC$, *with medial* $BM = i$.

CONCLUSION. $a^2 + c^2 = 2(\frac{1}{2}b)^2 + 2i^2$.

PROOF.—CASE I. If $\angle BMA = \angle BMC$, then

$$a^2 = (\tfrac{1}{2}b)^2 + i^2, \quad \text{and} \quad c^2 = (\tfrac{1}{2}b)^2 + i^2.$$

(243. In a right triangle, the square of the hypothenuse equals the sum of the squares of the other sides.)

CASE II. If $\angle BMA$ does not equal $\angle BMC$, one of them must be the greater. Call the greater BMC.

Then, in the obtuse-angled triangle BMC, by 305,

$$a^2 = (\tfrac{1}{2}b)^2 + i^2 + 2(\tfrac{1}{2}bj).$$

In $\triangle BMA$, by 306,

$$c^2 + 2(\tfrac{1}{2}bj) = (\tfrac{1}{2}b)^2 + i^2.$$

Adding,

$$a^2 + c^2 + 2(\tfrac{1}{2}bj) = 2(\tfrac{1}{2}b)^2 + 2i^2 + 2(\tfrac{1}{2}bj),$$

$$\therefore \quad a^2 + c^2 = 2(\tfrac{1}{2}b)^2 + 2i^2.$$

309. COROLLARY. The difference of the squares on the two sides is equivalent to twice the rectangle of the base and the projection of the medial on it.

THEOREM IV.

310. *The sum of the squares on the four sides of any quadrilateral is greater than the sum of the squares on the diagonals by four times the square on the sect joining the mid-points of the diagonals.*

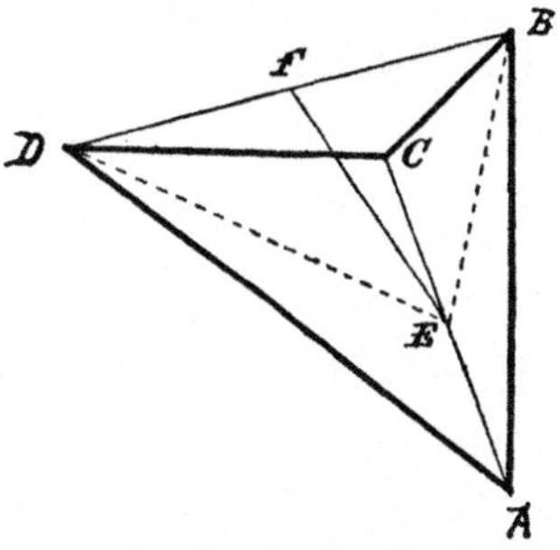

HYPOTHESIS. ***EF is the sect joining the mid-points of the diagonals AC and BD of the quadrilateral ABCD.***

CONCLUSION. $\overline{AB}^2 + \overline{BC}^2 + \overline{CD}^2 + \overline{DA}^2$
$= \overline{AC}^2 + \overline{BD}^2 + 4\overline{EF}^2.$

PROOF. Draw BE and DE.

By 308,

$$\overline{AB}^2 + \overline{BC}^2 = 2\left(\frac{AC}{2}\right)^2 + 2\overline{BE}^2,$$

and

$$\overline{CD}^2 + \overline{DA}^2 = 2\left(\frac{AC}{2}\right)^2 + 2\overline{DE}^2.$$

Adding,

$$\overline{AB}^2 + \overline{BC}^2 + \overline{CD}^2 + \overline{DA}^2 = 4\left(\frac{AC}{2}\right)^2 + 2\overline{BE}^2 + 2\overline{DE}^2.$$

But, by 308,

$$\overline{BE}^2 + \overline{DE}^2 = 2\left(\frac{BD}{2}\right)^2 + 2\overline{EF}^2,$$

$$\therefore \quad \overline{AB}^2 + \overline{BC}^2 + \overline{CD}^2 + \overline{DA}^2 = 4\left(\frac{AC}{2}\right)^2 + 4\left(\frac{BD}{2}\right)^2 + 4\overline{EF}^2$$
$$= \overline{AC}^2 + \overline{BD}^2 + 4\overline{EF}^2.$$

311. Corollary. The sum of the squares on the four sides of a parallelogram is equivalent to the sum of the squares on the diagonals, because the diagonals of a parallelogram bisect each other.

Problem I.

312. *To square any polygon.*

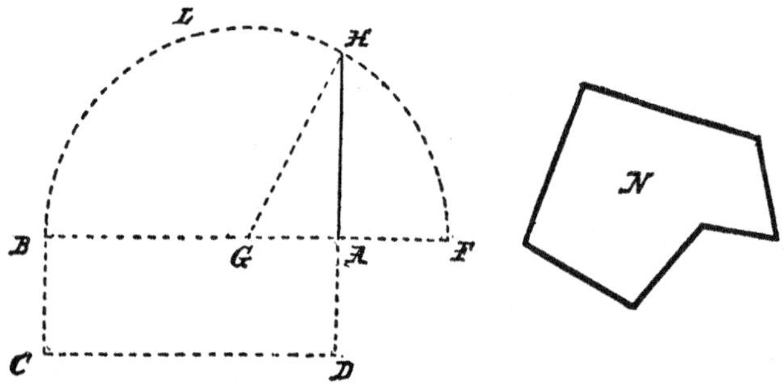

Given, **any polygon** *N*.

Required, *to describe a square equivalent to* N.

Construction. Describe, by 263, the rectangle $ABCD = N$.

Then, if $AB = BC$, the required square is *ABCD*.

If *AB* be not equal to *BC*, produce *BA*, and cut off $AF = AD$. Bisect *BF* in *G*, and, with center *G* and radius *GB*, describe *FLB*.

Produce *DA* to meet the circle in *H*.

The square on *AH* shall be equal to *N*.

Proof. Join *GH*. Then, by 296, because the sect *BF* is divided equally in *G* and unequally in *A*,

$$\therefore \text{ rectangle } BA,AF + \overline{AG}^2 = \overline{GF}^2 = \overline{GH}^2 = \overline{AH}^2 + \overline{AG}^2,$$

by 243;

$$\therefore \overline{AH}^2 = \text{rectangle } BA,AF = \text{rectangle } BA,AD = N.$$

Problem II.

313. *To divide a given sect into two parts so that the rectangle contained by the whole and one of the parts shall be equal to the square on the other part.*

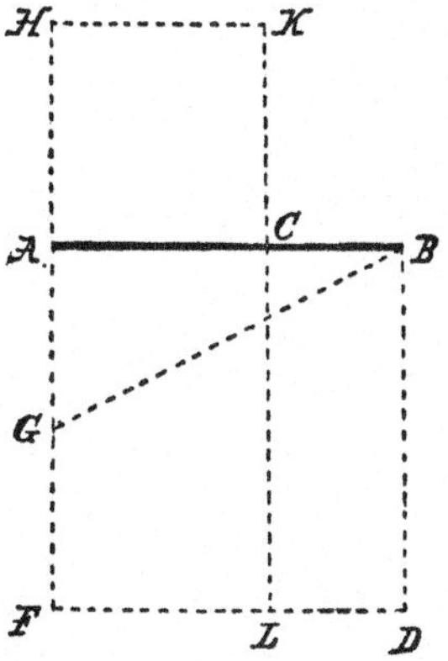

Given, *the sect AB.*

Required, *to find a point* C *such that rectangle* AB,BC $= \overline{AC}^2$.

Construction. On AB, by 241, describe the square $ABDF$.

By 136, bisect AF in G. Join BG.

Produce FA, and make $GH = GB$.

On AH describe the square $AHKC$.

Then $$AB,BC = \overline{AC}^2.$$

Proof. Produce KC to L. Then, by 298, because FA is bisected in G and produced to H, rectangle $FH,HA + \overline{AG}^2 = \overline{GH}^2 = \overline{GB}^2 = \overline{GA}^2 + \overline{AB}^2$.

$$\therefore\ FH,HA = \overline{AB}^2.$$

But, since $HK = HA$,

$$\therefore\ FH,HK = \overline{AB}^2.$$

Take from each the common part AL,

$$\therefore\ HC = CD;$$

that is,

$$\overline{AC}^2 = CB,BD = CB,BA.$$

PROBLEM III.

314. *To describe an isosceles triangle having each of the angles at the base double the angle at the vertex.*

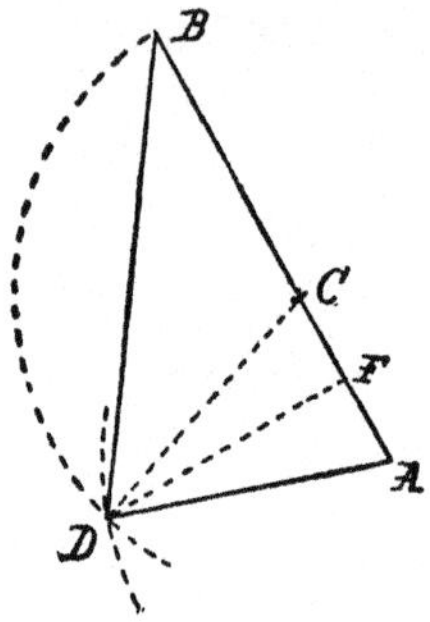

CONSTRUCTION. Take any sect, AB, and, by 313, divide it in C so that rectangle $AB,AC = \overline{BC}^2$. With center C and radius CB, describe a circle.

With same radius, but center A, describe a circle intersecting the preceding circle in D. Join AD, BD, CD.

ABD will be the triangle required.

PROOF. By construction, $CB = CD = DA$.

By 134, bisect $\measuredangle ADC$ by DF. By 137, DF is $\perp AC$, and bisects AC.

Then $$\overline{AB}^2 + \overline{AD}^2 = \overline{BD}^2 + 2AB,AF.$$

(306. The square on a side opposite an acute angle is less than the squares on the other two sides by twice the rectangle of either and the projection of the other on it.)

But $AC = 2AF$, $\quad \therefore \; AB,AC = AB,2AF = 2AB,AF$;

$$\therefore \text{ by our initial construction, } \overline{BC}^2 = 2AB,AF.$$

$$\therefore \; \overline{AB}^2 + \overline{AD}^2 = \overline{BD}^2 + \overline{BC}^2.$$

But, by construction, $AD = BC$, $\quad \therefore \; \overline{AB}^2 = \overline{BD}^2$, $\quad \therefore \; AB = BD$,

$$\therefore \; \measuredangle ADB = \measuredangle DAB = \measuredangle ACD = \measuredangle B + \measuredangle BDC = 2 \measuredangle B.$$

(173. The exterior angle of a triangle equals the sum of the two opposite interior angles.)

BOOK III.

THE CIRCLE.

I. Primary Properties.

315. If a sect turns about one of its end points, the other end point describes a curve called the *Circle.*

316. The fixed end point is called the *Center* of the circle.

317. The moving sect in any position is called a *Radius* of the circle.

318. As the motion of a sect does not enlarge or diminish it, all radii are equal.

319. Since the moving sect, after revolving through a perigon, returns to its original position, therefore the moving end point describes a *closed curve.*

This divides the plane into two surfaces, one of which is swept over by the moving sect. This finite plane surface is called the surface of the circle.

Any part of the circle is called an *Arc.*

THEOREM I.

320. *The sect to a point from the center of a circle is less than, equal to, or greater than, the radius, according as the point is within, on, or without the circle.*

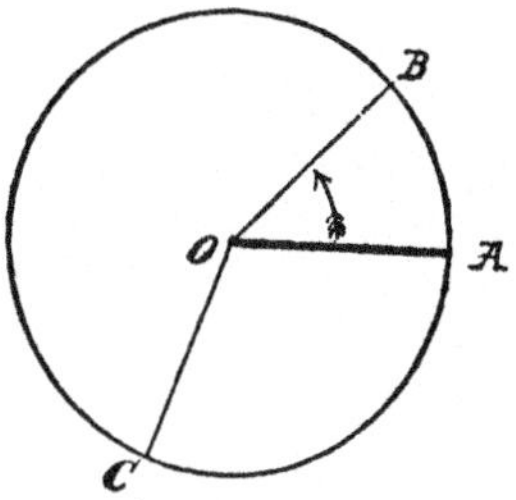

PROOF. If a point is on the circle, the sect drawn to it from the center is a radius, for it is one of the positions of the describing sect.

Any point, Q, within the circle lies on some radius, OQR,

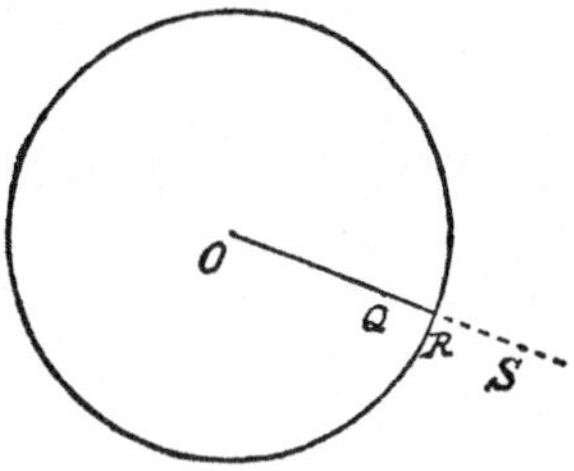

$$\therefore \quad OQ < OR.$$

If S is without the circle, then the sect OS contains a radius OR,

$$\therefore \quad OS > OR.$$

321. By 33, Rule of Inversion, a point is within, on, or without the circle according as its sect from the center is less than, equal to, or greater than, the radius.

322. A *Secant* is a line which passes through two points on the circle.

THEOREM II.

323. *A secant can meet the circle in only two points.*

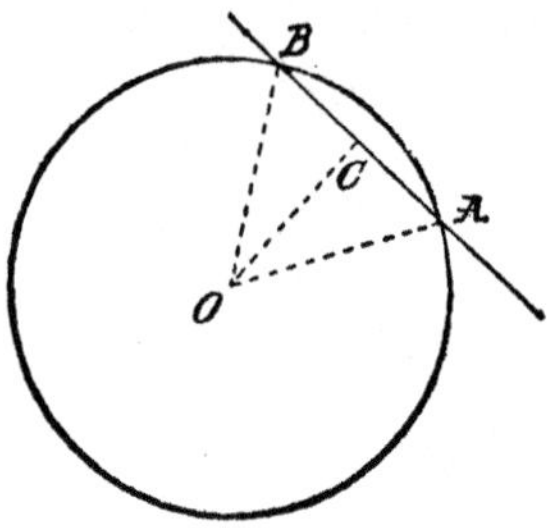

PROOF. By definition, all sects joining the center to points on the circle are equal, but from a point to a line there can be only two equal sects.

(155. No more than two equal sects can be drawn from a point to a line.)

324. A *Chord* is the part of a secant between the two points where it intersects the circle.

325. A *Segment* of a circle is the figure made by a chord and one of the two arcs into which the chord divides the circle.

326. When two arcs together make an entire circle, each is said to be the *Explement* of the other.

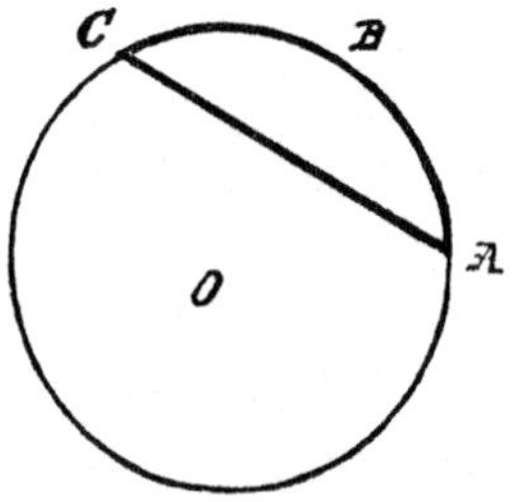

327. When two explemental arcs are equal, each is a *Semicircle.*

328. When two explemental arcs are unequal, the lesser is called the *Minor Arc*, and the greater is called the *Major Arc.*

329. A segment is called a *Major* or *Minor Segment* according as its arc is a major or minor arc.

THEOREM III.

330. *A circle has only one center.*

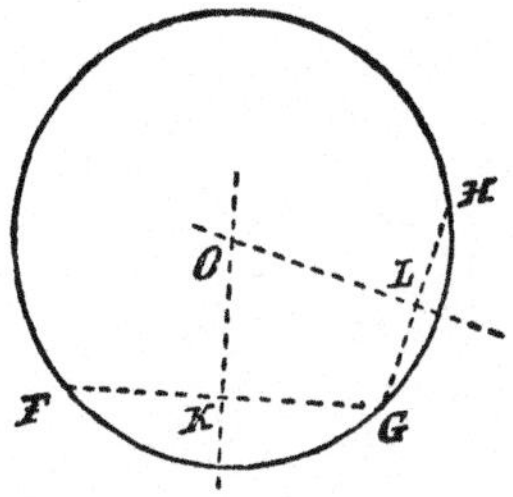

HYPOTHESIS. ***Let F, G, and H be points on a ⊙.***

CONCLUSION. ⊙ *FGH* has only one center.

PROOF. Join *FG* and *GH*.

Since, by definition, a center is a point from which all sects to the circle are equal, therefore any center of a circle through *F* and *G* is in the perpendicular bisector of *FG*, and any center of a circle through *G* and *H* is in the perpendicular bisector of *GH*.

(183. The locus of the point to which sects from two given points are equal is the perpendicular bisector of the sect joining them.)

But these two perpendicular bisectors can intersect in only one point,

∴ ⊙ *FGH* has only one center.

331. COROLLARY I. The perpendicular bisector of any chord passes through the center.

332. COROLLARY II. To find the center of any given circle, or of any given arc of a circle, draw two non-parallel chords and their perpendicular bisectors. The center is the point where these bisectors intersect.

333. A *Diameter* is a chord through the center.

334. A diameter is equal to two radii: so all diameters are bisected by the center of the circle, and are equal.

THEOREM IV.

335. *Circles of equal radii are congruent.*

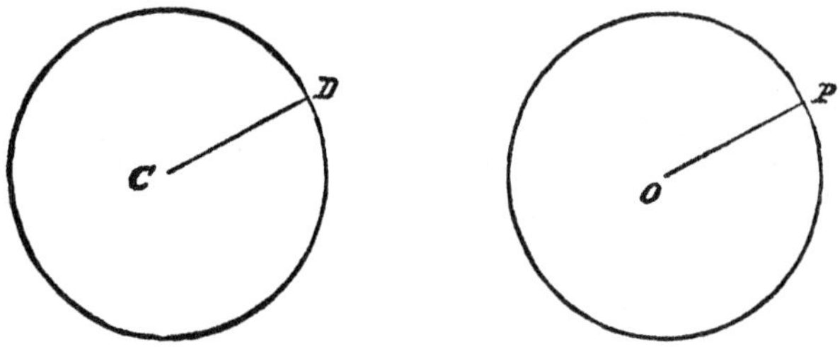

HYPOTHESIS. ***Two circles of which C and O are the centers, and radius CD = radius OP.***

CONCLUSION. The circles are congruent.

PROOF. Apply one circle to the other so that the center O shall coincide with center C, and sect OP fall upon line CD. Then, because $OP = CD$, the point P will coincide with the point D. Then every particular point in the one circle must coincide with some point in the other circle, because of the equality of radii.

(321. A point is on the circle when its sect from the center is equal to the radius.)

$$\therefore \quad \odot C \cong \odot O.$$

336. COROLLARY. After being applied, as above, the second circle may be turned about its center; and still it will coincide with the first, though the point P no longer falls upon D.

Hence, considering one circle as the trace of the other, —

A circle can be made to slide along itself by being turned about its center.

This fundamental property of this curve allows us to turn any figure connected with the circle about the center without changing its relation to the circle.

337. Circles which have the same center are called *Concentric.*

THEOREM V.

338. *Different concentric circles cannot have a point in common.*

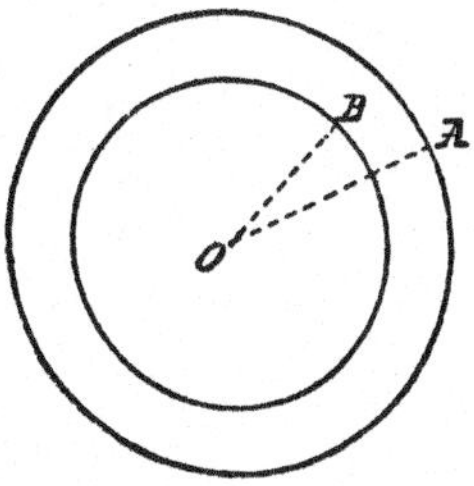

PROOF. The points of the circle with the lesser radius are all within the larger circle.

(321. A point is within the circle if its distance from the center is less than the radius.)

339. FIRST CONTRANOMINAL OF 338. Two different circles with a point in common are not concentric.

340. SECOND CONTRANOMINAL OF 338. Two concentric circles with a point in common coincide.

341. The center of a circle is a *Center of Symmetry*, the end points of any diameter being corresponding points.

This follows at once from the definition of Central Symmetry, and the fundamental property that the circle slides along itself when turned about its center, and so coincides with itself after turning about the center through *any angle.* The circle is the only closed curve which will slide upon its trace.

THEOREM VI.

342. *The perpendicular from the center of a circle to a secant bisects the chord; and, if a line through the center bisect a chord not passing through the center, it cuts it at right angles.*

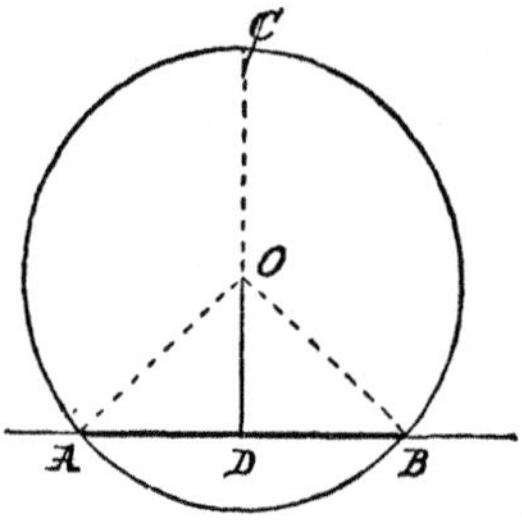

PROOF. For any chord, there is only one perpendicular from the center, only one line through the mid-point and center, only one perpendicular bisector; and these, by 331, are identical.

(331. The perpendicular bisector of any chord passes through the center.)

343. *Every diameter is an Axis of Symmetry.*

For if we fold over along a diameter, every point on the part of the circle turned over must fall on some point on the other part, since its sect from the center, which remains fixed, is a radius.

INVERSE. *Every line which is an axis of symmetry of a circle contains a diameter.*

For, if not, there would be a point symmetrical to the center, and this, too, would again be a center: the circle would thus have two centers.

(330. A circle has only one center.)

Each circle has therefore, besides its center of symmetry, an infinite number of axes of symmetry.

THEOREM VII.

344. *Every chord lies wholly within the circle.*

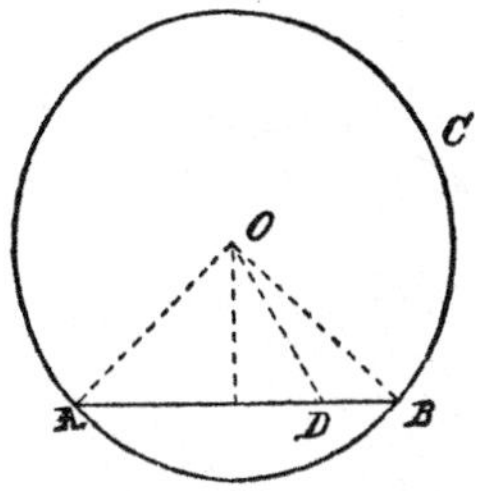

HYPOTHESIS. **Let** A **and** B **be any two points in** $\odot\ ABC$.

CONCLUSION. Every point on chord AB between A and B is within the $\odot\ ABC$.

PROOF. Take any point, D, in chord AB.

By 332, find O, the center of the circle.

Join OA, OD, OB.

Then OB makes a greater sect than OD from the foot of the perpendicular from O on the line AB,

(342. The perpendicular from center on line AB bisects chord AB.)

$$\therefore \quad OB > OD,$$

(154. The oblique which makes the greater sect from the foot of the perpendicular is the greater.)

$\therefore$ D is within the circle.

(320. A point is within the circle if its sect from the center is less than the radius.)

345. COROLLARY. If a line has a point within a circle, it is a secant, for the radius is greater than the perpendicular from the center to this line: so there will be two sects from the center to the line, each equal to the radius; that is, the line will pass through two points on the circle.

Thus, again, the circle is a closed curve.

THEOREM VIII.

346. *In a circle, two chords which are not both diameters do not mutually bisect each other.*

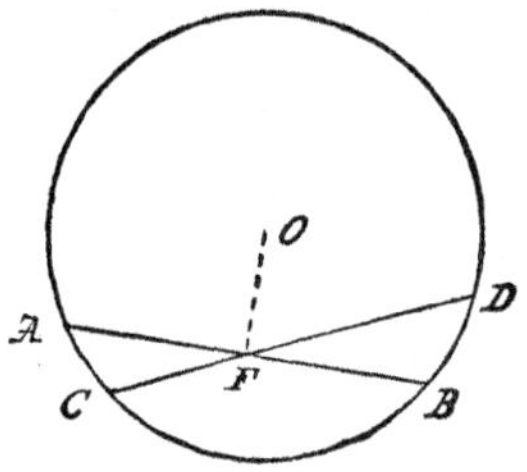

HYPOTHESIS. ***Let the chords** AB, CD, **which do not both pass through the center, cut one another in the point** F, **in the** $\odot$ $ACBD$.*

CONCLUSION. AB and CD do not mutually bisect each other.

PROOF. If one of them pass through the center, it is not bisected by the other, which does not pass through the center.

If neither pass through the center, find the center O, and join OF.

If F is the bisection point of one of the chords, as AB, then

$$\measuredangle\ OFB = \text{rt.}\ \measuredangle,$$

(342. If a line through the center bisect a chord not passing through the center, it cuts it at right angles.)

$$\therefore\quad \measuredangle\ OFD \text{ is oblique,}$$

$$\therefore\quad OF \text{ does not bisect } CD.$$

EXERCISES. 69. What is the locus of mid points of parallel chords?

70. Prove by symmetry that the diameter perpendicular to a chord bisects that chord, bisects the two arcs into which this chord divides the circle, and bisects the angles at the center subtended by these arcs.

THEOREM IX.

347. *If from any point not the center, sects be drawn to the different points of a circle, the greatest is that which meets the circle after passing through the center; the least is part of the same line.*

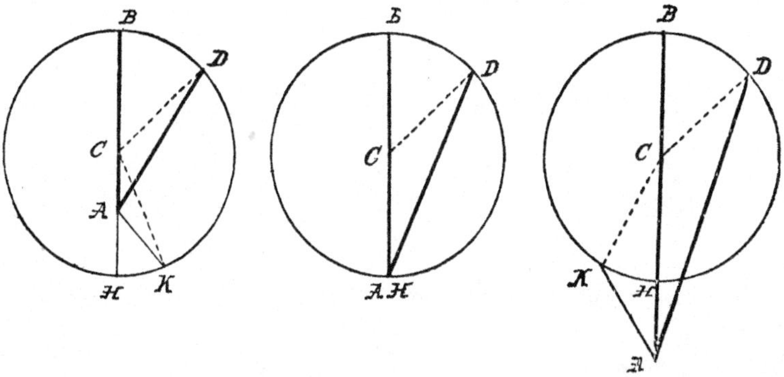

HYPOTHESIS. **From any point A, sects are drawn to a ⊙ $DBHK$, whose center is C.**

FIRST CONCLUSION. $ACB > AD$.

PROOF. Join CD.

Then, because $CB = CD$,

$$\therefore \quad AB = AC + CB = AC + CD > AD.$$

(156. Any two sides of a triangle are greater than the third.)

SECOND CONCLUSION. $AH < AK$.

PROOF. When A is within the circle,

$$HC = KC < KA + AC, \text{ by } 156.$$

Taking away AC from both sides, $\therefore \quad AH < AK$.

When A is on the circle, AH is a point.

When A is without the circle,

$$AC = AH + HC < AK + KC, \text{ by } 156.$$

Taking away $HC = KC$, $\therefore \quad AH < AK$.

348. COROLLARY. The diameter of a circle is greater than any other chord.

Theorem X.

349. *If from any point three sects drawn to a circle are equal, that point is the center.*

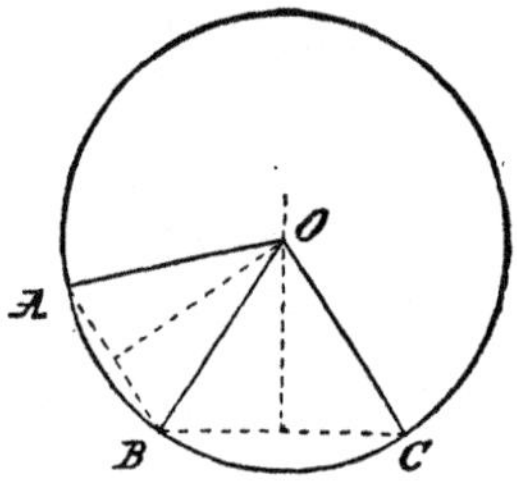

Hypothesis. ***From the point** O **to** $\odot ABC$, **let** $OA = OB = OC$.*

Conclusion. O is center of $\odot ABC$.

Proof. Join AB, and BC.

O is on the perpendicular bisector of chord AB, and also on that of chord BC,

(183. The locus of a point to which sects from two given points are equal is the perpendicular bisector of the sect joining them.)

$\therefore$ O is center of $\odot ABC$.

(332. The center is the intersection of perpendicular bisectors of two non-parallel chords.)

350. Contranominal of 349. From any point not the center, there cannot be drawn more than two equal sects to a circle.

351. Corollary I. If two circles have three points in common, they coincide.

Because from the center of one circle three equal sects can be drawn to points on the other circle,

$\therefore$ they are concentric, and they have a point in common,

$\therefore$ they coincide.

(340. Two concentric circles with a point in common coincide.)

352. Corollary II. Through three points not more than one circle can pass.

353. Corollary III. Two different circles cannot meet one another in more than two points.

354. A circle is *circumscribed about* a polygon when it passes through all the vertices of the polygon.

Then the polygon is said to be *inscribed in* the circle.

Problem I.

355. *Through any three points not in the same line, to describe a circle.*

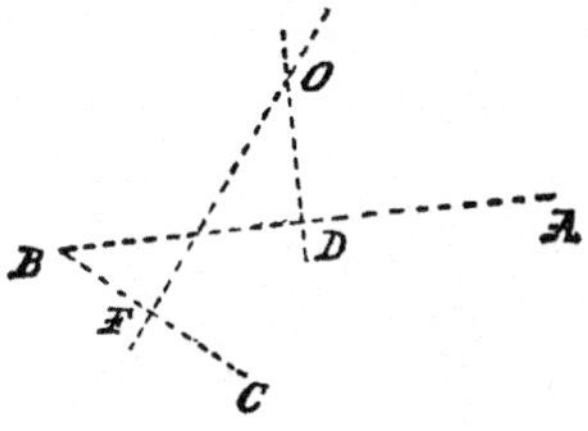

Given, **three points,** A, B, **and** C, **not in the same line.**

Required, *to describe a circle which shall pass through all of them.*

Construction. By 188, find the point O, such that

$$OA = OB = OC.$$

356. Corollary. To circumscribe a circle about any given triangle, by 355, pass a circle through its three vertices.

357. Four or more points which lie on the same circle are called *Concyclic.*

THEOREM XI.

358. *Perpendiculars from the center on equal chords are equal; and, on unequal chords, that on the greater is the lesser.*

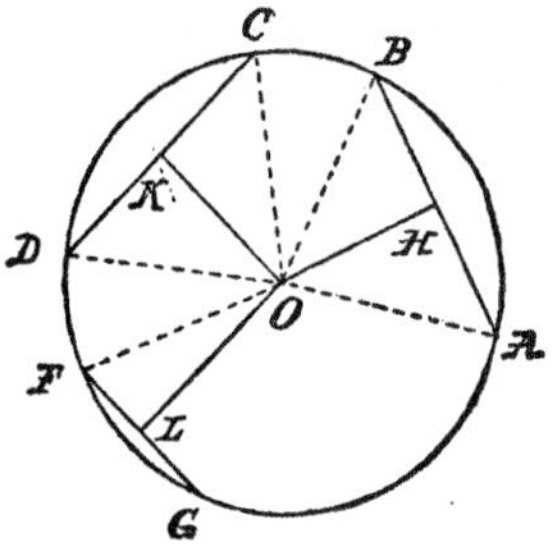

HYPOTHESIS. **From center** O,

$$OH, OK, OL, \perp \text{ chords } AB = CD > FG.$$

CONCLUSION. $OH = OK < OL$.

PROOF. Draw the radii OA, OB, OC, OD, OF.

By hypothesis, $AB = CD$,

$$\therefore \quad \triangle OAB \cong \triangle OCD,$$

(129. Triangles with three sides respectively equal are congruent.)

$\therefore$ the altitude $OH =$ corresponding altitude OK.

Again, because $CD > FG$,

$$\therefore \quad CK > FL.$$

But $\overline{CK}^2 + \overline{KO}^2 = \overline{OC}^2 = \overline{OF}^2 = \overline{FL}^2 + \overline{LO}^2$.

But $\overline{CK}^2 > \overline{FL}^2$.

$$\therefore \quad \overline{KO}^2 < \overline{LO}^2,$$

$$\therefore \quad OK < OL.$$

359. By 33, Rule of Inversion,

Chords having equal perpendiculars from the center are equal; and, of chords having unequal perpendiculars, the one with the lesser is the greater.

II. Angles at the Center.

360. The explemental angles at the center of a circle, whose arms are the same radii, are said to be *subtended by*, or to *stand upon*, the explemental arcs opposite them intercepted by the radii, the reflex angle upon the major arc.

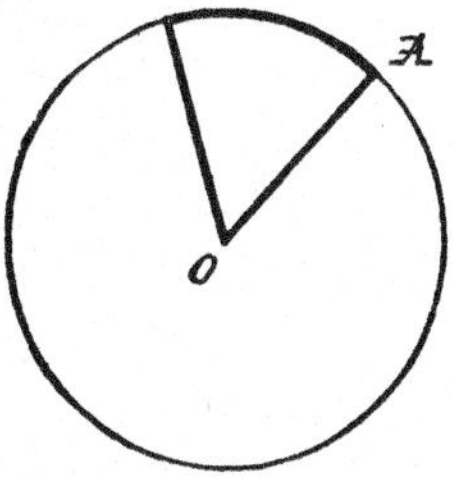

361. A *Sector* is the figure formed by two radii and the arc included between them.

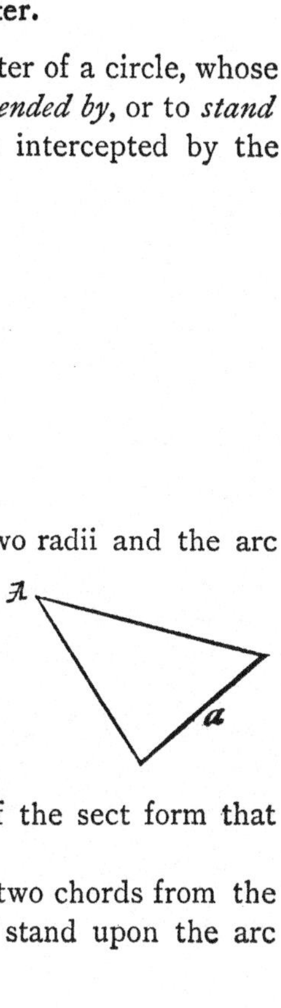

362. The angle of the sector is the angle at the center which stands upon the arc of the sector.

363. A given sect is said to subtend a certain angle from a given point when the lines drawn from the point to the ends of the sect form that angle.

364. An *inscribed angle* is formed by two chords from the same point on the circle, and is said to stand upon the arc between its arms.

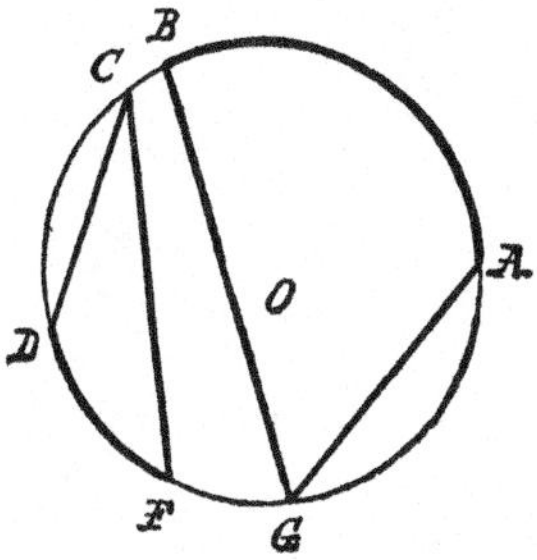

THEOREM XII.

365. *In the same or equal circles, equal arcs are congruent, subtend equal angles at the center, determine equal sectors, and are subtended by equal chords.*

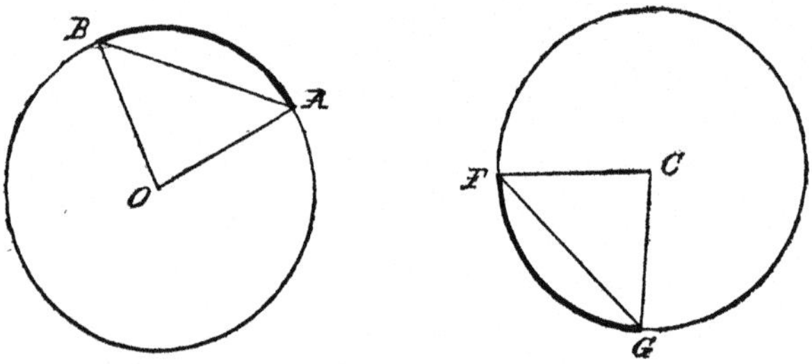

HYPOTHESIS. **Radius** OA = **radius** CF, **and arc** AB = **arc** FG.

CONCLUSIONS. I. $\measuredangle AOB = \measuredangle FCG$.

II. Sector AOB = sector FCG.

III. Chord AB = chord FG.

PROOF. Place the sector AOB over the sector FCG so that the center O shall fall on C, and the radius OA on the line CF. Then, because $OA = CF$,

$\therefore$ point A falls on point F.

Again, because the radii are equal, every point of the arc AB will fall on some part of the circle FG.

But arc AB = arc FG,

$\therefore$ point B falls on G,

$\therefore$ chord AB = chord FG,

$\measuredangle AOB = \measuredangle FCG$,

and sector AOB = sector FCG.

366. In the same or equal circles the *sum of two minor arcs* is the arc obtained by placing them on the same circle so as not to overlap, with one end point in common.

THEOREM XIII.

367. *A sum of two arcs of the same or equal circles subtends an angle at the center equal to the sum of the angles which each arc subtends separately.*

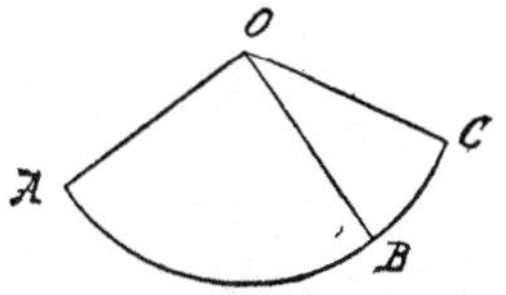

PROOF. Placing the arcs with two end points in common at B, join B and the other end points, A and C, to the center O.

Then the angles are in such a position, that, by definition 61,

$$\measuredangle AOB + \measuredangle BOC = \measuredangle AOC.$$

But $\measuredangle AOC$ is subtended by arc AC, which is the sum of arc AB subtending $\measuredangle AOB$, and arc BC subtending $\measuredangle BOC$.

368. COROLLARY. Two sectors of the same or equal circles may be so placed as to form a sector whose arc is the sum of their arcs, whose angle is the sum of their angles, and whose surface is the sum of their surfaces.

THEOREM XIV.

369. *In the same or equal circles, of two unequal arcs, the greater subtends the greater angle at the center, and determines the greater sector.*

PROOF. If the first arc is greater than the second, it is equal to the second plus a third arc; and so, by 367, the angle which the first subtends is greater than the angle which the second subtends by the angle which the third arc subtends at the center.

And like is true of the sectors.

370. From 365 and 369, by 33, Rule of Inversion,

In the same or equal circles, equal angles at the center intercept equal arcs, and determine equal sectors; and, of two unequal angles at the center, the greater intercepts the greater arc, and determines the greater sector.

371. Corollary. A diameter of a circle divides the enclosed surface into two equal parts.

372. Again, from 365 and 369, by 33, Rule of Inversion,

In the same or equal circles, equal sectors have equal arcs and equal angles; and, of two unequal sectors, the greater has the greater arc and the greater angle.

Theorem XV.

373. *In the same or equal circles, of two unequal minor arcs, the greater is subtended by the greater chord; of two unequal major arcs, the greater is subtended by the lesser chord.*

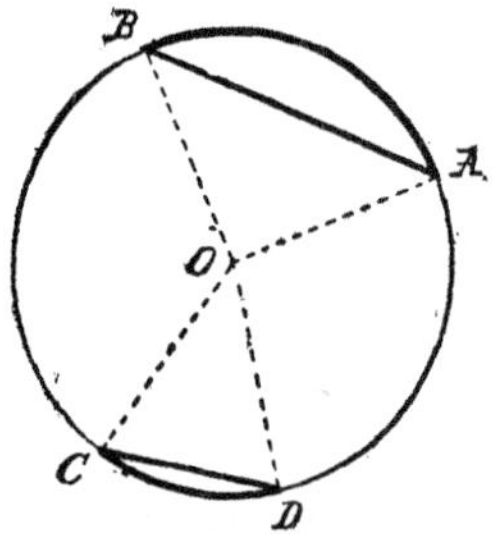

Hypothesis. ***Minor arc*** $AB >$ ***arc*** CD.

Conclusion. Chord $AB >$ chord CD.

Proof. $\measuredangle AOB > \measuredangle COD$,

(369. The greater arc subtends the greater angle at the center.)

$\therefore$ in $\triangle$s AOB and COD, side $AB >$ side CD.

(159. Two triangles with two sides equal, but the included angle greater in the first, have the third side greater in the first.)

Secondly, because the minor arc with its major arc together make up the entire circle, therefore to a greater major arc will correspond a lesser minor arc, and therefore a lesser chord.

374. From 365 and 373, by 33, Rule of Inversion,

In the same or equal circles, equal chords subtend equal major and minor arcs; and, of two unequal chords, the greater subtends the greater minor arc and the lesser major arc.

Theorem XVI.

375. *An angle at the center of a circle is double the inscribed angle standing upon the same arc.*

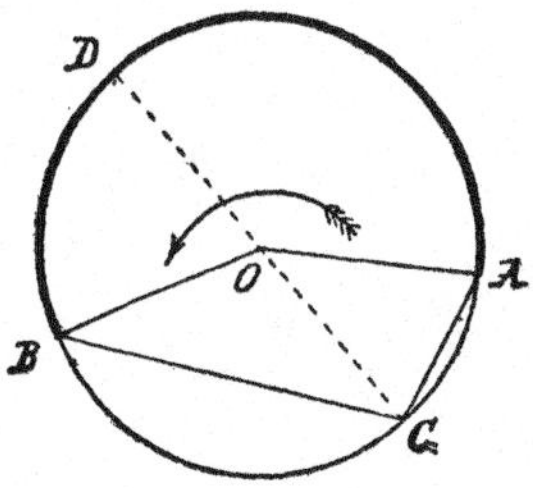

Fig. 1.

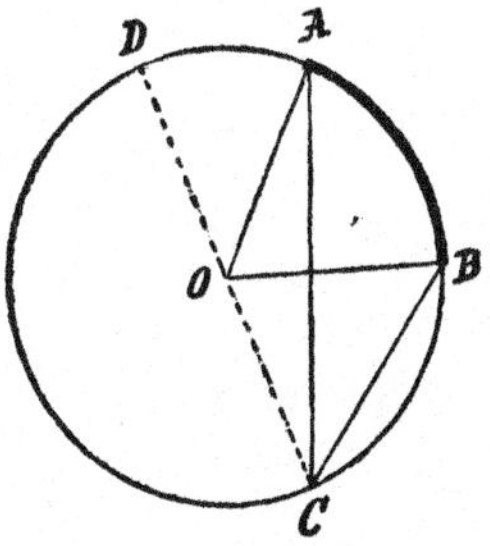

Fig. 2.

Hypothesis. ***Let*** AB ***be any arc,*** O ***the center,*** C ***any point on the circle not on arc*** AB.

Conclusion. $\measuredangle AOB = 2\measuredangle ACB$.

Proof. Join CO, and produce it to D.

Because, being radii, $OA = OC$,

$$\therefore \measuredangle OCA = \measuredangle OAC,$$

(126. The angles at the base of an isosceles triangle are equal.)

$$\therefore \measuredangle AOD = \measuredangle OCA + \measuredangle OAC = 2\measuredangle OCA.$$

(173. The exterior angle of a triangle equals the sum of the interior opposite angles.)

Similarly, $\measuredangle DOB = 2\measuredangle OCB$.

Hence (in Fig. 1) the sum or (in Fig. 2) the difference of the angles AOD, DOB, is double the sum or difference of OCA and OCB; that is, $\measuredangle AOB = 2\measuredangle ACB$.

III. Angles in Segments.

376. An angle made by two lines drawn from a point in the arc of a segment to the extremities of the chord is said to be inscribed in the segment, and is called the angle in the segment.

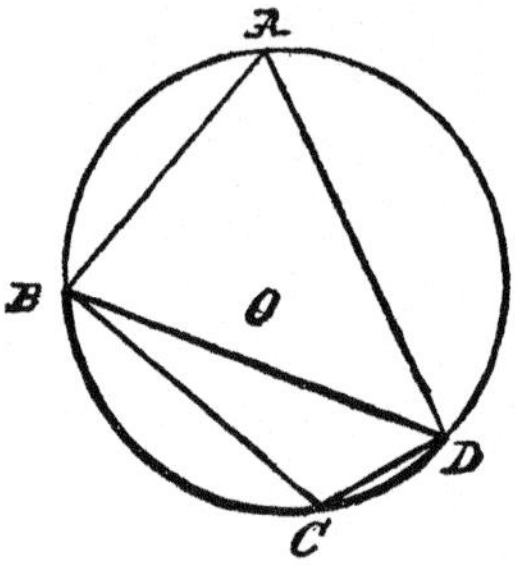

377. Corollary I. Angles inscribed in the same segment of a circle are equal.

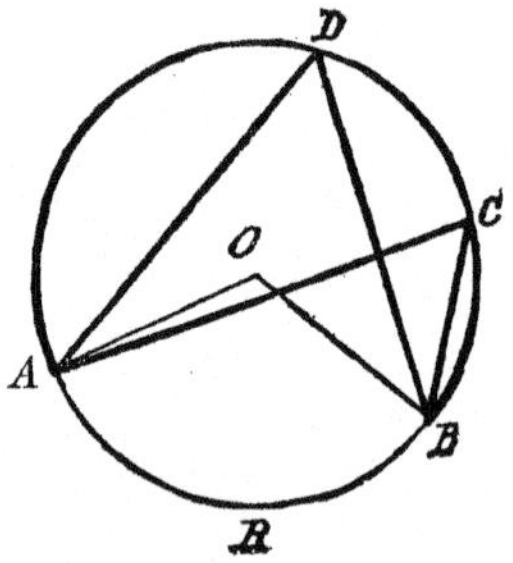

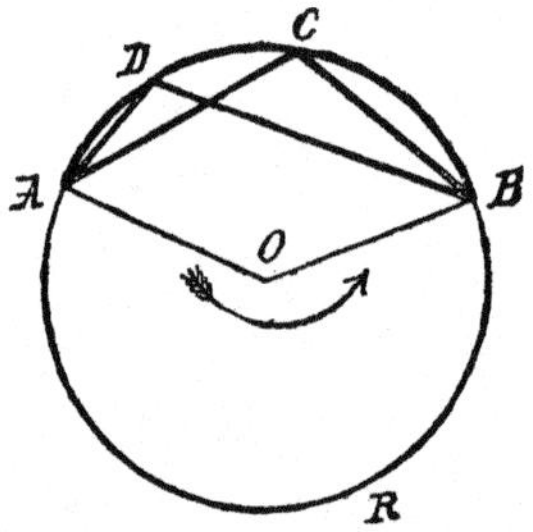

For each of the angles *ACB*, *ADB* is half of the angle subtended at the center by the arc *ARB*.

378. Corollary II. If a circle is divided into two segments by a chord, any pair of angles, one in each segment, will be supplemental.

For they are halves of the explemental angles at the center standing on the same explemental arcs.

379. Corollary III. The opposite angles of every quadrilateral inscribed in a circle are supplemental.

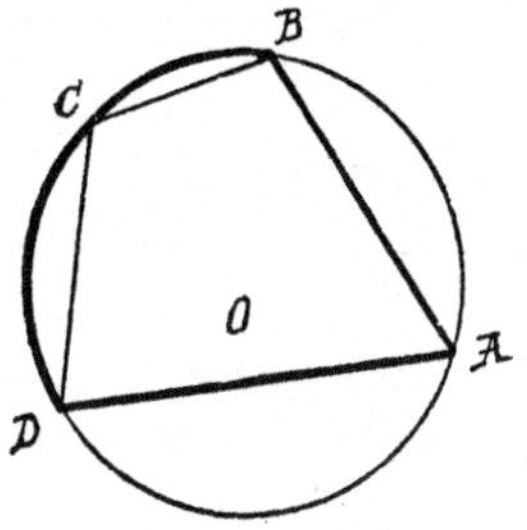

For they stand on explemental arcs, and so are halves of explemental angles at the center.

Theorem XVII.

380. *From a point on the side toward the segment, its chord subtends an angle less than, equal to, or greater than, the angle in the segment, according as the point is without, on, or within the arc of the segment.*

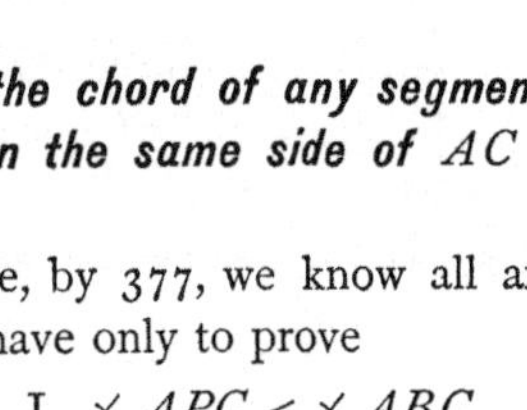

Hypothesis. ***AC* the chord of any segment, *ABC*; *P* any point without the segment on the same side of *AC* as *B*; *Q* any point within the segment.**

Conclusions. Since, by 377, we know all angles inscribed in the segment are equal, we have only to prove

I. $\measuredangle APC < \measuredangle ABC.$

II. $\measuredangle AQC > \measuredangle ABC.$

PROOF. I. Let R be the point where PC meets the circle, and join RA, making $\triangle APR$; then $\measuredangle APC < \measuredangle ARC$.

(142. An exterior angle of a triangle is greater than either interior opposite angle.)

II. For the same reason, if AQ is produced to meet the circumference at S, and SC joined, making $\triangle CSQ$, $\measuredangle AQC > \measuredangle CSA$.

381. By 33, Rule of Inversion,

On the side toward a segment, the vertex of a triangle, with its chord as base, will lie without, on, or within the arc of the segment, according as the vertical angle is less than, equal to, or greater than, an angle in the segment.

THEOREM XVIII.

382. *If two opposite angles of a quadrilateral are supplemental, a circle passing through any three of its vertices will contain the fourth.*

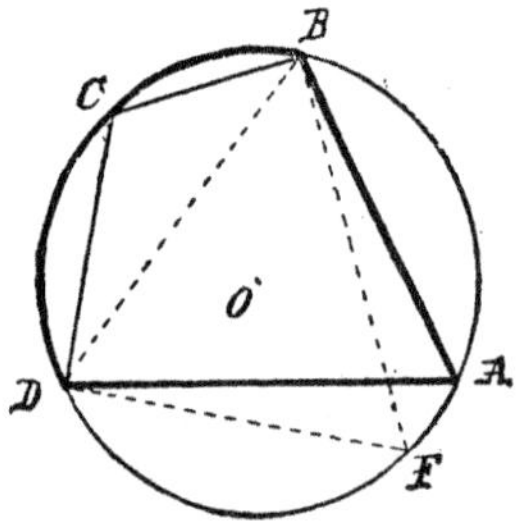

HYPOTHESIS. ***ABCD is a quadrilateral, with*** $\measuredangle A + \measuredangle C =$ st. $\measuredangle$.

CONCLUSION. The four vertices lie on the circle determined by any three of them.

PROOF. By 355, pass a circle through the three points B, C, D.

Take any point, F, on the arc DFB of the segment, on the same side of DB as A. Join FB, FD. Then $\measuredangle F + \measuredangle C =$ st. $\measuredangle$.

(379. The opposite angles of an inscribed quadrilateral are supplemental.)

From hypothesis, $\therefore$ $\measuredangle A = \measuredangle F$, $\therefore$ A is on the arc DFB.

(381. On the side toward a segment, the vertex of a triangle, with its chord as base, will lie on its arc if the vertical angle is equal to an angle in the segment.)

THEOREM XIX.

383. *The angle in a segment is greater than, equal to, or less than, a right angle, according as the arc of the segment is less than, equal to, or greater than, a semicircle.*

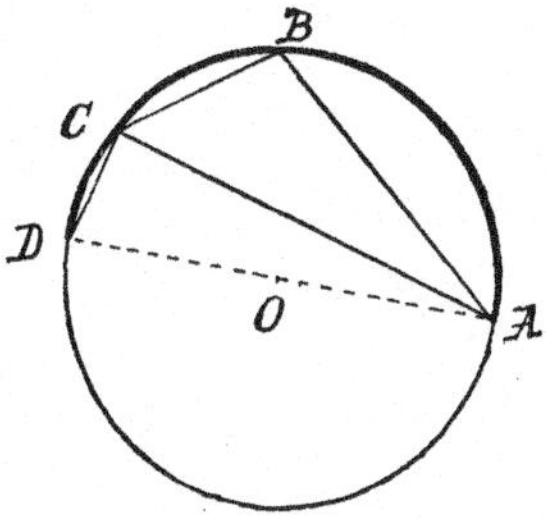

HYPOTHESIS. ***Let*** AD ***be a diameter of a circle whose center is*** O. ***Take*** B ***and*** C ***points on the same circle.***

CONCLUSIONS. I. $\measuredangle ACD =$ rt. $\measuredangle$, being half the straight angle AOD.

(375. An angle at the center is double the inscribed angle on the same arc.)

$$\text{II.} \quad \therefore \quad \measuredangle ADC < \text{rt.} \measuredangle.$$

(143. Any two angles of a triangle are together less than a straight angle.)

$$\text{III.} \quad \therefore \quad \measuredangle ABC > \text{rt.} \measuredangle,$$

since

$$\measuredangle ADC + \measuredangle ABC = \text{st.} \measuredangle.$$

(379. Opposite angles of an inscribed quadrilateral are supplemental.)

384. By 33, Rule of Inversion,

A segment is less than, equal to, or greater than, a semicircle, according as the angle in it is greater than, equal to, or less than, a right angle.

THEOREM XX.

385. *If two chords intersect within a circle, an angle formed and its vertical are each equal to half the angle at the center standing on the sum of the arcs they intercept.*

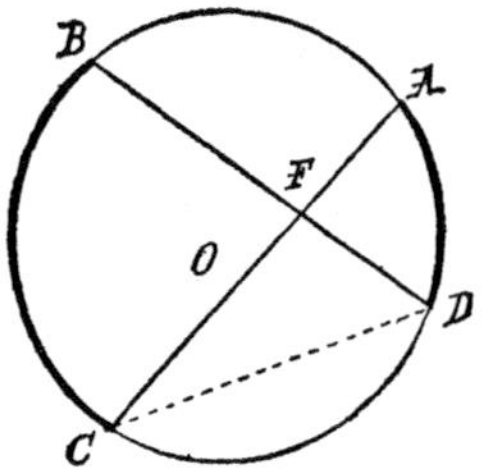

HYPOTHESIS. ***Let the chords AC, BD intersect at F within the circle.***

CONCLUSION. $\measuredangle BFC$ = half $\measuredangle$ at center standing on (arc BC + arc DA).

PROOF. Join CD.

$$\measuredangle BFC = \measuredangle BDC + \measuredangle DCA.$$

(173. The exterior angle of a triangle is equal to the sum of the opposite interior angles.)

But $2 \measuredangle BDC = \measuredangle$ at center on BC, and $2 \measuredangle DCA = \measuredangle$ at center on DA;

(375. An angle at the center is double the inscribed angle upon the same arc.)

$$\therefore \quad 2 \measuredangle BFC = \measuredangle \text{ at center on arc } (BC + DA).$$

(367. A sum of two arcs subtends an angle at the center equal to the sum of the angles subtended by the arcs.)

EXERCISES. 71. The end points of two equal chords of a circle are the vertices of a symmetrical trapezoid.

72. Every trapezoid inscribed in a circle is symmetrical.

THEOREM XXI.

386. *An angle formed by two secants is half the angle at the center standing on the difference of the arcs they intercept.*

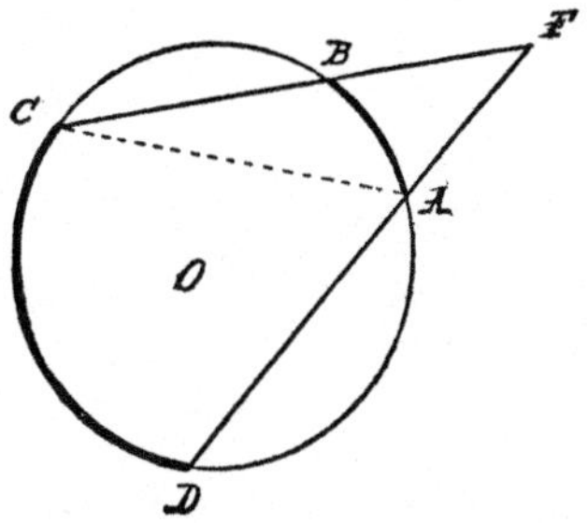

HYPOTHESIS. ***Let two lines from F cut the circle whose center is O, in the points A, B, C, and D.***

CONCLUSION. $\measuredangle F = \frac{1}{2} \measuredangle$ at O on difference between arc AB and arc CD.

PROOF. Join AC.

$$\measuredangle CAD = \measuredangle F + \measuredangle C.$$

Doubling both sides, $\measuredangle$ at O on arc $CD = 2 \measuredangle F + \measuredangle$ at O on arc AB;

$\therefore$ twice $\measuredangle F =$ difference of $\measuredangle$s at O on arcs CD and AB.

IV. Tangents.

387. A line which will meet the circle in one point only is said to be a *Tangent* to the circle.

388. The point at which a tangent touches the circle is called the *Point of Contact*.

THEOREM XXII.

389. *Of lines passing through the end of any radius, the perpendicular is a tangent to the circle, and every other line is a secant.*

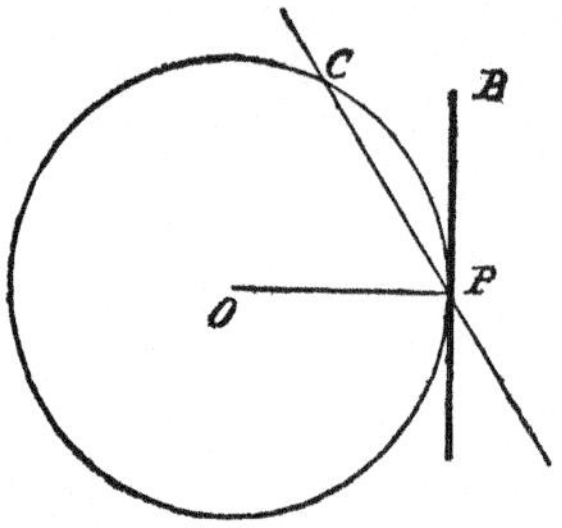

HYPOTHESIS. ***A radius OP perpendicular to PB, oblique to PC.***

CONCLUSIONS. I. PB is a tangent at P.

II. PC is a secant.

PROOF. (I.) The sect from O to any point on PB, except P, is $> OP$,

(150. The perpendicular is the least sect between a point and a line.)

$\therefore$ every point of PB except P is outside the circle.

(321. A point is without a circle if its sect from the center is greater than the radius.)

PROOF. (II.) A sect perpendicular to PC is less than the oblique OP,

$\therefore$ a point of PC is within the circle;

$\therefore$ PC is a secant.

390. COROLLARY I. One and only one tangent can be drawn to a circle at a given point on the circle.

391. COROLLARY II. To draw a tangent to a circle at a point on the circle, draw the perpendicular to the radius at the point.

392. COROLLARY III. The radius to the point of contact of any tangent is perpendicular to the tangent.

393. COROLLARY IV. The perpendicular to a tangent from the point of tangency passes through the center of the circle.

394. COROLLARY V. The perpendicular drawn from the center to the tangent passes through the point of contact.

ON THE THREE RELATIVE POSITIONS OF A LINE AND A CIRCLE.

395. COROLLARY VI. A line will be a secant, a tangent, or not meet the circle, according as its perpendicular from the center is less than, equal to, or greater than, the radius.

396. By 33, Rule of Inversion,

The perpendicular on a line from the center will be less than, equal to, or greater than, the radius, according as it is a secant, tangent, or non-meeter.

THEOREM XXIII.

397. *An angle formed by a tangent and a chord from the point of contact is half the angle at the center standing on the intercepted arc.*

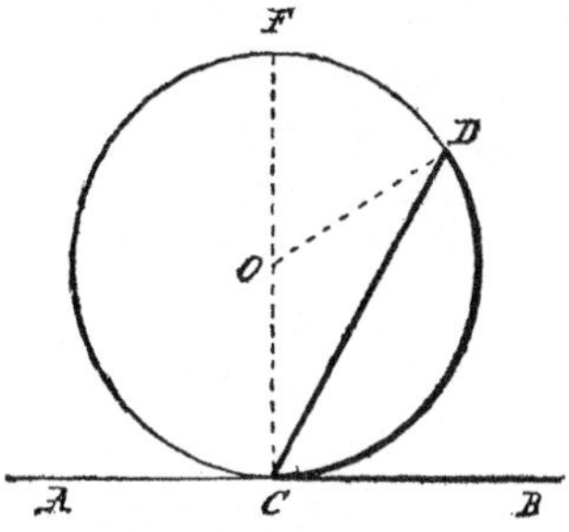

HYPOTHESIS. *AB* ***is tangent at*** *C*, ***and*** *CD* ***is a chord of*** ⊙ ***with center at*** *O*.

CONCLUSIONS. $\measuredangle DCB = \frac{1}{2} \measuredangle DOC.$

$\measuredangle DCA = \frac{1}{2}$ explement $DOC.$

PROOF. At C erect chord $CF \perp AB.$

CF is a diameter of the circle.

(393. The perpendicular to a tangent from the point of contact passes through the center of the circle.)

Join OD. Then

$$\text{rt.} \measuredangle OCA = \tfrac{1}{2} \text{st.} \measuredangle \text{ at } O,$$

also

$$\measuredangle OCD = \tfrac{1}{2} \measuredangle FOD.$$

(375. The angle at the center is double the inscribed angle on the same arc.)

Therefore, adding

$$\measuredangle DCA = \tfrac{1}{2} \text{reflex} \measuredangle DOC,$$

therefore the supplement of $\measuredangle DCA$, which is $\measuredangle DCB$, is half the explement of reflex $\measuredangle DOC$, which is $\measuredangle DOC$.

398. INVERSE. If the angle at the center standing on the arc intercepted by a chord equals twice the angle made by that chord and a line from its extremity on the same side as the arc, this line is a tangent.

PROOF. There is but one line which will make this angle; and we already know, from 397, that a tangent makes it.

EXERCISES. 73. The chord which joins the points of contact of parallel tangents to a circle is a diameter.

74. How may 397 be considered as a special case of 375?

75. A parallelogram inscribed in a circle must be a rectangle.

76. If a series of circles touch a given line at a given point, where will their centers all lie?

77. The angle of two tangents is double that of the chord of contact and the diameter through either point of contact.

THEOREM XXIV.

399. *The angle formed by a tangent and a secant is half the angle at the center standing on the difference of the intercepted arcs.*

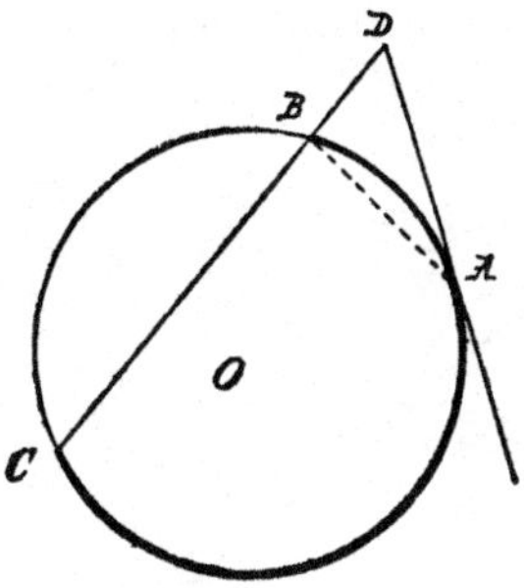

HYPOTHESIS. ***Of two lines from D, one cuts at B and C the ⊙ whose center is O, the other is tangent to the same ⊙ at A.***

CONCLUSION. $2 \measuredangle D =$ difference between $\measuredangle AOC$ and $\measuredangle AOB$.

PROOF. Join AB.

$$\measuredangle ABC = \measuredangle D + \measuredangle DAB,$$

(173. The exterior angle of a triangle is equal to the two interior opposite angles.)

$$\therefore \measuredangle AOC = 2 \measuredangle D + \measuredangle AOB,$$

(375. An angle at the center is double the inscribed angle on the same arc.)

and

(397. An angle formed by a tangent and chord is half the angle at the center on the intercepted arc.)

$\therefore$ twice $\measuredangle D$ is the difference between $\measuredangle AOC$ and $\measuredangle AOB$.

400. COROLLARY. The angle formed by two tangents is half the angle at the center standing on the difference of the intercepted arcs.

PROBLEM II.

401. *From a given point without a circle to draw a tangent to the circle.*

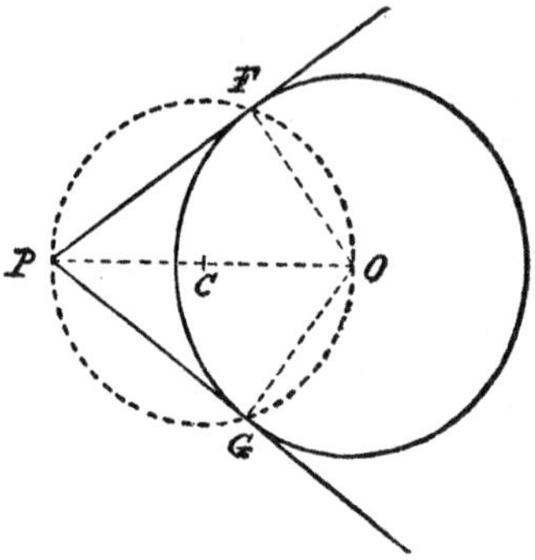

GIVEN, *a ⊙ with center O, and a point P outside it.*

REQUIRED, *to draw through* P *a tangent to the circle.*

CONSTRUCTION. Join *OP*. Bisect *OP* in *C*.

With center *C* and radius *CP* describe a circle cutting the given circle in *F* and *G*.

Join *PF* and *PG*.

These are tangents to ⊙ *FGB*.

PROOF. Join *OF*.

∡ *OFP* is a rt. ∡,

(383. The angle in a semicircle is a right angle.)

∴ *PF* is tangent at *F* to ⊙ *BFG*.

(389. A line perpendicular to a radius at its extremity is a tangent to the circle.)

402. COROLLARY. Two tangents drawn to a circle from the same external point are equal, and make equal angles with the line joining that point to the center.

V. Two Circles.

Theorem XXV.

403. *The line joining the centers of two circles which meet in two points is identical with the perpendicular bisector of the common chord.*

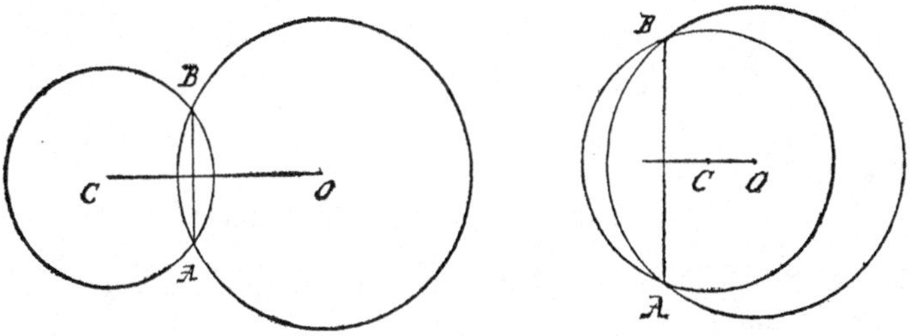

Proof. For the perpendicular bisector of the common chord must pass through the centers of the two circles.

(331. The perpendicular bisector of any chord passes through the center.)

404. Corollary. Since any common chord is bisected by the line joining the centers, therefore if the two circles meet at a point on the line of centers, there is no common chord, and these circles have no second point in common.

405. Two circles which meet in one point only are said to touch each other, or to be tangent to one another, and the point at which they meet is called their point of contact.

406. By 343,

Two circles, not concentric, have always one, and only one, common axis of symmetry; namely, their line of centers.

For this is the only line which contains a diameter of each.

THEOREM XXVI.

OBVERSE OF 404.

407. *If two circles have one common point not on the line through their centers, they have also another common point.*

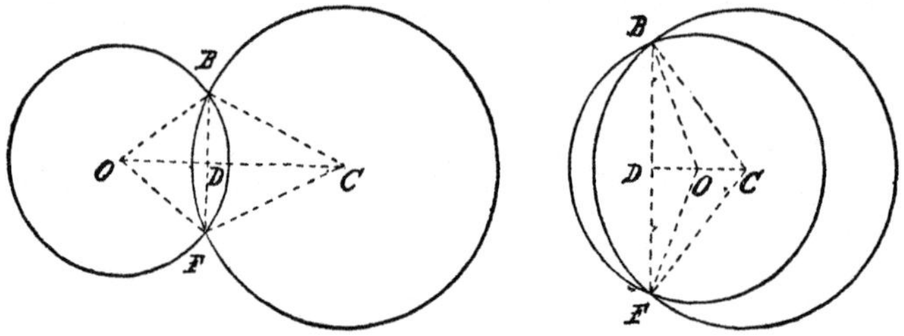

HYPOTHESIS. **⊙ with center O and ⊙ with center C, having a common point B not on OC.**

CONCLUSION. They have another common point.

PROOF. Join OC, and from B drop a line perpendicular to OC at D, and prolong it, making $DF = BD$.

F is the second common point.

For $\triangle ODB \cong \triangle OFD$, and $\triangle CBD \cong \triangle CFD$;

(124. Triangles having two sides and the included angle equal in each are congruent.)

$\therefore \quad OF = OB, \quad$ and $\quad CF = CB$;

$\therefore \quad F$ is on both ⊙s.

(321. A point is on the circle if its distance from the center is equal to the radius.)

408. CONTRANOMINAL OF 407. If two circles touch one another, the line through their centers passes through the point of contact.

409. COROLLARY. Two circles which touch one another have a common tangent at their point of contact; namely, the perpendicular through that point to the line joining their centers.

410. Calling the sect joining the centers of two circles their center-sect c, and calling their radii r_1 and r_2, we have, in regard to the relative positions of two circles, —

1. If $c > r_1 + r_2$, therefore the $\odot$s are wholly exterior.

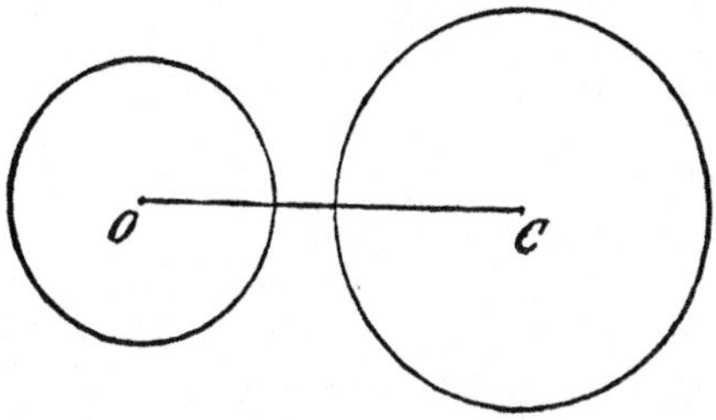

2. If $c = r_1 + r_2$, therefore the $\odot$s touch externally.

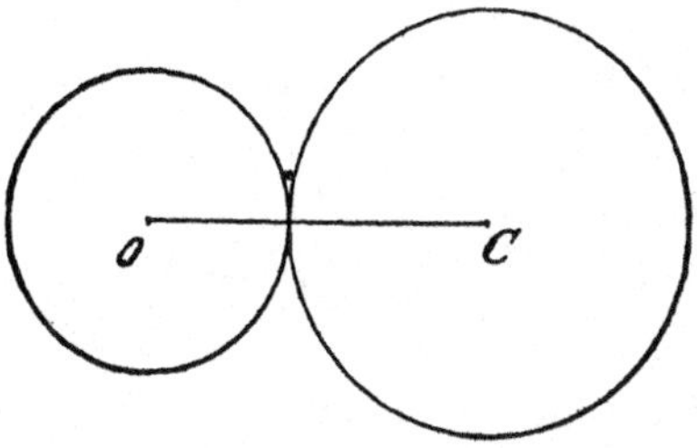

3. If $c < r_1 + r_2$, but $c >$ the difference of radii, therefore the $\odot$s cut each other.

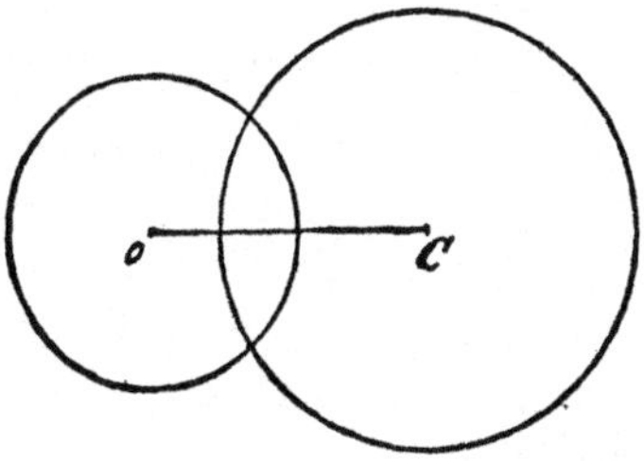

4. If $c =$ the difference of radii, therefore the ⊙s touch internally.

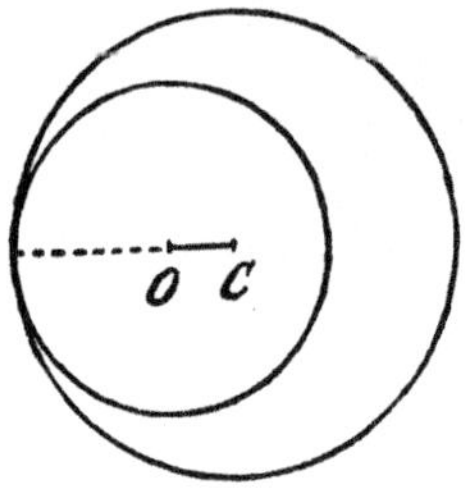

5. If $c <$ the difference of radii, therefore one ⊙ is wholly interior to the other.

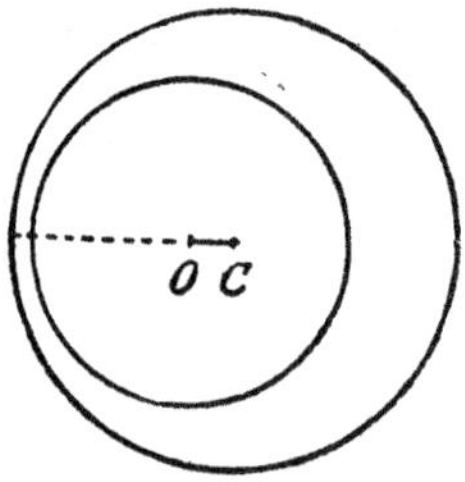

411. By 33, Rule of Inversion, the five inverses to the above are true.

Exercises. 78. How must a line through one of the common points of two intersecting circles be drawn in order that the two circles may intercept equal chords on it?

79. Through one of the points of intersection of two circles draw the line on which the two circles intercept the greatest sect.

80. If any two lines be drawn through the point of contact of two circles, the lines joining their second intersections with each circle will be parallel.

VI. Problems.

Problem III.

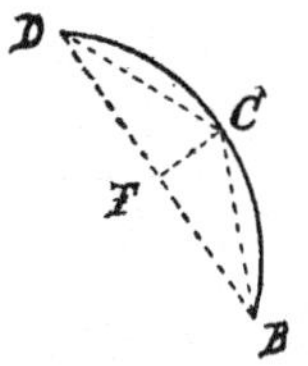

412. *To bisect a given arc.*

Given, **the arc** *BD*.

Required, *to bisect it.*

Construction. Join *BD*, and bisect the sect *BD* in *F*; at *F* erect a perpendicular cutting the arc in *C*.

C is the mid point of the arc.

Proof. Join *BC*, *CD*.

$$\triangle BCF \cong \triangle DFC,$$

(124. Triangles having two sides and the included angle in each equal are congruent.)

$\therefore$ chord $BC =$ chord CD,

$\therefore$ arc $BC =$ arc CD.

(374. In the same circle, equal chords subtend equal minor arcs.)

413. A polygon is said to be *circumscribed about* a circle when all its sides are tangents to the circle.

The circle is then said to be *inscribed in* the polygon.

414. A circle which touches one side of a triangle and the other two sides produced is called an *Escribed Circle.*

PROBLEM IV.

415. *To describe a circle touching three given lines which are not all parallel, and do not all pass through the same point.*

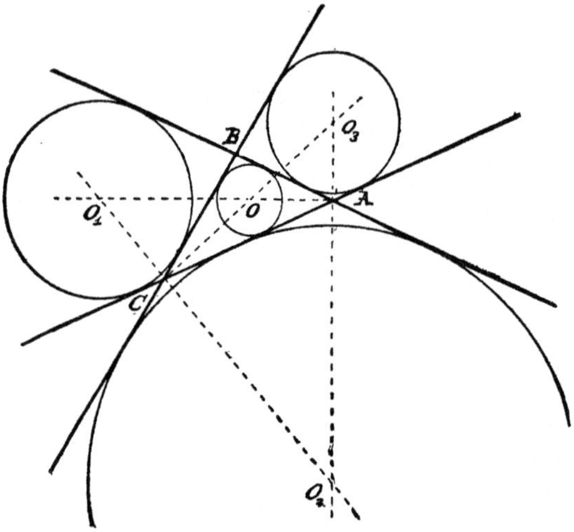

GIVEN, ***three lines intersecting in** A, B, **and** C.*

REQUIRED, *to describe a circle touching them.*

CONSTRUCTION. Draw the bisectors of the angles at A and C.

These four bisectors will intersect in four points, O, O_1, O_2, O_3.

A circle described with any one of these points as center, and its perpendicular on any one of the three given lines as radius, will touch all three.

PROOF. By 186, every point in the bisector of an angle is equally distant from its arms;

Therefore, since O is on the bisector of $\measuredangle A$, the perpendicular from O on AB equals the perpendicular from O on AC, which also equals the perpendicular from O on CB, since O is also on the bisector of $\measuredangle C$.

416. The four tangents common to two circles occur in two pairs intersecting on the common axis of symmetry.

PROBLEM V.

417. *In a given circle to inscribe a triangle equiangular to a given triangle.*

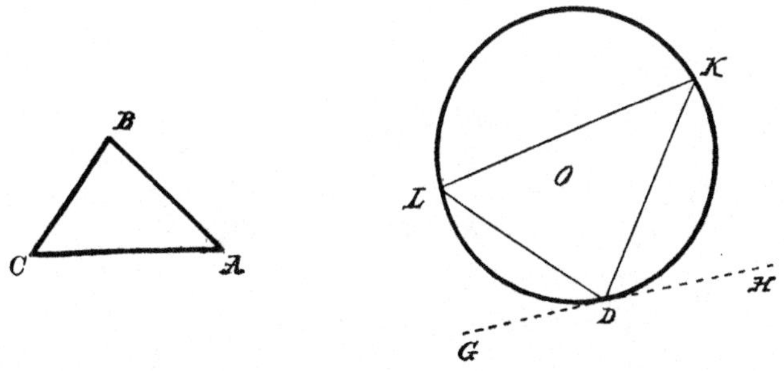

GIVEN, *a* ⊙ *and* △ ABC.

REQUIRED, *to describe in the* ⊙ *a* △ *equiangular to* △ ABC.

CONSTRUCTION. Draw a tangent GH touching the circle at the point D.

Make ∡ HDK = ∡ C, and ∡ GDL = ∡ A.

K and L being on the circumference, join KL.

DKL is the required triangle.

PROOF. ∡ KDL = ∡ B.

(174. The three angles of a triangle are equal to a straight angle.)

∡ K = ∡ A, and ∡ L = ∡ C.

(375. An inscribed angle is half the angle at the center on the intercepted arc.)
(397. An angle formed by a tangent and a chord is half the angle at the center on the intercepted arc.)

BOOK IV.

REGULAR POLYGONS.

I. Partition of a Perigon.

PROBLEM I.

418. *To bisect a perigon.*

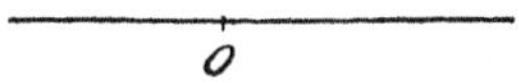

SOLUTION. To bisect the perigon at the point O, draw any line through O.

This divides the perigon into two straight angles, and all straight angles are equal.

419. COROLLARY. By drawing a second line through O at right angles to the first, we cut the perigon into four equal parts; and as we can bisect any angle, so we can cut the perigon into 8, 16, 32, 64, etc., equal parts.

PROBLEM II.

420. *To trisect a perigon.*

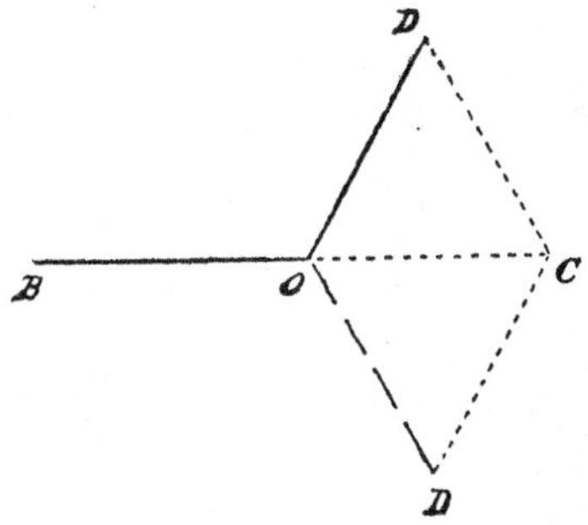

SOLUTION. To trisect the perigon at the point O, to O draw any line BO; on BO produced take a sect OC; on OC construct, by 132, an equilateral triangle CDO.

∡ DOB is one-third of a perigon.

For ∡ DOC is one-third of a st. ∡,

(174. The three angles of a triangle are equal to a straight angle.)

∴ ∡ DOB is two-thirds of a st. ∡.

421. COROLLARY. Since we can bisect any angle, so we may cut the perigon into 6, 12, 24, 48, etc., equal parts.

422. REMARK. To trisect *any* given angle is a problem beyond the power of strict Elementary Geometry, which allows the use of only the compasses and an unmarked ruler. There is an easy solution of it, which oversteps these limits only by using two marks on the straight-edge. The trisection of the angle, the duplication of the cube, and the quadrature of the circle, are the three famous problems of antiquity.

PROBLEM III.

423. *To cut a perigon into five equal parts.*

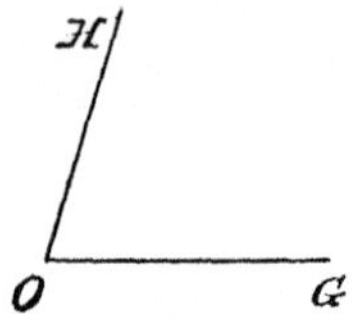

SOLUTION. By 314, describe an isosceles triangle ABC having

$$\measuredangle A = \measuredangle C = 2 \measuredangle B.$$

Then $\measuredangle A$ is two-fifths of a st. $\measuredangle$.

(174. The three angles of any triangle are equal to a straight angle.)

Therefore, to get a fifth of a perigon at a point O, construct, by 164,

$$\measuredangle GOH = \measuredangle A.$$

424. COROLLARY. Since we can bisect any angle, we may cut a perigon into 10, 20, 40, 80, etc., equal parts.

PROBLEM IV.

425. *To cut a perigon into fifteen equal parts.*

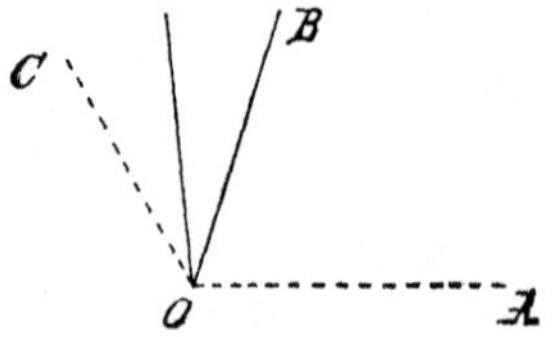

SOLUTION. At the perigon point O, by 420, construct the $\measuredangle AOC$ = one-third of a perigon.

By 423, make $\measuredangle AOB$ = one-fifth of a perigon.

Then of such parts, as a perigon contains fifteen, $\measuredangle AOC$ contains five, and $\measuredangle AOB$ contains three, therefore $\measuredangle BOC$ contains two.

So bisecting $\measuredangle BOC$ gives one-fifteenth of a perigon.

426. COROLLARY. Hence a perigon may be divided into 30, 60, 120, etc., equal parts.

II. Regular Polygons and Circles.

PROBLEM V.

427. *To inscribe in a circle a regular polygon having a given number of sides.*

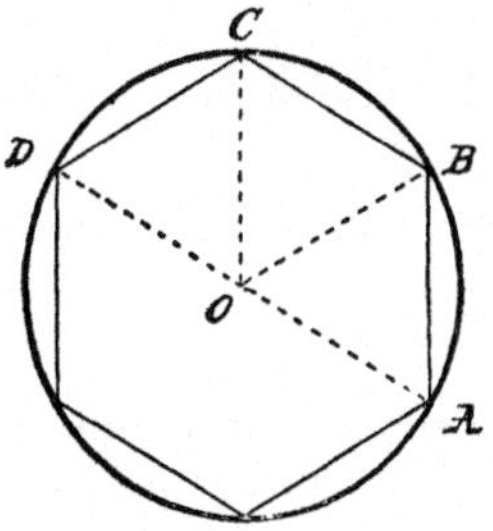

This problem can be solved if a perigon can be divided into the given number of equal parts.

For let the perigon at O, the center of the circle, be divided into a number of equal parts, and extend their arms to meet the circle in A, B, C, D, etc. Draw the chords, AB, BC, CD, etc.

Then shall $ABCD$, etc., be a regular polygon.

For if the figure be turned about its center O, until OA coincides with the trace of OB, therefore, because the angles are all equal, OB will coincide with the trace of OC, and OC with the trace of OD, etc.; then AB will coincide with the trace of BC, and BC with the trace of CD, etc.;

$$\therefore \quad AB = BC = CD = \text{etc.},$$

$\therefore$ the polygon is equilateral.

Moreover, since then ABC will coincide with the trace of BDC,

$$\therefore \quad \measuredangle ABC = \measuredangle BCD = \text{etc.},$$

$\therefore$ the polygon is equiangular.

Therefore $ABCD$, etc., is a regular polygon, and it is inscribed in the given circle.

428. REMARK. From the time of Euclid, about 300 B.C., no advance was made in the inscription of regular polygons until GAUSS, in 1796, found that a regular polygon of 17 sides was inscriptible, and in his abstruse Arithmetic, published in 1801, gave the following: —

In order that the geometric division of the circle into n parts may be possible, n must be 2, or a higher power of 2, or else a prime number of the form $2^m + 1$, or a product of two or more different prime numbers of that form, or else the product of a power of 2 by one or more different prime numbers of that form.

In other words, it is necessary that n should contain no odd divisor not of the form $2^m + 1$, nor contain the same divisor of that form more than once.

Below 300, the following 38 are the only possible values of n: 2, 3, 4, 5, 6, 8, 10, 12, 15, 16, 17, 20, 24, 30, 32, 34, 40, 48, 51, 60, 64, 68, 80, 85, 96, 102, 120, 128, 136, 160, 170, 192, 204, 240, 255, 256, 257, 272.

EXERCISES. 81. The square inscribed in a circle is double the square on the radius, and half the square on the diameter.

82. Prove that each diagonal is parallel to a side of the regular pentagon.

83. An inscribed equilateral triangle is equivalent to half a regular hexagon inscribed in the same circle.

84. An equilateral triangle described on a given sect is equivalent to one-sixth of a regular hexagon described on the same sect.

85. If a triangle is equilateral, show that the radius of the circumscribed circle is double that of the inscribed; and the radius of an escribed, triple.

86. The end points of a sect slide on two lines at right angles: find the locus of its mid-point.

Problem VI.

429. *To circumscribe about a given circle a regular polygon having a given number of sides.*

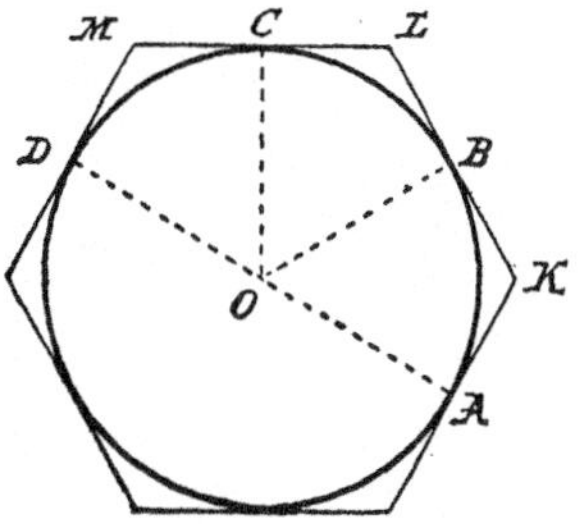

This problem can be solved if a perigon can be divided into the given number of equal parts.

For let the perigon at O, the center of the circle, be divided into a number of equal angles, and extend their arms to meet the circle in A, B, C, D, etc. Draw perpendiculars to these arms at A, B, C, D, etc.

These will be tangents.

Call their points of intersection K, L, M, etc.

Then shall KLM, etc., be a regular polygon.

For if the figure be turned about its center O until OA coincides with the trace of OB, then, because the angles are all equal, OB will coincide with the trace of OC, and OC with the trace of OD, etc.

Therefore the tangents at A, B, C, etc., will coincide with the traces of the tangents at B, C, D, etc.

Hence the polygon will coincide with its trace;

$$\therefore \quad KL = LM = \text{etc.},$$

also

$$\measuredangle K = \measuredangle L = \measuredangle M = \text{etc.};$$

therefore the polygon is regular, and it is circumscribed about the given circle.

430. Corollary. Hence we can circumscribe about a circle regular polygons of 3, 4, 5, 6, 8, 10, 12, 15, 16, 17, etc., sides.

Problem VII.

431. *To circumscribe a circle about a given regular polygon.*

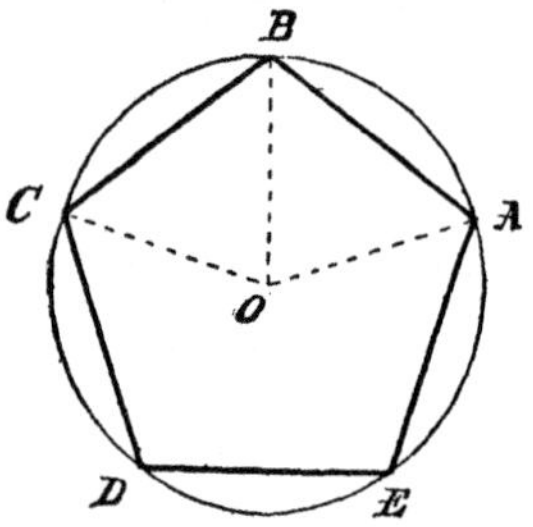

Given, *a regular polygon, as* $ABCDE$.

Required, *to describe a circle about it.*

Construction. Bisect $\measuredangle EAB$ and $\measuredangle ABC$ by lines intersecting in O. With center O and radius OA describe a circle.

This shall be the required circle.

Proof. Join OC. Then

$$\triangle OBC \cong \triangle OBA,$$

(124. Triangles having two sides and the included angle in each equal are congruent.)

$$\therefore \quad \measuredangle OCB = \measuredangle OAB = \text{half one } \measuredangle \text{ of the regular polygon,}$$

$$\therefore \quad \measuredangle BCD \text{ is bisected.}$$

Similarly prove each $\measuredangle$ of the polygon bisected,

$$\therefore \quad OA = OB = OC = OD = OE,$$

(148. In a triangle, sides opposite equal angles are equal.)

therefore a circle with radius OA passes through B, C, D, E, and is circumscribed about the given polygon.

PROBLEM VIII.

432. *To inscribe a circle in a given regular polygon.*

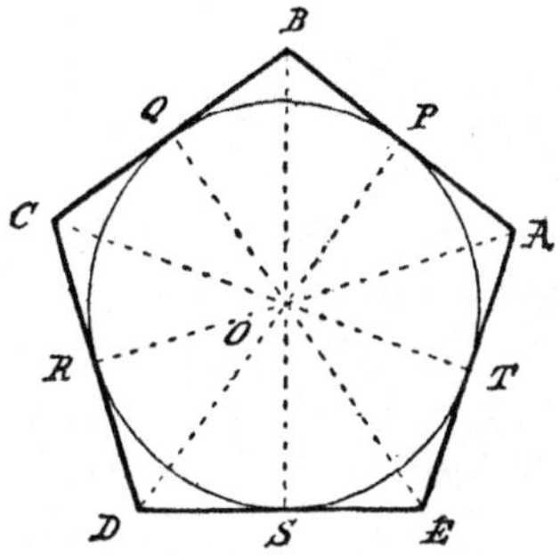

GIVEN, ***a regular polygon***, *ABCDE*.

REQUIRED, *to inscribe a circle in it.*

CONSTRUCTION. Bisect ∡ *EAB* and ∡ *ABC* by lines intersecting in *O*. From *O* drop *OP* perpendicular to *AB*.

The ⊙ with center *O* and radius *OP* shall be the ⊙ required.

PROOF. Join *OC*, *OD*, *OE*, and draw $OQ \perp BC$, $OR \perp CD$, $OS \perp DE$, and $OT \perp EA$. Then

$$\triangle\ OBC \cong \triangle\ OBA,$$

(124. Triangles having two sides and the included angle in each equal are congruent.)

$\therefore$ ∡ *OCB* = ∡ *OAB* = half one ∡ of the regular polygon,

$\therefore$ ∡ *BCD* is bisected.

Similarly prove each ∡ of the polygon bisected.

Again,

$$\triangle\ OBP \cong \triangle\ OBQ,$$

(176. Triangles having two angles and a corresponding side in each equal are congruent.)

$$\therefore\ OP = OQ.$$

Similarly,

$$OQ = OR = OS = OT,$$

therefore a circle with radius OP will touch AB, BC, CD, DE, EA, at points P, Q, R, S, T;

$\therefore$ it is inscribed in the given polygon.

433. COROLLARY I. The inscribed and circumscribed circles of a regular polygon are concentric.

434. COROLLARY II. The bisectors of the angles of a regular polygon all meet in a point which is the center both of the circumscribed and inscribed circles, and is called the center of the regular polygon.

435. COROLLARY III. The perpendicular bisectors of the sides of a regular polygon all pass through its center.

436. The radius of its circumscribed circle is called the *radius* of a regular polygon. The radius of its inscribed circle is called its *apothem.*

437. The side of a regular hexagon inscribed in a circle is equal to the radius. For the sects from the center to the ends of a side make an isosceles triangle, one of whose angles is one-third a straight angle; therefore it is equilateral.

III. Least Perimeter in Equivalent Figures. — Greatest Surface in Isoperimetric Figures.

438. Any two figures are called *Isoperimetric* when their perimeters are equal.

THEOREM I.

439. *Of all equivalent triangles having the same base, that which is isosceles has the least perimeter.*

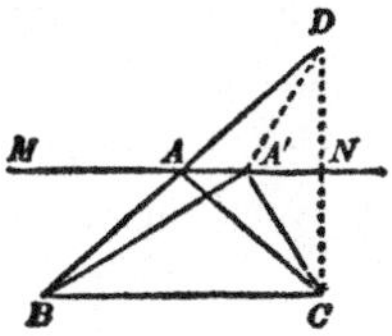

HYPOTHESIS. ***Let ABC be an isosceles triangle, and $A'BC$ any equivalent triangle having the same base.***

CONCLUSION. $AB + AC < A'B + A'C$.

PROOF. $AA' \parallel BC$.

(253. INVERSE. Equivalent triangles on the same base, and on the same side of it, are between the same parallels.)

Draw $CND \perp AA'$, meeting BA produced in D. Join $A'D$.

$$\measuredangle NAC = \measuredangle ACB,$$

(168. If a transversal cuts two parallels, the alternate angles are equal.)

$$\measuredangle ACB = \measuredangle ABC,$$

(126. In an isosceles triangle the angles opposite the equal sides are equal.)

$$\measuredangle ABC = \measuredangle DAN,$$

(169. If a transversal cuts two parallels, the corresponding angles are equal.)

$$\therefore \quad \triangle ACN \cong \triangle ADN,$$

(128. Triangles having two angles and the included side equal in each are congruent.)

$\therefore$ AN is the perpendicular bisector of CD,

$\therefore$ $AD = AC$, and $A'D = A'C$.

(183. The locus of the point to which sects from two given points are equal, is the perpendicular bisector of the sect joining them.)

But

$$BD < A'B + A'D,$$

(156. Any two sides of a triangle are together greater than the third.)

$$\therefore \quad AB + AC < A'B + A'C.$$

440. COROLLARY. Of all equivalent triangles, that which is equilateral has the least perimeter.

For the triangle having the least perimeter enclosing a given surface must be isosceles whichever side is taken as the base.

THEOREM II.

441. *Of all isoperimetric triangles having the same base, that which is isosceles has the greatest surface.*

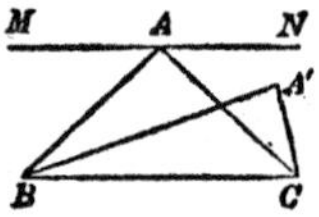

HYPOTHESIS. ***Let*** ABC ***be an isosceles triangle; and let*** $A'BC$ ***standing on the same base*** BC***, have an equal perimeter; that is,***

$$A'B + A'C = AB + AC.$$

CONCLUSION. $\triangle ABC > \triangle A'BC.$

PROOF. The vertex A' must fall between BC and the parallel AN; since, if it fell upon AN, by the preceding proof, $A'B + A'C > AB + AC$; and, if it fell beyond AN, the sum $A'B + A'C$ would be still greater.

Therefore the altitude of $\triangle ABC$ is greater than the altitude of $\triangle A'BC$, and hence also its surface.

442. COROLLARY. Of all isoperimetric triangles, that which is equilateral is the greatest.

For the greatest triangle having a given perimeter must be isosceles whichever side is taken as the base.

THEOREM III.

443. *Of all triangles formed with the same two given sides, that in which these sides are perpendicular to each other has the greatest surface.*

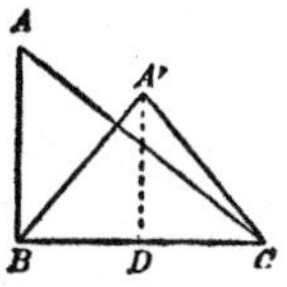

HYPOTHESIS. **Let *ABC*, *A'BC*, be two triangles having the sides *AB*, *BC*, respectively equal to *A'B*, *BC*; and let ∡ *ABC* be right.**

CONCLUSION. △ *ABC* > △ *A'BC*.

PROOF. Taking *BC* as the common base, the altitude *AB* of △ *ABC* is greater than the altitude *A'D* of △ *A'BC*.

THEOREM IV.

444. *Of all isoperimetric plane figures, the circle contains the greatest surface.*

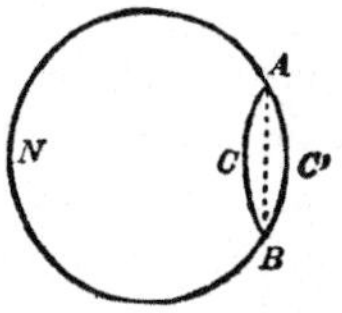

PROOF. With a given perimeter, there may be an indefinite number of figures differing in form and size. The surface may be as small as we please, but cannot be increased indefinitely.

Therefore, among all the figures of the same perimeter, there must be one greatest figure, or several equivalent greatest figures of different forms.

Every closed figure, if the greatest of a given perimeter, must be *convex;* that is, such that any sect joining two points of the perimeter lies wholly within the figure. For let *ACBNA* be a non-convex figure, the sect *AB*, joining two of the points in its perimeter, lying without the figure; then if the re-entrant portion *ACB* be revolved about the line *AB* into the position *AC'B*, the figure *AC'BNA* has the same perimeter as the first figure, but a greater surface.

Now let *AFBCA* be a figure of greatest surface formed with a given perimeter; then, taking any point *A* in its perimeter, and drawing

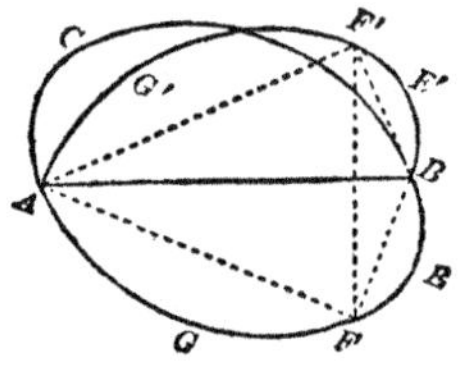

AB to bisect the perimeter, it also bisects the surface. For if the surface of one of the parts, as *AFB*, were greater than that of the other part, *ACB*, then if the part *AFB* were revolved upon the line *AB* into the position *AF'B*, the surface of the figure *AF'BFA* would be greater than that of the figure *AFBCA*, and yet would have the same perimeter.

Now the angles *AFB* and *AF'B* must be right angles, else the triangles *AFB* and *AF'B* could be increased, by 443, without varying the chords *AF*,*FB*, *AF'*,*F'B*, and then (the segments *AGF*, *FEB*, *AG'F'*, *F'E'B*, still standing on these chords) the whole figure would have increased without changing its perimeter.

But *F* is any point in the curve *AFB*; therefore this curve is a semicircle.

(384. The arc of a segment is a semicircle if the angle in it is right.)

Therefore the whole figure is a circle.

THEOREM V.

445. *Of all equivalent plane figures, the circle has the least perimeter.*

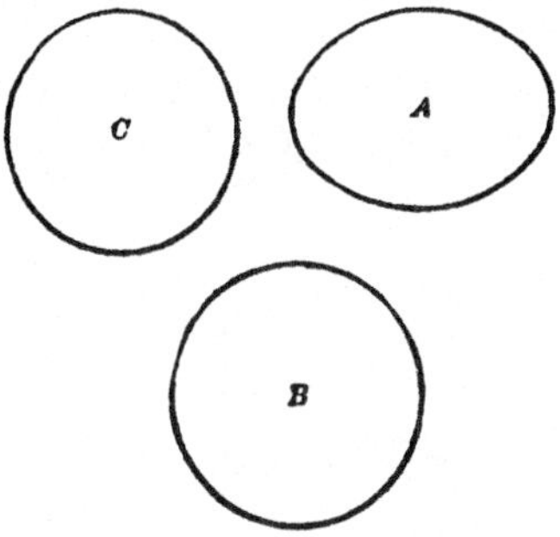

HYPOTHESIS. ***Let C be a circle, and A any other figure having the same surface as C.***

CONCLUSION. The perimeter of C is less than that of A.

PROOF. Suppose B a circle with the same perimeter as the figure A; then, by 444, $A < B$ $\therefore$ $C < B$.

But, of two circles, that which has the less surface has the less perimeter;

$\therefore$ perimeter of $C <$ perimeter of B, or of A.

THEOREM VI.

446. *Of all the polygons constructed with the same given sides, that is the greatest which can be inscribed in a circle.*

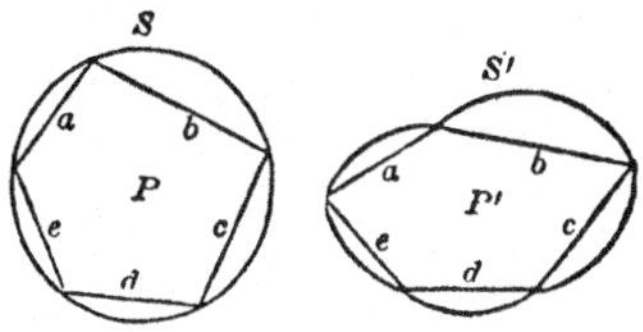

HYPOTHESIS. ***Let P be a polygon constructed with the sides a, b, c, d, e, and inscribed in a circle S, and let P' be any other***

polygon constructed with the same sides, and not inscriptible in a circle.

CONCLUSION. $P > P'$.

PROOF. Upon the sides a, b, c, etc., of the polygon P' construct circular segments equal to those standing on the corresponding sides of P. The whole figure S' thus formed has the same perimeter as the circle S, therefore, by 444, surface of $S > S'$; subtracting the circular segments from both, we have

$$P > P'.$$

THEOREM VII.

447. *Of all isoperimetric polygons having the same number of sides, the regular polygon is the greatest.*

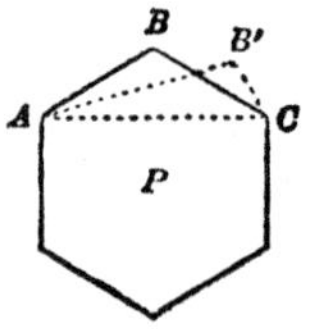

PROOF. I. The greatest polygon P, of all the isoperimetric polygons of the same number of sides, must have its sides equal; for if two of its sides, as AB', $B'C$, were unequal, we could, by 439, increase its surface by substituting for the triangle $AB'C$ the isoperimetric isosceles triangle ABC.

II. The greatest polygon constructed with the same number of equal sides must, by 446, be inscriptible in a circle. Therefore it is a regular polygon.

EXERCISES. 87. Of all triangles that can be inscribed in a given triangle, that whose vertices are the feet of the altitudes of the original triangle has the least perimeter.

Theorem VIII.

448. *Of all equivalent polygons having the same number of sides, the regular polygon has the least perimeter.*

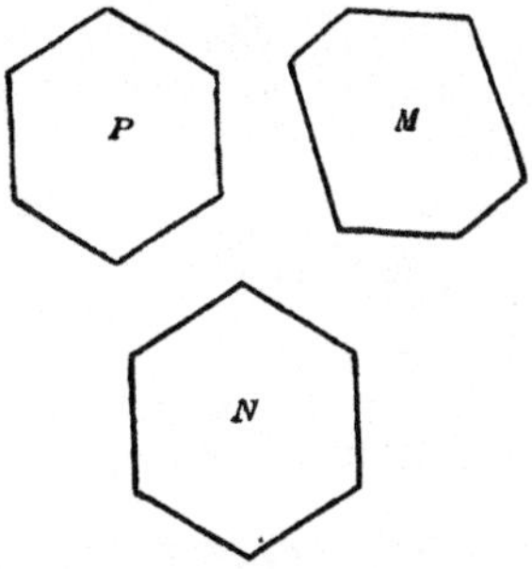

Hypothesis. ***Let P be a regular polygon, and M any equivalent irregular polygon having the same number of sides as P.***

Conclusion. The perimeter of P is less than that of M.

Proof. Let N be a regular polygon having the same perimeter and the same number of sides as M; then, by 447,

$$M < N, \quad \text{or} \quad P < N.$$

But, of two regular polygons having the same number of sides, that which has the less surface has the less perimeter; therefore the perimeter of P is less than that of N or of M.

Theorem IX.

449. *If a regular polygon be constructed with a given perimeter, its surface will be the greater, the greater the number of its sides.*

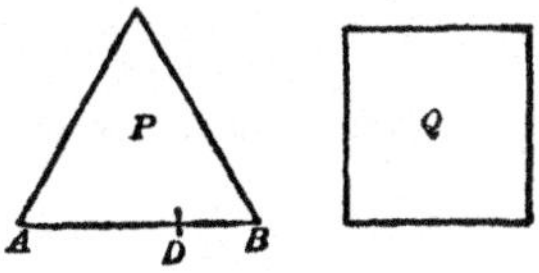

PROOF. Let P be the regular polygon of three sides, and Q the regular polygon of four sides, constructed with the same given perimeter.

In any side AB of P take any arbitrary point D; the polygon P may be regarded as an irregular polygon of four sides, in which the sides AD, DB, make a straight angle with each other; then, by 447, the irregular polygon P of four sides is less than the regular isoperimetric polygon Q of four sides.

In the same manner it follows that Q is less than the regular isoperimetric polygon of five sides, and so on.

THEOREM X.

450. *Of equivalent regular polygons, the perimeter will be the less, the greater the number of sides.*

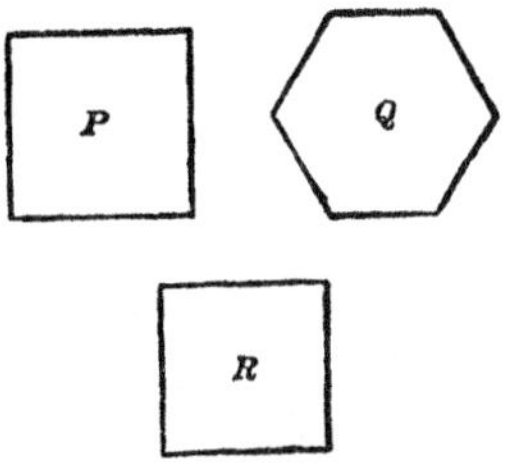

HYPOTHESIS. ***Let P and Q be equivalent regular polygons, and let Q have the greater number of sides.***

CONCLUSION. The perimeter of P will be greater than that of Q.

PROOF. Let R be a regular polygon having the same perimeter as Q and the same number of sides as P; then, by 449,

$$Q > R, \qquad \text{or} \qquad P > R;$$

therefore the perimeter of P is greater than that of R or of Q.

BOOK V.

RATIO AND PROPORTION.

Multiples.

451. NOTATION. In Book V., capital letters denote magnitudes.

Magnitudes which are or may be of different kinds are denoted by letters taken from different alphabets.

$A + A$ is abbreviated into $2A$.

$$A + A + A = 3A.$$

The small Italic letters m, n, p, q, denote whole numbers.

mA means A taken m times;

nA means A taken n times;

$$\therefore \quad mA + nA = (m + n)A.$$

nA taken m times is mnA.

452. A greater magnitude is said to be a *Multiple* of a lesser magnitude when the greater is the sum of a number of parts each equal to the less; that is, when the greater contains the less an exact number of times.

453. A lesser magnitude is a *Submultiple*, or *Aliquot Part*, of a greater magnitude when the less is contained an exact number of times in the greater.

454. When each of two magnitudes is a multiple of, or exactly contains, a third magnitude, they are said to be *Commensurable.*

455. If there is no magnitude which each of two given magnitudes will contain an exact number of times, they are called *Incommensurable.*

456. REMARK. It is important the student should know, that of two magnitudes of the same kind taken at hazard, or one being given, and the other deduced by a geometrical construction, it is very much more likely that the two should be incommensurable than that they should be commensurable.

To treat continuous magnitudes as commensurable would be to omit the normal, and give only the exceptional case. This makes the arithmetical treatment of ratio and proportion radically incomplete and inadequate for geometry.

PROBLEM I.

457. *To find the greatest common submultiple or greatest common divisor of two given magnitudes, if any exists.*

Let AB and CD be the two magnitudes.

From AB, the greater, cut off as many parts as possible, each equal to CD, the less. If there be a remainder FB, set it off in like manner as often as possible upon CD. Should there be a second remainder HD, set it off in like manner upon the first remainder, and so on.

The process will terminate only if a remainder is obtained which is an aliquot part of the preceding one; and, should it so terminate, the two given magnitudes will be commensurable, and have the last remainder for their greatest common divisor.

For suppose *HD* the last remainder.

Then *HD* is an aliquot part of *FB*, and so of *CH*, and therefore of *CD*, and therefore of *AF*. Thus being a submultiple of *AF* and *FB*, it is contained exactly in *AB*. And, moreover, it is the greatest common divisor of *AB* and *CD*.

For since every divisor of *CD* and *AB* must divide *AF*, it must divide *FB* or *CH*, and therefore also *HD*.

Hence the common divisor cannot be greater than *HD*.

458. INVERSE OF 457. If two magnitudes be commensurable, the above process will terminate.

For now, by hypothesis, we have a greatest common divisor *G*.

But *G* is contained exactly in every remainder.

For *G*, being a submultiple of *CD*, is also an aliquot part of *AF*, a multiple of *CD*; and therefore, to be a submultiple of *AB*, it must be an aliquot part of *FB* the first remainder. Similarly, *CD* and the first remainder *FB* being divisible by *G*, the second remainder *HD* must be so, and in the same way the third and every subsequent remainder.

But the alternate remainders decrease by more than half, and so the process must terminate at *G*; for otherwise a remainder would be reached which, being less than *G*, could not be divisible by *G*.

459. CONTRANOMINAL OF 457. The above process applied to incommensurable magnitudes is interminable.

460. OBVERSE OF 457. If the above process be interminable, the magnitudes are incommensurable.

On this depends the demonstration, given in 461, of a remarkable theorem proved in the tenth book of Euclid's "Elements."

THEOREM I.

461. *The side and diagonal of a square are incommensurable.*

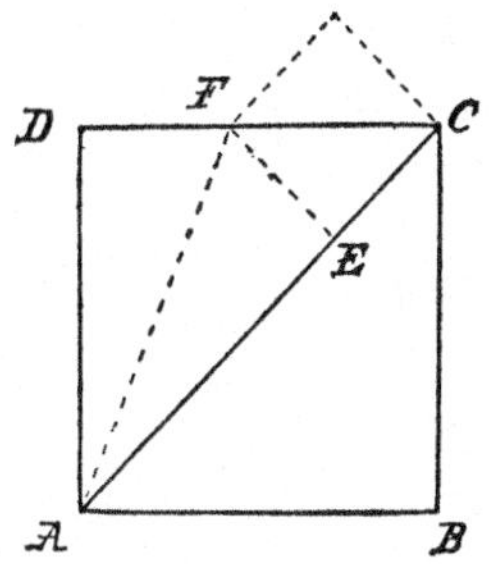

HYPOTHESIS. ***Let** ABCD **be a square;** AB, **a side;** AC, **a diagonal.***

CONCLUSION. Then will AB and AC be incommensurable.

PROOF. $AC > AB$, but $< 2AB$,

Therefore a first remainder EC is obtained by setting off on AC a part $AE = AB$.

Erect EF perpendicular to AC and meeting CD in F. Join AF.

$$\triangle ADF \cong \triangle AEF,$$

(179. Right triangles are congruent when the hypothenuse and one side are equal respectively in each.)

$$\therefore \quad DF = FE.$$

Again, rt. $\triangle$ CEF is isosceles, because one of the complemental angles, ECF, is half a rt. $\measuredangle$,

$$\therefore \quad CE = EF = FD.$$

Hence a common divisor of EC and DC would be also a common divisor of EC and FC.

But EC and FC are again the side and diagonal of a square, therefore the process is interminable.

462. *Another demonstration that the side and diagonal of any square are incommensurable.*

If you suppose them commensurable, let P represent their common measure; that is, the aliquot part which is contained an exact number of times in the side, and also an exact number of times in the diagonal.

Let m represent the number of times P is contained in the side S, and n the number of times P is contained in the diagonal D; so that $S = mP$, and $D = nP$, where m and n are whole numbers.

Then, by 243, the square on D = 2 sq. on S, and

$$\therefore \quad n^2 = 2m^2,$$

therefore n^2 is an even number,
therefore n is an even number, since the square of an odd number is odd; therefore n may be represented by $2q$,

$$\therefore \quad 4q^2 = 2m^2,$$

$$\therefore \quad 2q^2 = m^2,$$

therefore m must be even, and can be represented by $2t$, and therefore q must be even, and so on forever.

That is, m and n must be such whole numbers that they will divide by 2, and give quotients which will again divide by 2, and so on forever.

This is impossible; for even if m and n were high powers of 2, they would by constant division reduce to 1, an odd number.

SCALES OF MULTIPLES.

463. By taking magnitudes each equal to A, one, and two, three, four, etc., of them together we obtain a set of magnitudes depending upon A, and all known when A is known; namely, A, $2A$, $3A$, $4A$, $5A$, $6A$, . . . and so on, each being obtained by putting A to the preceding one.

This we shall call the scale of multiples of A.

464. If m be a whole number, mA and mB are called equimultiples of A and B, or the same multiples of A and B.

465. We assume: As A is greater than, equal to, or less than B, so is mA greater than, equal to, or less than mB respectively.

466. By 33, Rule of Inversion,

As mA is greater than, equal to, or less than mB, so is A greater than, equal to, or less than B respectively.

Theorem II.

467. *Commensurable magnitudes have also a common multiple.*

If A and B are commensurable magnitudes, there is some multiple of A which is also a multiple of B.

Proof. Let C be a common divisor of A and B.

The scale of multiples of C is

$$C, 2C, 3C \ldots .$$

Now, by hypothesis, one of the multiples of this scale, suppose pC, is equal to A, and one, suppose qC, is equal to B.

Hence, by 465, the multiple pqC is equal to qA, and the same multiple is equal to pB; therefore,

$$qA = pB.$$

468. Inverse of 467. Magnitudes which have a common multiple are commensurable.

Proof. If $pA = qB$, then $\frac{A}{q}$ will go p times into B, and q times into A.

469. Any whole number or fraction is commensurable with every whole number and fraction, being each divisible by unity over the product of the denominators.

To find a common multiple, we have only to multiply together the whole numbers and the numerators of the fractions.

470. Multiplication by a whole number or fraction is distributive,

$$m(A + B + \ldots) = mA + mB + \ldots$$

471. Multiplication being commutative for whole numbers or fractions,

$$\therefore \quad m(nA) = mnA = nmA = n(mA).$$

472. Magnitudes which are of the same kind can, being multiplied, exceed each the other.

Scale of Relation.

473. The *Scale of Relation* of two magnitudes of the same kind is a list of the multiples of both, all arranged in ascending order of magnitude; so that, any multiple of either magnitude being assigned, the scale of relation points out between which multiples of the other it lies.

474. If we call the side of any square S, and its diagonal D, their scale of relation will commence thus, —

S, D, $2S$, $2D$, $3S$, $4S$, $3D$, $5S$, $4D$, $6S$, $7S$, $5D$, $8S$, $6D$, $9S$, $7D$, $10S$, $11S$, $8D$, $12S$, $9D$, $13S$, $14S$, $10D$, $15S$, $11D$, $16S$, $12D$, $17S$, $18S$, $13D$, $19S$, $14D$, $20S$, $21S$, $15D$, etc.

$141S$, $100D$, $142S$, etc.

$1414213S$, $1000000D$, $1414214S$, etc.

$141421356S$, $100000000D$, $141421357S$, etc.

And we have proved that no multiple of S will ever equal any multiple of D.

475. If A be less than B, one multiple at least of the scale of A will lie between each two consecutive multiples of the scale of B.

Moreover, if A and B are two finite magnitudes of the same kind, however small A may be, we may, by continuing the scale of multiples of A sufficiently far, at length obtain a multiple of A greater than B.

So we are justified in saying,

I. We can always take mA greater than B or pB.

II. We can always take nA such that it is greater than pB, but not greater than qB, provided that A is less than B, and p than q.

476. The scale of relation of two magnitudes will be changed if one is altered in size ever so little; for some multiple of the altered magnitude can be found which will exceed, or fall short of, the same multiple of it before alteration, by more than the other original magnitude, and consequently the interdistribution of the multiples of the two original magnitudes will differ from the interdistribution of the multiples of one of the original magnitudes and the second altered.

Hence, when two magnitudes are known, the order of their multiples is fixed and known.

Inversely, if by any means the order of the multiples is known, and also one of the original magnitudes, the other is of fixed size, even though we may not yet be in a condition to find it.

477. The order in which the multiples of A lie among the multiples of B, when all are arranged in ascending order of magnitude, and the series of multiples continued indefinitely, determines what is called the *Ratio* of A to B, and written $A : B$, in which the first magnitude is called the *Antecedent*, and the second the *Consequent*.

478. Hitherto we have compared one magnitude to another, with respect to quantity, only in general, according to the logical division greater than, equal to, less than.

But double, triple, quadruple, quintuple, sextuple, are special cases of a more subtile relation which exists between every two magnitudes of the same kind.

Ratio is the relation of one magnitude to another with respect to quantuplicity.

479. If the multiples of A and B interlie in the same order as the multiples of C and D, then the ratio A to B is the same as the ratio C to D, and the four magnitudes are said to be *Proportional*, or to form a *Proportion.*

This is written $A : B :: C : D$, and read "A to B is the same as C to D."

The second pair of magnitudes may be of a different kind from the first pair.

A and D are called the *Extremes*, B and C the *Means*, and D is said to be a *Fourth Proportional* to A, B, and C.

480. In the special case when A and B are commensurable, we can estimate their quantuple relation by considering what multiples they are of some common standard; and so we can get two numbers whose ratio will be the same as the ratio of A to B.

481. The ratio of two magnitudes is the same as the ratio of two other magnitudes, when, any equimultiples whatsoever of the antecedents being taken, and likewise any equimultiples whatsoever of the consequents, the multiple of one antecedent is greater than, equal to, or less than, that of its consequent, according as the multiple of the other antecedent is greater than, equal to, or less than, that of its consequent.

482. Thus, the ratio of A to B is the same as the ratio of C to D when mA is greater than, equal to, or less than nB, according as mC is greater than, equal to, or less than nB, whatever whole numbers m and n may be.

483. Three magnitudes (A, B, C) of the same kind are said to be proportionals when the ratio of the first to the second is the same as the ratio of the second to the third; that is, when

$$A : B :: B : C.$$

In this case, C is said to be the *Third Proportional* to A and B, and B the *Mean Proportional* between A and C.

THEOREM III.

484. *Ratios which are the same as the same ratio are the same as one another.*

HYPOTHESIS. $A : B :: C : D$, and $A : B :: X : Y$.

CONCLUSION. $C : D :: X : Y$.

PROOF. By the inverse of 479, the multiples of C and D have the same interorder as those of A and B; and, thc same being true of the multiples of X and Y, therefore the multiples of X and Y have the same interorder as those of C and D.

485. REMARK. Thus ratios come under the statement, "Things equal to the same thing are equal to each other;" and a proportion may be spoken of as an equality of ratios.

486. Since the inverse of a complete definition is true, therefore if a pair of numbers p and q can be found such that pA is greater than, equal to, or less than qB when pC is not respectively greater than, equal to, or less than qD, then the ratio of A to B is unequal to the ratio of C to D.

487. The first is to be to the second in a greater (or less) ratio than the third to the fourth when a multiple of the first takes a more (or less) advanced position among the multiples of the second than the same multiple of the third takes among those of the fourth.

488. The ratio of A to B is greater than that of C to D when two whole numbers m and n can be found such that mA is greater than nB, while mC is not greater than nD; or such that mA is equal to nB, while mC is less than nD.

THEOREM IV.

489. *Equal magnitudes have the same ratio to the same magnitude, and the same has the same ratio to equal magnitudes.*

For if $A = B$,

$$\therefore \quad mA = mB.$$

So if $mA > nC$,

$$\therefore \quad mB > nC;$$

and if equal, equal; if less, less;

$$\therefore \quad A : C :: B : C.$$

And also if $pT > mA$,

$$\therefore \quad pT > mB;$$

and if equal, equal; if less, less;

$$\therefore \quad T : A :: T : B.$$

Theorem V.

490. *Of two unequal magnitudes, the greater has a greater ratio to any other magnitude than the less has; and the same magnitude has a greater ratio to the less, of two other magnitudes, than it has to the greater.*

If $A > B$, m can be found such that mB is less than mA by a greater magnitude than C.

Hence, if $mA = nC$, or if mA is between nC and $(n + 1)C$, mB will be less than nC; therefore, by 488,

$$A : C > B : C.$$

Also, since $nC > mB$, while nC is not $> mA$,

$$\therefore \quad C : B > C : A.$$

491. From 489 and 490, by 33, Rule of Inversion,

If $A : C :: B : C$, or if $T : A :: T : B$,

$$\therefore \quad A = B.$$

If $A : C > B : C$, or if $T : A < T : B$,

$$\therefore \quad A > B.$$

If $A : C < B : C$, or if $T : A > T : B$,

$$\therefore \quad A < B.$$

THEOREM VI.

492. *If any number of magnitudes be proportionals, as one of the antecedents is to its consequent, so will all the antecedents taken together be to all the consequents.*

Let

$$A : B :: C : D :: E : F,$$

then

$$A : B :: A + C + E : B + D + F.$$

For as $mA >, =,$ or $< nB$, so, by inverse of 482, is $mC >, =,$ or $< nD$; and so also is $mE >, =,$ or $< nF$;

$\therefore$ so also is $mA + mC + mE >, =,$ or $< nB + nD + nF$,

and therefore so is $m(A + C + E) >, =,$ or $< n(B + D + F)$;

whence

$$A : B :: A + C + E : B + D + F.$$

THEOREM VII.

493. *The ratio of the equimultiples of two magnitudes is the same as the ratio of the magnitudes themselves.*

$$mA : mB :: A : B.$$

For as $pA >, =,$ or $< qB$, so is $mpA >, =,$ or $< mqB$;

but $mpA = pmA$, and $mqB = qmB$,

$\therefore$ as $pA >, =,$ or $< qB$,

so is $pmA >, =,$ or $< qmB$, whatever be the values of p and q;

$$\therefore\ A : B :: mA : mB.$$

THEOREM VIII.

494. *If four magnitudes be proportionals, then, if the first be greater than the third, the second will be greater than the fourth; and if equal, equal; and if less, less.*

GIVEN, $A : B :: C : D$.

If $A > C$,

$$\therefore \quad A : B > C : B;$$

therefore, from our hypothesis,

$$C : D > C : B;$$

therefore, by 491,

$$B > D.$$

If $A = C$,

$$\therefore \quad A : B :: C : B,$$

$$\therefore \quad C : D :: C : B,$$

$$\therefore \quad B = D.$$

If $A < C$,

$$\therefore \quad A : B < C : B,$$

$$\therefore \quad C : D < C : B,$$

$$\therefore \quad B < D.$$

THEOREM IX.

495. *If four magnitudes of the same kind be proportionals, they will also be proportionals when taken alternately.*

Let $A : B :: C : D$, the four magnitudes being of the same kind, then alternately,

$$A : C :: B : D.$$

For, by 493,

$$mA : mB :: A : B :: C : D :: nC : nD,$$

$$\therefore \quad mA : mB :: nC : nD;$$

therefore, by 494, $mA >, =$, or $< nC$, as $mB >, =$, or $< nD$; and, this being true for all values of m and n,

$$\therefore \quad A : C :: B : D.$$

Theorem X.

496. *If two ratios are equal, the sum of the antecedent and consequent of the first has to the consequent the same ratio as the sum of the antecedent and consequent of the other has to its consequent.*

If $A : B :: C : D$,
then

$$A + B : B :: C + D : D.$$

Proof. Whatever multiple of $A + B$ we choose to examine, take the same multiple of A, say $17A$, and let it lie between some two multiples of B, say $23B$ and $24B$; then, by hypothesis, $17C$ lies between $23D$ and $24D$.

Add $17B$ to all the first, and $17D$ to all the second; then $17(A + B)$ lies between $40B$ and $41B$, and $17(C + D)$ between $40D$ and $41D$; and in the same manner for any other multiples.

BOOK VI.

RATIO APPLIED.

I. Fundamental Geometric Proportions.

THEOREM I.

497. *If two lines are cut by three parallel lines, the intercepts on the one are to one another in the same ratio as the corresponding intercepts on the other.*

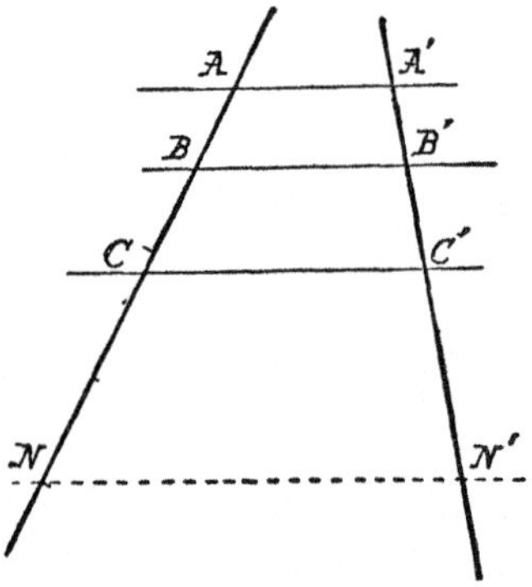

HYPOTHESIS. ***Let the three parallels*** AA', BB', CC', ***cut two other lines in*** A, B, C, ***and*** A', B', C', ***respectively.***

CONCLUSION. $AB : BC :: A'B' : B'C'$.

PROOF. On the line ABC, by laying off m sects $= AB$, take $BM = mAB$, and, in the same way, $BN = n \,.\, BC$, taking M and N on the same side of B. From M and N draw lines $\parallel AA'$, cutting $A'B'C'$ in M' and N'.

$$\therefore \; B'M' = m \,.\, A'B', \quad \text{and} \quad B'N' = n \,.\, B'C'.$$

(227. If three or more parallels intercept equal sects on one transversal, they intercept equal sects on every transversal.)

But whatever be the numbers m and n, as BM (or $m \,.\, AB$) is $>$, $=$, or $<$ BN (or $n \,.\, BC$),
so is $B'M'$ (or $m \,.\, A'B'$) respectively $>$, $=$, or $<$ $B'N'$ (or $n \,.\, B'C'$);

$$\therefore \; AB : BC :: A'B' : B'C'.$$

498. REMARK. Observe that the reasoning holds good, whether B is between A and C, or beyond A, or beyond C.

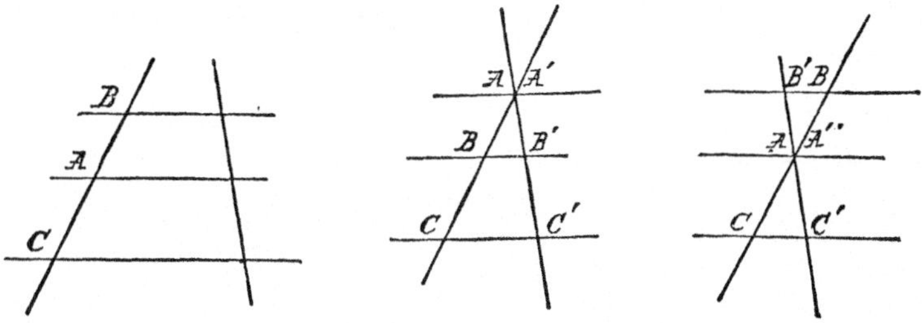

499. COROLLARY I. If the points A and A' coincide, the figure ACC' will be a triangle; therefore a line parallel to one side of a triangle divides the other two sides proportionally.

500. COROLLARY II. If two lines are cut by four parallel lines, the intercepts on the one are to one another in the same ratio as the corresponding intercepts on the other.

501. If a sect AB is produced, and the line cut at a point P outside the sect AB, the sect AB is said to be *divided externally* at P, and AP and BP are called *External Segments* of AB.

In distinction, if the point *P* is on the sect *AB*, it is said to be *divided internally*.

THEOREM II.

502. *A given sect can be divided internally into two segments having the same ratio as any two given sects, and also externally unless the ratio be one of equality; and, in each case, there is only one such point of division.*

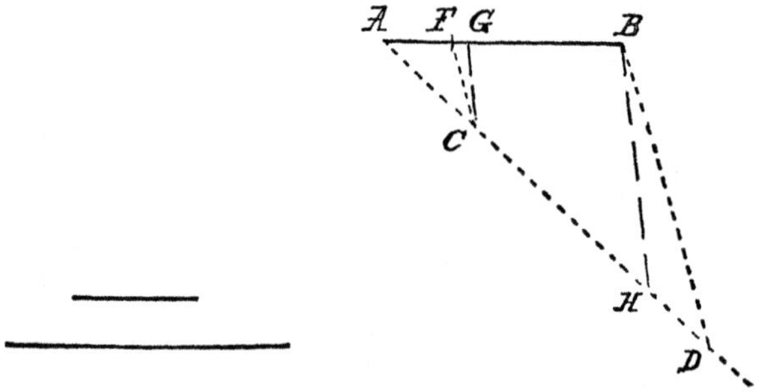

GIVEN, *the sect* *AB*.

On a line from *A* making any angle with *AB*, take *AC* and *CD* equal to the two given sects. Join *BD*.

Draw *CF* ‖ *DB* and meeting *AB* in *F*. By 497, *AB* is divided internally at *F* in the given ratio.

If it could be divided internally at *G* in the same ratio, *BH* being drawn ‖ *CG* to meet *AD* in *H*,

AG would be to *GB* as *AC* to *CH*, and therefore not as *AC* to *CD*.

(476. The scale of relation of two magnitudes will be changed if one is altered in size ever so little.)

Hence *F* is the only point which divides *AB* internally in the given ratio.

Again, if CD be taken so that A and D are on the same side of C, the like construction will determine the external point of division.

In this case the construction will fail if $CD = AC$, for D would coincide with A.

As above, we may also prove that there can be only one point of external division in the given ratio.

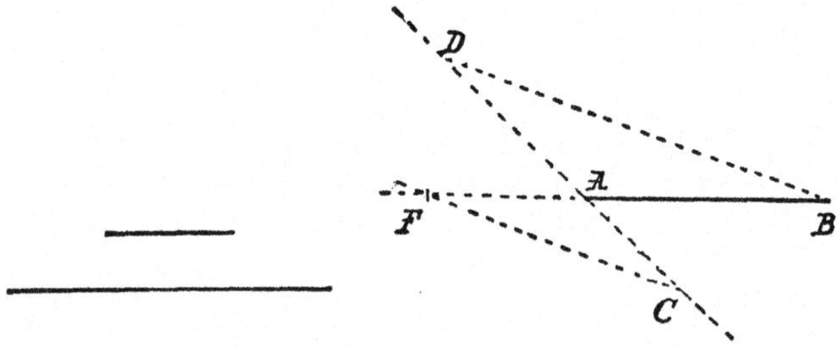

503. Inverse of 499. A line which divides two sides of a triangle proportionally is parallel to the third.

For a parallel from one of the points would divide the second side in the same ratio, but there is only one point of division of a given sect in a given ratio.

Theorem III.

504. *Rectangles of equal altitude are to one another in the same ratio as their bases.*

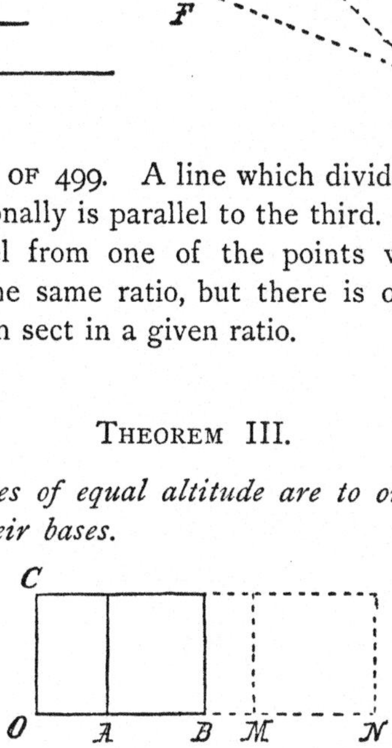

Let AC, BC, be two rectangles having the common side OC, and their bases OA, OB, on the same side of OC.

In the line OAB take $OM = m.OA$, and $ON = n.OB$, and complete the rectangles MC and NC.

Then $MC = m \, . \, AC$, and $NC = n \, . \, BC$; and as OM is >, =, or < ON, so is MC respectively >, =, or < NC;

$\therefore$ rectangle AC : rectangle BC :: base OA : base OB.

505. Corollary. Parallelograms or triangles of equal altitude are to one another as their bases.

Theorem IV.

506. *In the same circle, or in equal circles, angles at the center and sectors are to one another as the arcs on which they stand.*

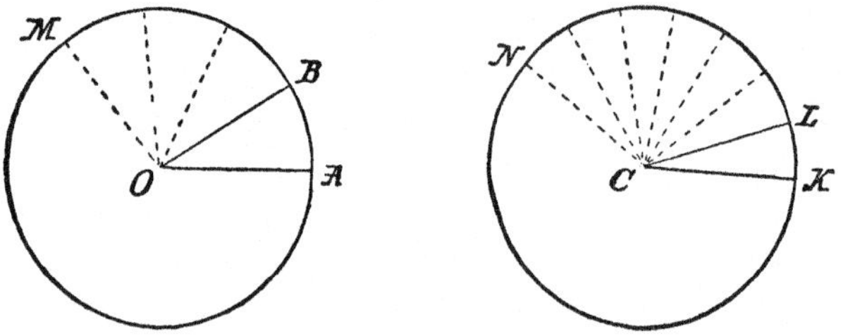

Let O and C be the centers of two equal circles; AB, KL, any two arcs in them.

Take an arc $AM = m \, . \, AB$;

then the angle or the sector between OA and OM equals $m \, . \, AOB$.

(365. In equal circles, equal arcs subtend equal angles at the center.)

Also take an arc $KN = n \, . \, KL$; then the angle or sector between CK and CN equals $n \, . \, KCL$.

But as AOM >, =, or < KCN, so respectively is arc AM >, =, or < arc KN;

(370 and 372. In equal circles, equal angles at the center or equal sectors intercept equal arcs, and of two unequal angles or sectors the greater has the greater arc.)

$\therefore$ AOB : KCL :: arc AB : arc KL.

II. Similar Figures.

507. *Similar figures are those of which the angles taken in the same order are equal, and the sides between the equal angles proportional.*

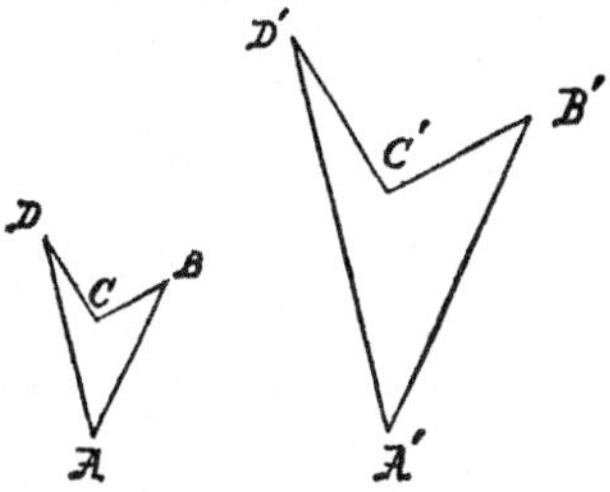

The figure $ABCD$ is similar to $A'B'C'D'$ if $\measuredangle A = \measuredangle A'$, $\measuredangle B = \measuredangle B'$, etc., and also

$$AB : A'B' :: BC : B'C' :: CD : C'D' :: \text{etc.}$$

THEOREM V.

508. *Mutually equiangular triangles are similar.*

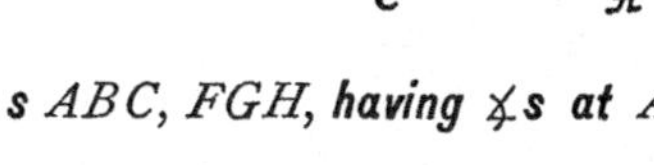

HYPOTHESIS. △s ABC, FGH, **having** $\measuredangle$**s at** A, B, **and** C, $=$ $\measuredangle$**s at** F, G, **and** H.

CONCLUSION. $AB : FG :: BC : GH :: CA : HF$.

PROOF. Apply △ FGH to △ ABC so that the point G coincides with B, and GH falls on BC.

Then, because $\measuredangle G = \measuredangle B$,

$\therefore$ GF falls on BA.

Now, since $\measuredangle BF'H' = \measuredangle A$ by hypothesis,

$$\therefore \quad F'H' \parallel AC,$$

(166. If corresponding angles are equal, the lines are parallel.)

therefore, by 498, $AB : BF' :: CB : BH'$.

In the same way, by applying the $\triangle FGH$ so that the $\measuredangle$s at H and C coincide, we may prove that $BC : GH :: CA : HF$.

509. COROLLARY. A triangle is similar to any triangle cut off by a line parallel to one of its sides.

THEOREM VI.

510. *Triangles having their sides taken in order proportional are similar.*

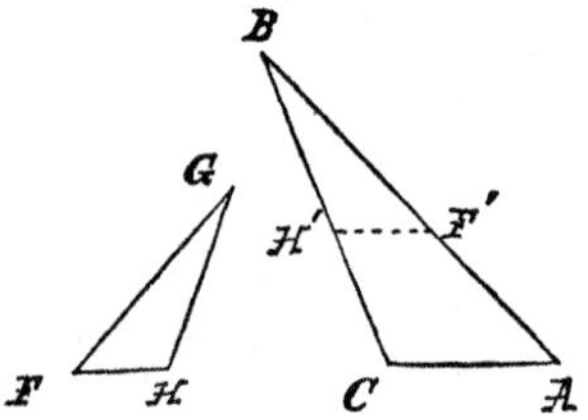

HYPOTHESIS. $AB : FG :: BC : GH :: CA : HF$.

CONCLUSION. $\measuredangle C = \measuredangle H$, and $\measuredangle A = \measuredangle F$.

PROOF. On BA take $BF' = GF$, and draw $F'H' \parallel CA$;

$$\therefore \quad AB : F'B :: BC : BH' :: CA : H'F'.$$

(508. Mutually equiangular triangles are similar.)

Since $FG = F'B$, $\quad \therefore \quad BC : GH :: BC : BH'$,

(484. Ratios equal to the same ratio are equal.)

therefore, by 491, $\quad GH = BH'$.

In the same way $\quad HF = H'F'$,

$$\therefore \quad \triangle FGH \cong \triangle F'BH'.$$

But $F'BH'$ is similar to ABC.

Theorem VII.

511. *Two triangles having one angle of the one equal to one angle of the other, and the sides about these angles proportional, are similar.*

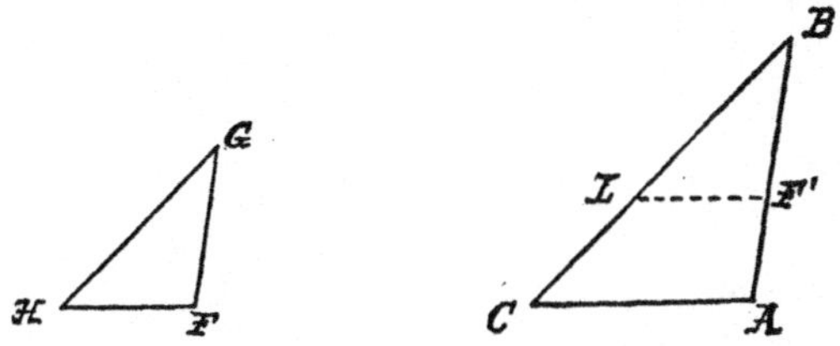

Hypothesis. $\measuredangle B = \measuredangle G$, and $AB : BC :: FG : GH.$
Conclusion. $\triangle ABC \sim \triangle FGH$ (using $\sim$ for the word "similar").
Proof. In BA take $BF' = GF$, and draw $F'L \parallel AC$,

$$\therefore \quad \triangle F'BL \sim \triangle ABC,$$

(508. Equiangular triangles are similar.)

$$\therefore \quad AB : BC :: F'B : BL.$$

But $F'B = FG$ by construction; therefore, from our hypothesis,

$$AB : BC :: F'B : GH,$$

$$\therefore \quad F'B : BL :: F'B : GH,$$

therefore, by 491,

$$BL = GH,$$

$$\therefore \quad \triangle F'BL \cong \triangle FGH,$$

(124. Triangles having two sides and the included angle respectively equal are congruent.)

$$\therefore \quad \triangle FGH \sim \triangle ABC.$$

THEOREM VIII.

512. *If two triangles have two sides of the one proportional to two sides of the other, and an angle in each opposite one corresponding pair of these sides equal, the angles opposite the other pair are either equal or supplemental.*

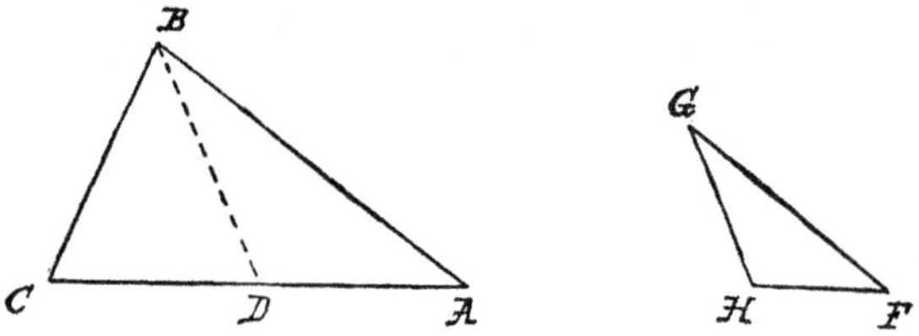

The angles included by the proportional sides are either equal or unequal.

CASE I. If they are equal, then the third angles are equal.

(174. The sum of the angles of a triangle is a straight angle.)

CASE II. If the angles included by the proportional sides are unequal, one must be the greater.

HYPOTHESIS. ABC *and* FGH $\triangle$s *with* $AB : BC :: FG : GH$, $\measuredangle A = \measuredangle F$, $\measuredangle B > \measuredangle G$.

CONCLUSION. $\measuredangle C + \measuredangle H =$ st. $\measuredangle$.

PROOF. Make $\measuredangle ABD = \measuredangle G$;

$$\therefore \quad \triangle ABD \sim \triangle FGH;$$

(508. Equiangular triangles are similar.)

$$\therefore \quad AB : BD :: FG : GH;$$

therefore, from our hypothesis,

$$AB : BD :: AB : BC;$$

therefore, by 491, $BD = BC$,

$$\therefore \quad \measuredangle C = \measuredangle BDC.$$

But $\measuredangle BDC + \measuredangle BDA =$ st. $\measuredangle$,

$$\therefore \quad \measuredangle C + \measuredangle H = \text{st. } \measuredangle.$$

513. COROLLARY. If two triangles have two sides of the one proportional to two sides of the other, and an angle in each opposite one corresponding pair of these sides equal, then if one of the angles opposite the other pair is right, or if they are oblique, but not supplemental, or if the side opposite the given angle in each triangle is not less than the other proportional side, the triangles are similar.

514. In similar figures, sides between equal angles are called *Homologous*, or corresponding. The ratio of a side of one polygon to its homologous side in a similar polygon is called the *Ratio of Similitude* of the polygons. Similar figures are said to be *similarly placed* when each side of the one is parallel to the corresponding side of the other.

THEOREM IX.

515. *If two unequal similar figures are similarly placed, all lines joining a vertex of one to the corresponding vertex of the other are concurrent.*

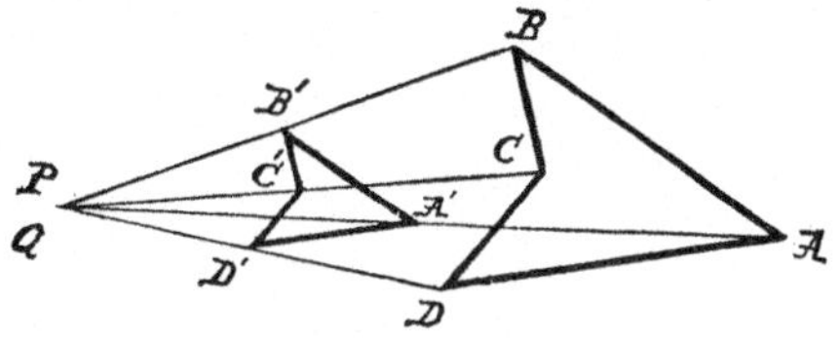

HYPOTHESIS. *ABCD and A'B'C'D' two similar figures similarly placed.*

CONCLUSION. The lines AA', BB', CC', etc., meet in a point P.

PROOF. Since AB and $A'B'$ are unequal,

$$\therefore \quad AA' \text{ and } BB' \text{ are not } \parallel.$$

Call their point of intersection P. Then

$$AP : A'P :: BP : B'P :: AB : A'B'.$$

(508. Equiangular triangles are similar.)

In the same way, if BB' and CC' meet in Q, then

$$BQ : B'Q :: BC : B'C'.$$

But, by hypothesis,

$$AB : A'B' :: BC : B'C',$$

$$\therefore BP : B'P :: BQ : B'Q,$$

$\therefore$ the point Q coincides with P.

(502. A line can be cut only at a single point into external segments having a given ratio.)

516. Corollary. Similar polygons may be divided into the same number of triangles similar and similarly placed.

For if, with their corresponding sides parallel, one of the polygons were placed inside the other, the lines joining corresponding vertices would so divide them.

517. The point of concurrence of the lines joining the equal angles of two similar and similarly placed figures is called the *Center of Similitude* of the two figures.

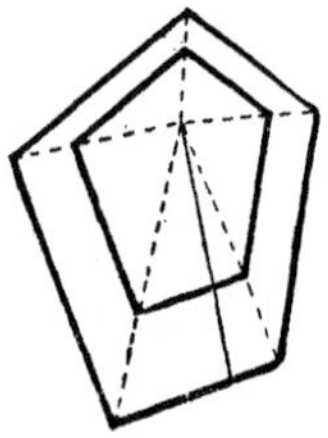

518. Corollary. The sects from the center of similitude along any line to the points where it meets corresponding sides of the similar figures are in the ratio of those sides.

Exercises. 87. Construct a polygon similar to a given polygon, the ratio of similitude of the two polygons being given.

THEOREM X.

519. *A perpendicular from the right angle to the hypothenuse divides a right-angled triangle into two other triangles similar to the whole and to one another.*

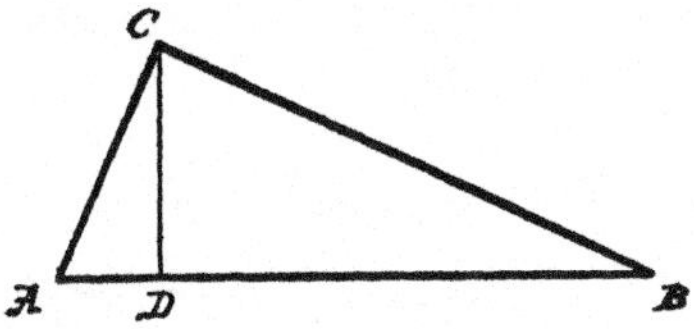

HYPOTHESIS. $\triangle ABC$ **right-angled at** C.

$CD \perp AB$.

CONCLUSION. $\triangle ACD \sim \triangle ABC \sim \triangle CBD$.

PROOF. $\measuredangle CAD = \measuredangle BAC$, and rt. $\measuredangle ADC =$ rt. $\measuredangle ACB$,

$\therefore \quad \measuredangle ACD = \measuredangle ABC$,

(174. The three angles of any triangle are equal to a straight angle.)

$\therefore \quad \triangle ACD \sim \triangle ABC$.

In the same way we may prove $\triangle CBD \sim \triangle ABC$;

$\therefore \quad \triangle ACD \sim \triangle CBD$.

520. COROLLARY I. Each side of the right triangle is a mean proportional between the hypothenuse and its adjacent segment.

For since $\triangle ACD \sim \triangle ABC$,

$\therefore \quad AB : AC :: AC : AD$.

521. COROLLARY II. The perpendicular is a mean proportional between the segments of the hypothenuse.

522. COROLLARY III. Since, by 383, lines from any point in a circle to the ends of a diameter form a right angle, therefore, if from any point of a circle a perpendicular be dropped

upon a diameter, it will be a mean proportional between the segments of the diameter.

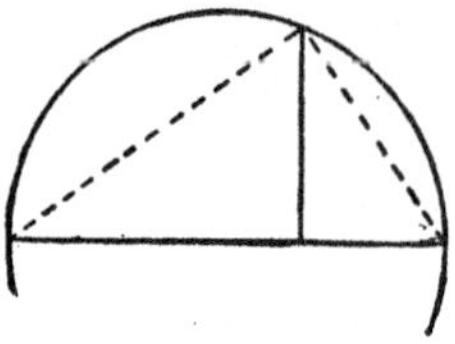

Theorem XI.

523. *The bisector of an interior or exterior angle of a triangle divides the opposite side internally or externally in the ratio of the other two sides of the triangle.*

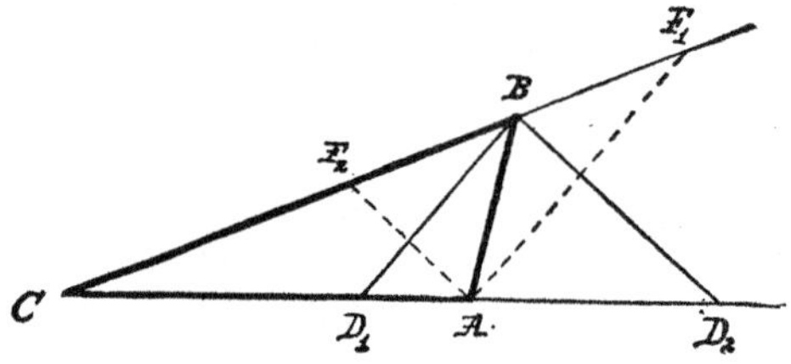

Hypothesis. ABC any △. BD the bisector of ∡ at B.
Conclusion. $AB : BC :: AD : DC$.
Proof. Draw $AF \parallel BD$.

Then, of the two angles at B given equal by hypothesis, one equals the corresponding interior angle at F, and the other the corresponding alternate angle at A,

(168. A line cutting two parallels makes alternate angles equal, and (169) corresponding angles equal.)

$$\therefore \quad AB = BF.$$

(126. If two angles of a triangle are equal, the sides opposite are equal.)

But $\quad BF : BC :: AD : DC,$

(499. If a line be parallel to a side of a triangle, it cuts the other sides proportionally.)

$$\therefore \quad AB : BC :: AD : DC.$$

524. INVERSE. Since, by 502, a sect can be cut at only one point internally in a given ratio, and at only one point externally in a given ratio, therefore, by 32, Rule of Identity, if one side of a triangle is divided internally or externally in the ratio of the other sides, the line drawn from the point of division to the opposite vertex bisects the interior or exterior angle.

525. When a sect is divided internally and externally into segments having the same ratio, it is said to be *divided harmonically.*

526. COROLLARY. The bisectors of an interior and exterior angle at one vertex of a triangle divide the opposite side harmonically.

THEOREM XII.

527. *If a sect,* AB, *is divided harmonically at the points* P *and* Q, *the sect* PQ *will be divided harmonically at the points* A *and* B.

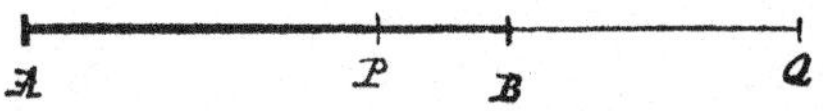

HYPOTHESIS. ***The sect AB divided internally at P, and externally at Q, so that***

$$AP : BP :: AQ : BQ;$$

therefore, by inversion,

$$BP : AP :: BQ : AQ;$$

therefore, by alternation,

$$BP : BQ :: AP : AQ.$$

528. The points *A, B,* and *P, Q,* of which each pair divide harmonically the sect terminated by the other pair, are called four *Harmonic Points.*

III. Rectangles and Polygons.

THEOREM XIII.

529. *If four sects are proportional, the rectangle contained by the extremes is equivalent to the rectangle contained by the means.*

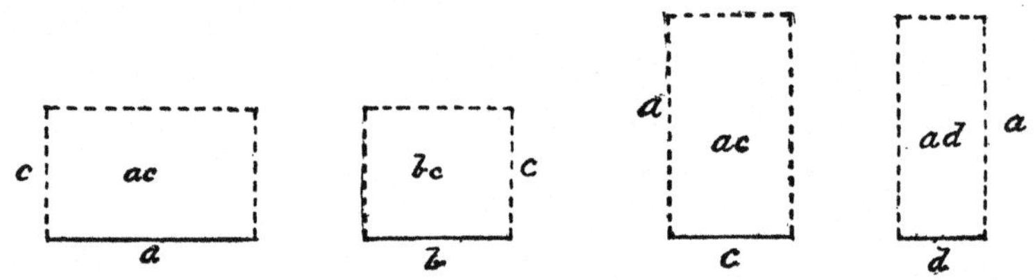

Let the four sects a, b, c, d, be proportional.

Then rectangle $ad = bc$.

PROOF. On a and on b construct rectangles with altitude $= c$. On c and on d construct rectangles of altitude a.
Then

$a : b :: ac : bc$, and $c : d :: ac : ad$.

(504. Rectangles of equal altitudes are to each other as their bases.)

But, by hypothesis,

$a : b :: c : d$,

$\therefore \; ac : bc :: ac : ad$;

therefore, by 491,

$bc = ad$.

530. INVERSE. If two rectangles are equivalent, the sides of the one will form the extremes, and the sides of the other the means, of a proportion.

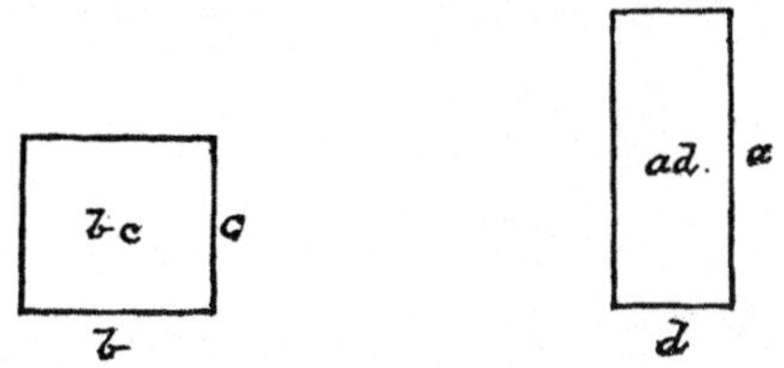

HYPOTHESIS. **Rectangle** $ad = bc$.

CONCLUSION. $a : b :: c : d$.

PROOF. Since $ad = bc$, therefore, by 489, $ac : bc :: ac : ad$; but $ac : bc :: a : b$, and $ac : ad :: c : d$,

$$\therefore a : b :: c : d.$$

531. COROLLARY. If three sects are proportional, the rectangle of the extremes is equivalent to the square on the mean.

THEOREM XIV.

532. *If two chords intersect either within or without the circle, the rectangle contained by the segments of the one is equivalent to the rectangle contained by the segments of the other.*

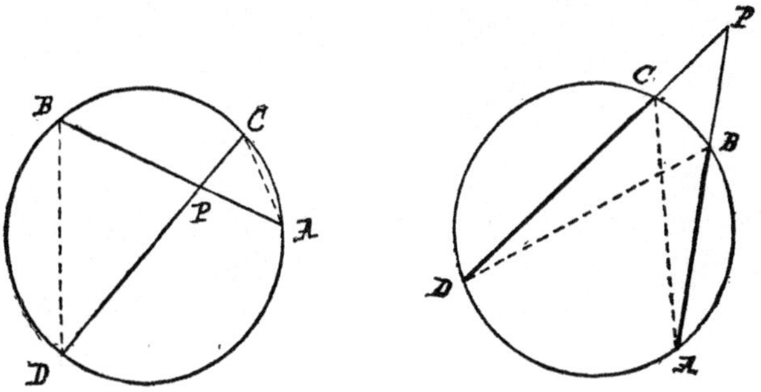

HYPOTHESIS. ***Let the chords** AB **and** CD **intersect in** P.*

CONCLUSION. Rectangle $AP.PB$ = rectangle $CP.PD$.

PROOF. $\measuredangle PAC = \measuredangle PDB$,

(376. Angles in the same segment of a circle are equal.)

and

$$\measuredangle APC = \measuredangle BPD,$$

$$\therefore \quad \triangle APC \sim \triangle BPD,$$

(508. Equiangular triangles are similar.)

$$\therefore \quad AP : CP :: PD : PB,$$

$$\therefore \quad AP.PB = CP.PD.$$

533. COROLLARY. Let the point P be without the circle, and suppose DCP to revolve about P until C and D coincide; then the secant DCP becomes a tangent, and the rectangle

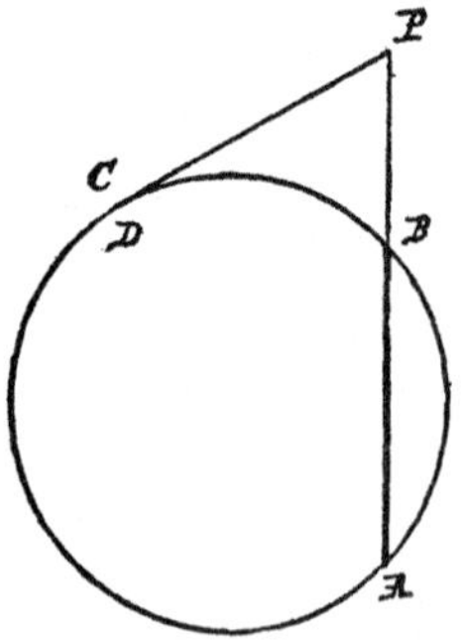

$CP.PD$ becomes the square on PC. Therefore, if from a point without a circle a secant and tangent be drawn, the rectangle of the whole secant and part outside the circle is equivalent to the square of the tangent.

Theorem XV.

534. *The rectangle of two sides of a triangle is equivalent to the rectangle of two sects drawn from that vertex so as to make equal angles with the two sides, and produced, one to the base, the other to the circle circumscribing the triangle.*

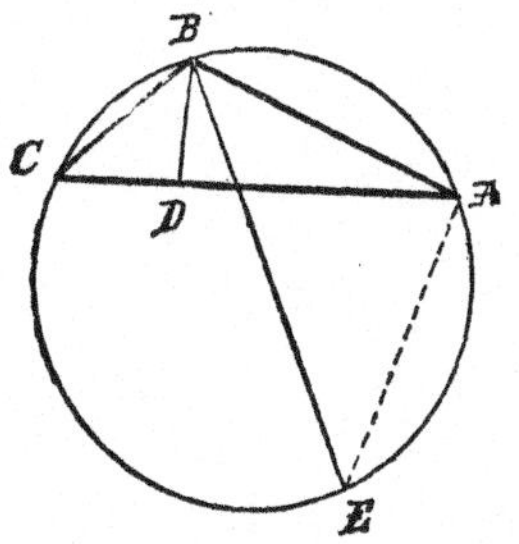

Hypothesis. $\measuredangle ABE = \measuredangle CBD$.
Conclusion. Rectangle $AB.BC = DB.BE$.
Proof. Join AE. Then

$$\measuredangle C = \measuredangle E,$$

(376. Angles in the same segment of a circle are equal.)

$\measuredangle CBD = \measuredangle ABE$, by hypothesis,

$$\therefore \triangle CBD \sim \triangle ABE,$$

$$\therefore AB : BE :: DB : BC,$$

$$\therefore AB.BC = DB.BE.$$

535. Corollary I. If BD and BE coincide, they bisect the angle B; therefore rectangle $AB.BC = DB.BE = DB(BD + DE) = BD^2 + BD.DE = BD^2 + CD.DA$ (by 532). Therefore, when the bisector of an angle of a triangle

meets the base, the rectangle of the two sides is equivalent to the rectangle of the segments of the base, together with the square of the bisector.

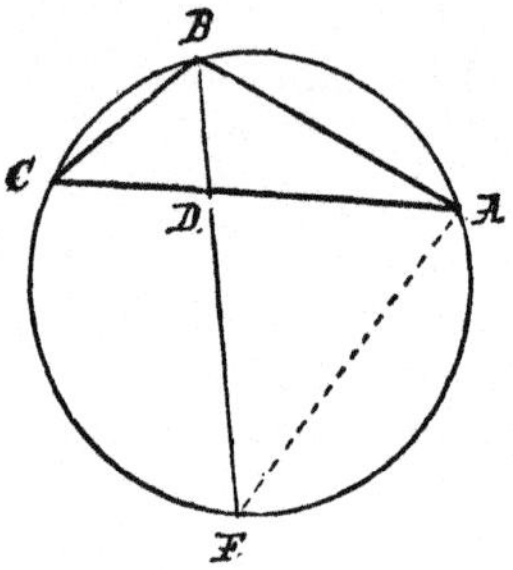

536. Corollary II. If BD be a perpendicular, BE is a diameter, for angle BAE is then right; therefore in any triangle

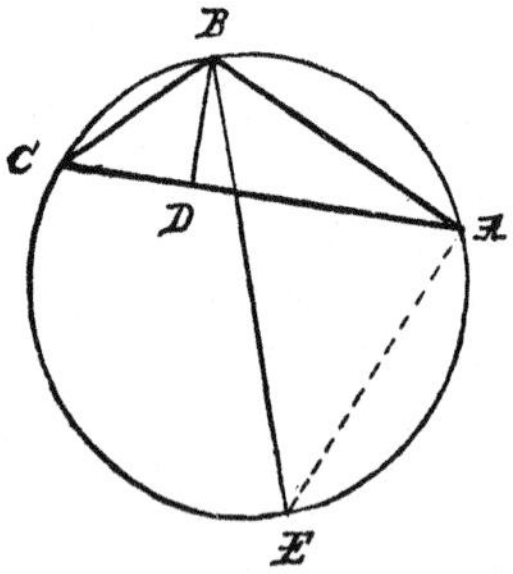

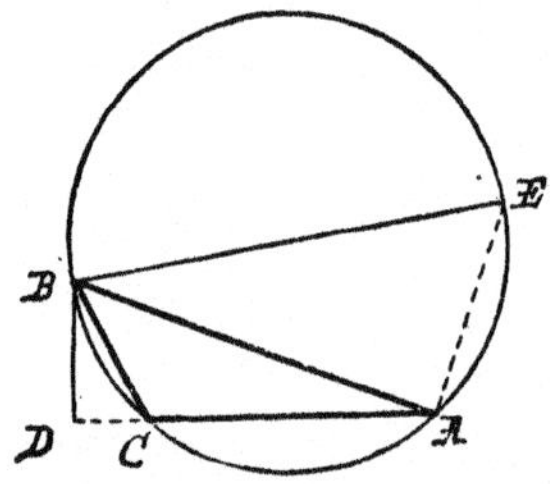

the rectangle of two sides is equivalent to the rectangle of the diameter of the circumscribed circle by the perpendicular to the base from the vertex.

Exercises. 88. Prove that the inverse of 535 does not hold when $AB = BC$.

89. Prove 535 when it is an exterior angle which is bisected.

Theorem XVI.

537. *The rectangle of the diagonals of a quadrilateral inscribed in a circle is equivalent to the sum of the two rectangles of its opposite sides.*

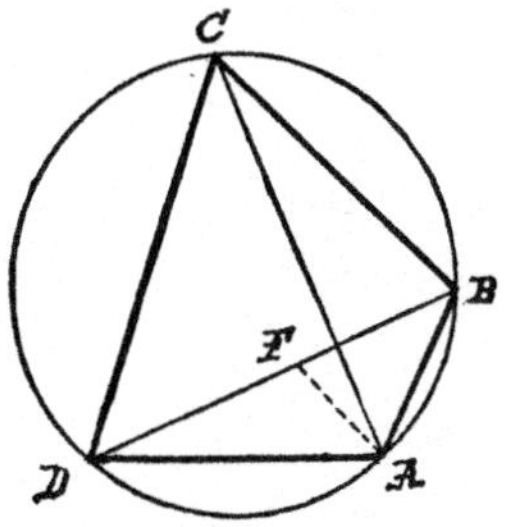

Hypothesis. *ABCD* ***an inscribed quadrilateral.***

Conclusion. Rectangle $AC \cdot BD = AB \cdot CD + BC \cdot DA$.

Proof. By 164, make $\measuredangle DAF = \measuredangle BAC$.

To each add $\measuredangle FAC$.

Then, in $\triangle$s *ACD* and *ABF*,

$$\measuredangle DAC = \measuredangle BAF;$$

also

$$\measuredangle ACD = \measuredangle ABD,$$

(376. Angles inscribed in the same segment are equal.)

$$\therefore \quad \triangle ACD \sim \triangle ABF,$$

(508. Equiangular triangles are similar.)

$$\therefore \quad AC : AB :: CD : BF;$$

therefore, by 529,

$$AC \cdot BF = AB \cdot CD.$$

Again, by construction, $\measuredangle DAF = \measuredangle BAC$.

Moreover, $\measuredangle ADF = \measuredangle ACB$,

(376. Angles inscribed in the same segment are equal.)

$$\therefore \triangle ADF \sim \triangle ABC,$$

(508. Equiangular triangles are similar.)

$$\therefore AC : AD :: CB : DF;$$

therefore, by 529,

$$AC . FD = AD . BC,$$

$$\therefore AB . CD + BC . DA = AC . BF + AC . FD$$
$$= AC(BF + FD) = AC . BD.$$

Problem I.

538. *To alter a given sect in a given ratio.*

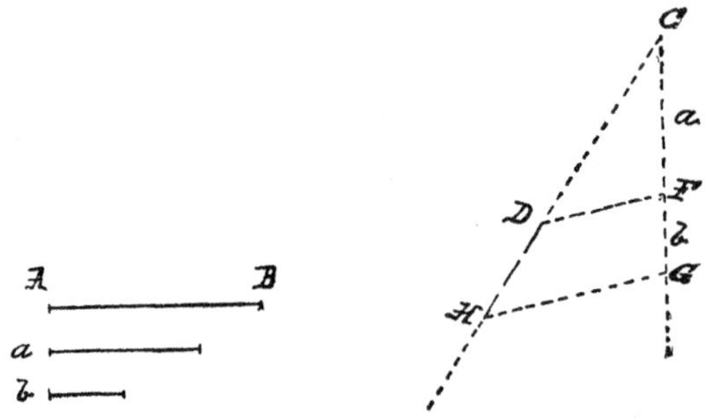

Given, ***the ratio as that of sect*** *a* ***to sect*** *b*, ***and given the sect*** *AB*.

Required, *to find a sect to which* AB *shall have the same ratio as* a *to* b.

Construction. Make any angle C.

On one arm cut off $CD = AB$. On the other arm cut off $CF = a$, and $FG = b$.

Join DF, and through G draw $GH \parallel DF$.

$$\therefore AB : DH :: a : b.$$

539. This is the same as finding a fourth proportional to three given sects.

To find a third proportional to a and b, make $AB = b$ in the above construction.

540. Every alteration of a magnitude is an alteration in some ratio.

Two or more alterations are jointly equivalent to some one alteration, and then this single alteration which produces the joint effect of two is said to be *compounded* of those two.

The composition of the ratios of a to b and C to D is performed by assuming F, altering it into G, so that $F : G :: a : b$, then altering G into H, so that $G : H :: C : D$.

The joint effect turns F into H; and the ratio of F to H is the ratio compounded of the two ratios, $a : b$ and $C : D$.

541. A ratio arising from the composition of two equal ratios is called the *Duplicate Ratio* of either.

Theorem XVII.

542. *Mutually equiangular parallelograms have to one another the ratio which is compounded of the ratios of their sides.*

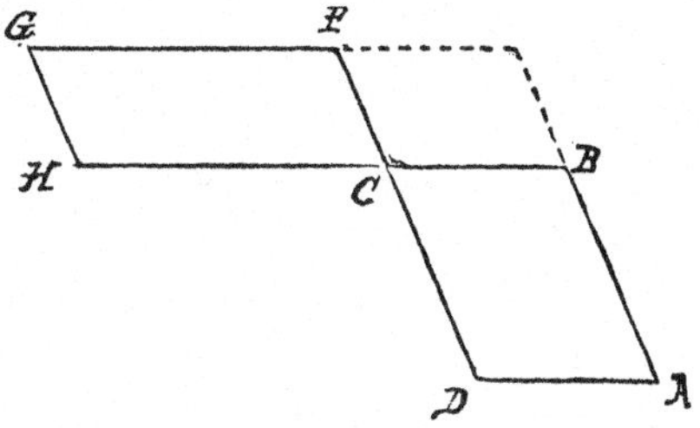

Hypothesis. *In* ▱ AC, ∡ BCD = ∡ HCF *of* ▱ CG.

Conclusion. ▱ AC : ▱ CG = ratio compounded of DC : CF, and BC : CH.

PROOF. Place the ▱s so that HC and CB are in one line; then, by 109, DC and CF are in one line. Complete the ▱ BF.

Then

$$▱ AC : ▱ BF :: DC : CF,$$

and

$$▱ BF : ▱ CG :: BC : CH,$$

(505. Parallelograms of equal altitude are as their bases.)

∴ ▱ AC has to ▱ CG the ratio compounded of $DC : CF$ and $BC : CH$.

543. COROLLARY I. Triangles which have one angle of the one equal or supplemental to one angle of the other, being halves of equiangular parallelograms, are to one another in the ratio compounded of the ratios of the sides about those angles.

544. COROLLARY II. Since all rectangles are equiangular parallelograms, therefore the ratio compounded of two ratios between sects is the same as the ratio of the rectangle contained by the antecedents to the rectangle of the consequents.

If the ratio compounded of $a : a'$, and $b : b'$, be written $\frac{a}{a'} \cdot \frac{b}{b'}$, this corollary proves $\frac{a}{a'} \cdot \frac{b}{b'} = \frac{ab}{a'b'}$; and the composition of ratios obeys the same laws as the multiplication of fractions. Thus $\frac{a}{b} \cdot \frac{a}{b} = \frac{a^2}{b^2}$, and so the duplicate ratio of two sects is the same as the ratio of the squares on those sects.

It will be seen hereafter that the special case of 542, when the parallelograms are rectangles, is made the foundation of all mensuration of surfaces.

EXERCISES. 90. If mutually equiangular parallelograms are equivalent, so are rectangles with the same sides.

91. Equivalent parallelograms having the same sides as equivalent rectangles are mutually equiangular.

THEOREM XVIII.

545. *Similar triangles are to one another as the squares on their corresponding sides.*

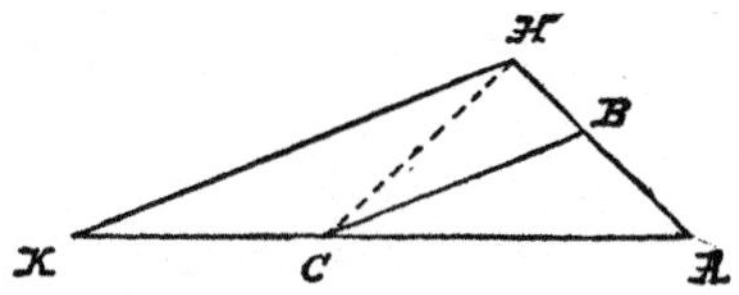

Let the similar triangles ABC, AHK, be placed so as to have the sides AB, AC, along the corresponding sides AH, AK, and therefore $BC \parallel HK$. Join CH.

Since, by 505, triangles of equal altitude are as their bases,

$$\therefore \quad \triangle ABC : \triangle AHC :: AB : AH,$$

and $\triangle AHC : \triangle AHK :: AC : AK = AB : AH$, by hypothesis;

$$\therefore \quad \triangle ABC : \triangle AHK = \frac{AB}{AH} \cdot \frac{AB}{AH} = \frac{AB^2}{AH^2}.$$

EXERCISES. 92. The ratio of the surfaces of two similar triangles is the square of the ratio of similitude of the triangles.

93. If the bisector of an angle of a triangle also bisect a side, the triangle is isosceles.

94. In every quadrilateral which cannot be inscribed in a circle, the rectangle contained by the diagonals is less than the sum of the two rectangles contained by the opposite sides.

95. The rectangles contained by any two sides of triangles inscribed in equal circles are proportional to the perpendiculars on the third sides.

96. Squares are to one another in the duplicate ratio of their sides.

THEOREM XIX.

546. *Similar polygons are to each other as the squares on their corresponding sides.*

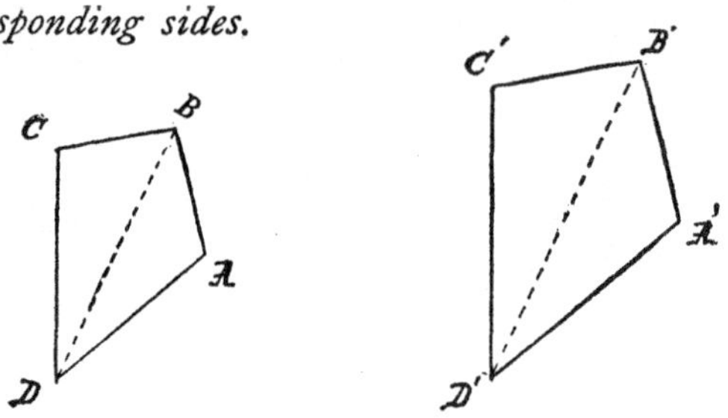

HYPOTHESIS. ***$ABCD$ and $A'B'C'D'$ two similar polygons, of which BC and $B'C'$ are corresponding sides.***

CONCLUSION. $ABCD : A'B'C'D' :: AB^2 : A'B'^2$.

PROOF. By 516, the polygons may be divided into similar triangles.

By 545, any pair of corresponding triangles are as the squares on corresponding sides,

$$\therefore \quad \frac{\triangle ABD}{\triangle A'B'D'} = \frac{BD^2}{B'D'^2} = \frac{\triangle BCD}{\triangle B'C'D'} = \frac{BC^2}{B'C'^2};$$

therefore, by 496,

$$\frac{\triangle ABD + \triangle BCD}{\triangle A'B'D' + \triangle B'C'D'} = \frac{BC^2}{B'C'^2}.$$

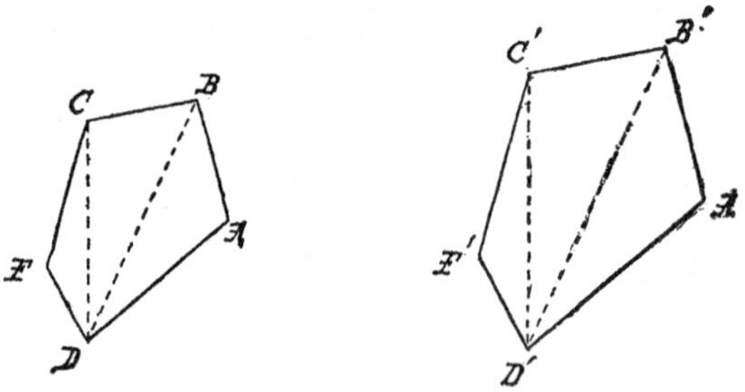

In the same way for a third pair of similar triangles, etc.

547. Corollary. If three sects form a proportion, a polygon on the first is to a similar polygon similarly described on the second as the first sect is to the third.

Theorem XX.

548. *The perimeters of any two regular polygons of the same number of sides have the same ratio as the radii of their circumscribed circles.*

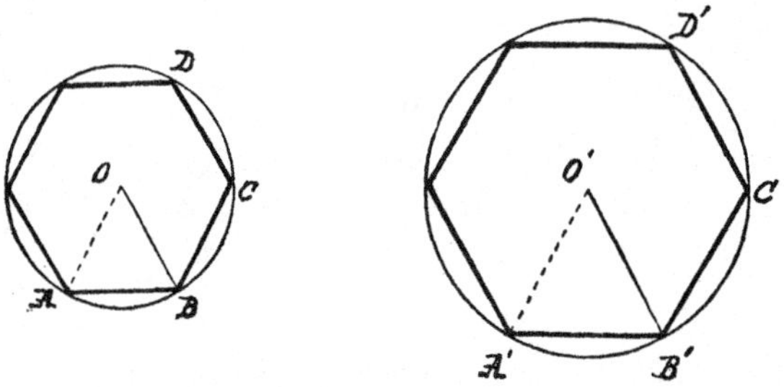

Proof. The angles of two regular polygons of the same number of sides are all equal, and the ratio between any pair of sides is the same, therefore the polygons are similar.

But lines drawn from the center to the extremities of any pair of sides are radii of the circumscribed circles, and make similar triangles,

$$\therefore \quad \frac{AB}{A'B'} = \frac{BC}{B'C'} = \frac{r}{r'};$$

therefore, by 496,

$$\frac{AB + BC}{A'B' + B'C'} = \frac{r}{r'} = \frac{CD}{C'D'};$$

$$\therefore \quad \frac{AB + BC + CD}{A'B' = B'C' + C'D'} = \frac{r}{r'}, \text{ etc.}$$

Problem II.

549. *To find a mean proportional between two given sects.*

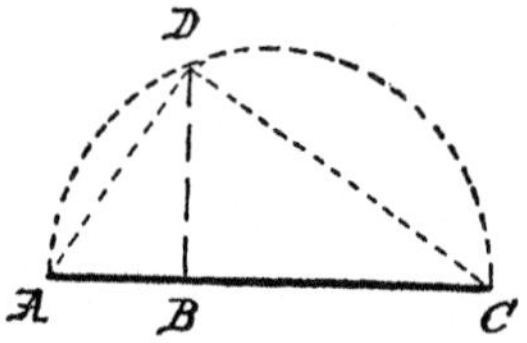

Let AB, BC, be the two given sects. Place AB, BC, in the same line, and on AC describe the semicircle ADC. From B draw BD perpendicular to AC.

BD is the mean proportional.

(383. An angle inscribed in a semicircle is a right angle.)

(521. If from the right angle a perpendicular be drawn to the hypothenuse, it will be a mean proportional between the segments of the hypothenuse.)

Exercises. 97. If the given sects were AC and BC, placed as in the above figure, how would you find a mean proportional between them?

98. Half the sum of two sects is greater than the mean proportional between them.

99. The rectangle contained by two sects is a mean proportional between their squares.

100. The sum of perpendiculars drawn from any point within an equilateral triangle to the three sides equals its altitude.

101. The bisector of an angle of a triangle divides the triangle into two others, which are proportional to the sides of the bisected angle.

102. Lines which trisect a side of a triangle do not trisect the opposite angle.

103. In the above figure, if $AF \perp AD$ meet DB produced at F, then $\triangle ABD = \triangle FCB$.

PROBLEM III.

550. *On a given sect to describe a polygon similar to a given polygon.*

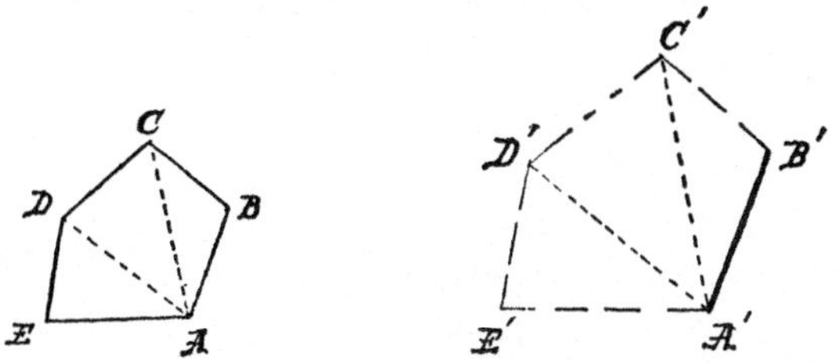

Let $ABCDE$ be the given polygon, and $A'B'$ the given sect.

Join AD, AC.

Make

$$\measuredangle A'B'C' = \measuredangle ABC,$$
$$\measuredangle B'A'C' = \measuredangle BAC,$$
$$\therefore \quad \measuredangle B'C'A' = \measuredangle BCA,$$

and

$$\triangle A'B'C' \sim \triangle ABC.$$

Then make

$$\measuredangle A'C'D' = \measuredangle ACD,$$
$$\measuredangle C'A'D' = \measuredangle CAD,$$
$$\therefore \quad \measuredangle C'D'A' = \measuredangle CDA,$$

and

$$\triangle A'C'D' \sim \triangle ACD, \text{ etc.}$$

Therefore, from the first pair of similar triangles,

$$A'B' : B'C' :: AB : BC,$$

and

$$B'C' : C'A' :: BC : CA.$$

From the second pair of similar triangles,

$$C'A' : C'D' :: CA : CD,$$
$$\therefore \quad B'C' : C'D' :: BC : CD,$$

and so on.

Thus all the $\measuredangle$s in one polygon are = the corresponding $\measuredangle$s in the other, and the sides about the corresponding $\measuredangle$s are proportional.

$\therefore$ the polygons are similar.

THEOREM XXI.

551. *In a right-angled triangle, any polygon upon the hypothenuse is equivalent to the sum of the similar and similarly described polygons on the other two sides.*

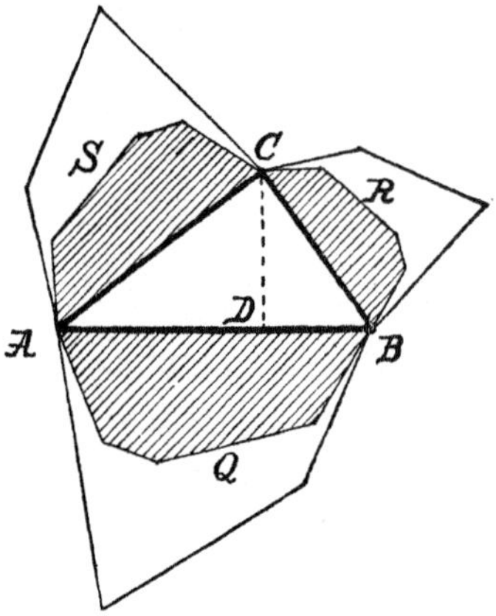

Draw $CD \perp AB$.

Therefore, by 520,

$$AB : AC :: AC : AD;$$

therefore, by 547,

$$AB : AD :: Q : S.$$

Similarly,

$$AB : BD :: Q : R.$$

Therefore, by 496,

$$AB : AD + DB :: Q : R + S,$$

$$\therefore \quad Q = R + S.$$

EXERCISES. 104. If 551 applies to semicircles, show the triangle equivalent to two crescent-shaped figures, called the lunes of Hippocrates of Chios (about 450 B.C.).

PROBLEM IV.

552. *To describe a polygon equivalent to one, and similar to another, given polygon.*

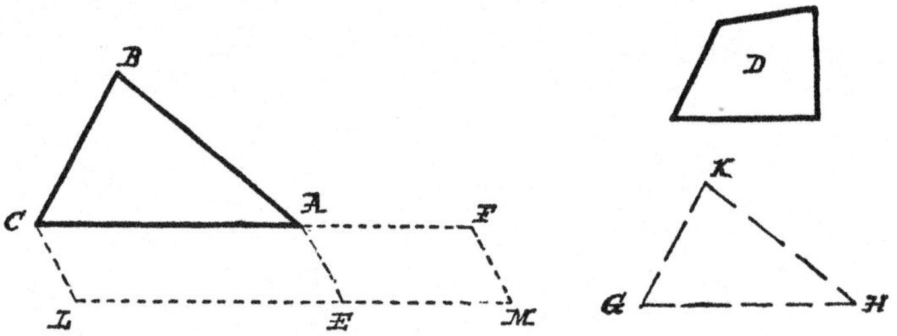

Let D be the one, and ABC the other, given polygon.

By 262, on AC describe any ▱ $CE = ABC$, and on AE describe ▱ $AM = D$, and having ∡ AEM = ∡ CAE.

By 549, between AC and AF find a mean proportional GH.

By 550, on GH construct the figure KGH similar and similarly described to the figure ABC.

KGH is the figure required.

It may be proved, as in 542, that AC and AF are in one line, and also LE and EM;
therefore, by 505,

$$AC : AF :: \text{▱ } CE : \text{▱ } AM :: ABC : D.$$

But

$$AC : AF :: ABC : KGH,$$

(547. If three sects form a proportion, a polygon on the first is to a similar polygon on the second as the first sect is to the third.)

$$\therefore \; ABC : D :: ABC : KGH;$$

therefore, by 491,

$$KGH = D.$$

GEOMETRY OF THREE DIMENSIONS.

BOOK VII.

OF PLANES AND LINES.

553. Already, in 50, a plane has been defined as the surface generated by the motion of a line always passing through a fixed point while it slides along a fixed line.

554. Already, in 97, the theorem has been assumed, that, if two points of a line are in a plane, the whole line lies in that plane.

Two other assumptions will now be made:—

555. Any number of planes may be passed through any line.

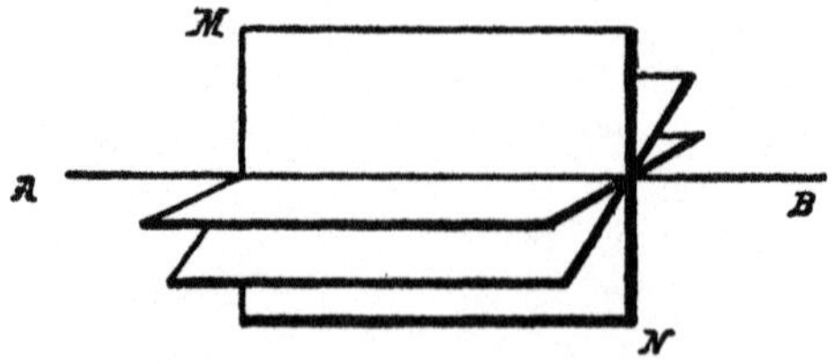

556. A plane may be revolved on any line lying in it.

THEOREM I.

557. *Through two intersecting lines, one plane, and only one, passes.*

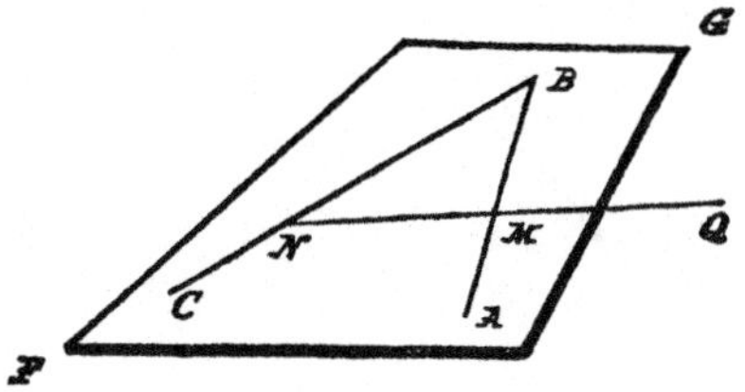

HYPOTHESIS. **Two lines AB, BC, meeting in B.**

CONCLUSION. One plane, and only one, passes through them.

PROOF. By 555, let any plane FG be passed through AB, and, by 556, be revolved around on AB as an axis until it meets any point C of the line BC.

The line BC then has two points in the plane FG; and therefore, by 554, the whole line BC is in this plane.

Also, any plane containing AB *and* BC *must coincide with* FG.

For let Q be any point in a plane containing AB and BC.

Draw QMN in this plane to cut AB, BC, in M and N. Then, since M and N are points in the plane EF, therefore, by 554, Q is a point in the plane EF.

Similarly, any point in a plane containing AB, BC, must lie in EF; therefore any plane containing AB, BC, must coincide with EF.

558. COROLLARY I. Two lines which intersect lie in one plane, and a plane is completely determined by the condition that it passes through two intersecting lines.

559. COROLLARY II. Any number of lines, each of which intersects all the others at different points, lie in the same plane; but a line may pass through the intersection of two others without being in their plane.

560. COROLLARY III. *A line, and a point without that line, determine a plane.*

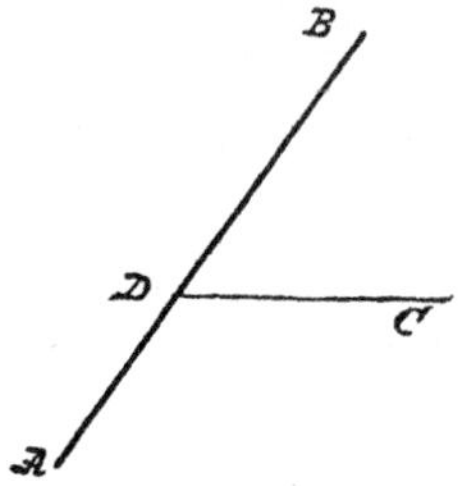

PROOF. Suppose *AB* the line, and *C* the point without *AB*.

Draw the line *CD* to any point *D* in *AB*. Then one plane contains *AB* and *CD*, therefore one plane contains *AB* and *C*.

Again, any plane containing *AB* must contain *D*; therefore, any plane containing *AB* and *C* must contain *CD* also.

But there is only one plane that can contain *AB* and *CD*.

Therefore there is only one plane that can contain *AB* and *C*.

Hence the plane is completely determined.

561. COROLLARY IV. Three points not in the same line determine a plane.

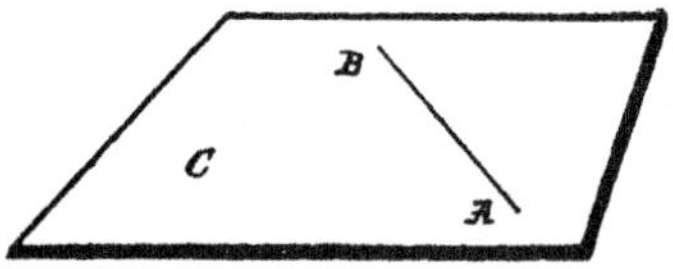

For let *A*, *B*, *C* be three such points. Draw the line *AB*.

Then a plane which contains *A*, *B*, and *C* must contain *AB* and *C*; and a plane which contains *AB* and *C* must contain *A*, *B*, and *C*. Now, *AB* and *C* are contained by one plane, and one only; therefore *A*, *B*, and *C* are contained by one plane, and one only. Hence the plane is completely determined.

562. Two parallel lines determine a plane.

For, by the definition of parallel lines, the two lines are in the same plane; and, as only one plane can be drawn to contain one of the lines and any point in the other line, it follows that only one plane can be drawn to contain both lines.

Theorem II.

563. *If two planes cut one another, their common section must be a straight line.*

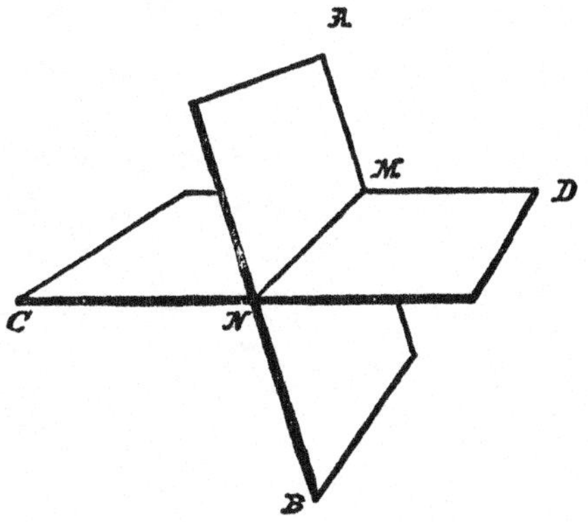

Hypothesis. ***Let** AB **and** CD **be two intersecting planes.***

Conclusion. Their common section is a straight line.

Proof. Let M and N be two points common to both planes. Draw the straight line MN. Therefore, by 554, since M and N are in both planes, the straight line MN lies in both planes.

And no point out of this line can be in both planes; because then two planes would each contain the same line and the same point without it, which, by 560, is impossible.

Hence every point in the common section of the planes lies in the straight line MN.

564. CONTRANOMINAL OF 554. A line which does not lie altogether in a plane may have no point, and cannot have more than one point, in common with the plane.

Therefore three planes which do not pass through the same line cannot have more than one point in common; for, by 563, the points common to two planes lie on a line, and this line can have only one point in common with the third plane.

565. All planes are congruent; hence properties proved for one plane hold for all. A plane will slide upon its trace.

PRINCIPLE OF DUALITY.

566. When any figure is given, we may construct a *reciprocal* figure by taking planes instead of points, and points instead of planes, but lines where we had lines.

The figure reciprocal to four points which do not lie in a plane will consist of four planes which do not meet in a point.

From any theorem we may infer a reciprocal theorem.

Two points determine a line.	Two planes determine a line.
Three points which are not in a line determine a plane.	Three planes which do not pass through a line determine a point.
A line and a point without it determine a plane.	A line and a plane not through it determine a point.
Two lines in a plane determine a point.	Two lines through a point determine a plane.

There is also a more special principle of duality, which, in the plane, takes points and lines as reciprocal elements; for they have this fundamental property in common, that two elements of one kind determine one of the other. Thus, from a proposition relating to lines or angles in axial symmetry, we get a proposition relating to points or sects in central symmetry.

The angle between two corresponding lines is bisected by the axis.	The sect between two corresponding points is bisected by the center.

THEOREM III.

567. *If a line be perpendicular to two lines lying in a plane, it will be perpendicular to every other line lying in the plane and passing through its foot.*

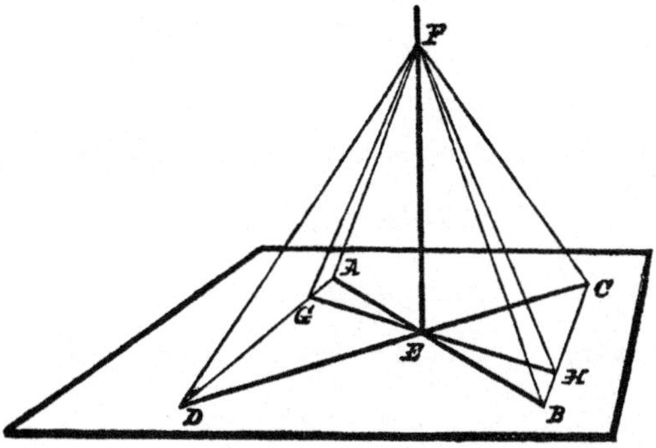

HYPOTHESIS. ***Let the line EF be perpendicular to each of the lines AB, CD, at E, the point of their intersection.***

CONCLUSION. $EF \perp GH$, any other line lying in the plane $ABCD$, and passing through E.

PROOF. Take $AE = BE$, and $CE = DE$.

Join AD and BC, and let G and H be the points in which the joining lines intersect the third line GH.

Take any point F in EF. Join FA, FB, FC, FD, FG, FH.

Then

$$\triangle AED \cong \triangle BEC,$$

(124. Triangles having two sides and the included angle equal are congruent.)

$$\therefore \quad AD = BC, \quad \text{and} \quad \measuredangle DAE = \measuredangle CBE.$$

Then

$$\triangle AEG \cong \triangle BEH,$$

(128. Triangles having two angles and the included side equal are congruent.)

$$\therefore \quad GE = HE, \quad \text{and} \quad AG = BH.$$

Then

$$\triangle AEF \cong \triangle BEF,$$

(124. Triangles having two sides and the included angle equal are congruent.)

$$\therefore \quad AF = BF.$$

In the same way

$$CF = DF.$$

Then

$$\triangle ADF \cong \triangle BCF,$$

(129. Triangles having three sides in each respectively equal are congruent.)

$$\therefore \quad \angle DAF = \angle CBF.$$

Then

$$\triangle AFG \cong \triangle BFH,$$

(124. Triangles having two sides and the included angle equal are congruent.)

$$\therefore \quad FG = FH.$$

Then

$$\triangle FEG \cong \triangle FEH,$$

(129. Triangles having three sides in each respectively equal are congruent.)

$$\therefore \quad \angle FEG = \angle FEH,$$

$$\therefore \quad FE \perp GH.$$

As GH is any line whatever lying in the plane $ABCD$, and passing through E,

$$\therefore \quad EF \perp \text{ every such line.}$$

568. A line meeting a plane so as to be perpendicular to every line lying in the plane and passing through the point of intersection, is said to be perpendicular to the plane. Then also the plane is said to be perpendicular to the line.

569. Corollary. At a given point in a plane, only one perpendicular to the plane can be erected; and, from a point without a plane, only one perpendicular can be drawn to the plane.

570. Inverse of 567. All lines perpendicular to another line at the same point lie in the same plane.

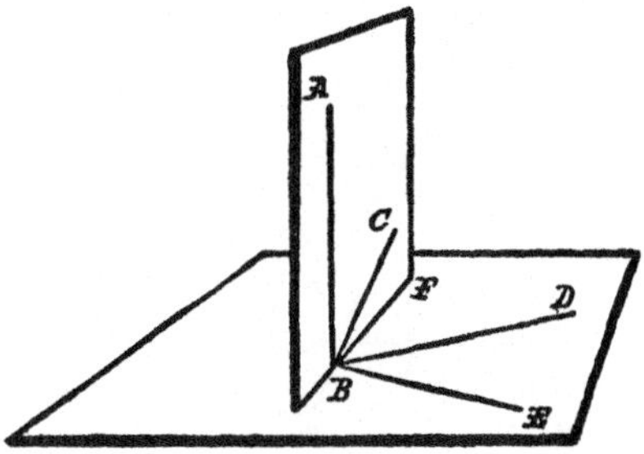

Hypothesis. *AB any line ⊥ BD and BE. BC any other line ⊥ AB.*

Conclusion. *BC* is in the plane *BDE.*

Proof. For if not, let the plane passing through *AB*, *BC*, cut the plane *BDE* in the line *BF*. Then *AB*, *BC*, and *BF* are all in one plane; and because *AB* ⊥ *BD* and *BE*, therefore, by 567,

AB ⊥ *BF*.

But, by hypothesis, *AB* ⊥ *BC*;
therefore, in the plane *ABF* we have two lines *BC*, *BF*, both ⊥ *AB* at *B*, which, by 105, is impossible;

∴ *BC* lies in the plane *DBE*.

571. Corollary. If a right angle be turned round one of its arms as an axis, the other arm will generate a plane; and when this second arm has described a perigon, it will have passed through every point of this plane.

572. Through any point *D* without a given line *AB* to pass a plane perpendicular to *AB*. In the plane determined by the line *AB* and the point *D*, draw *DB* ⊥ *AB*; then revolve the rt. ∡ *ABD* about *AB*.

Theorem IV.

573. *Lines perpendicular to the same plane are parallel to each other.*

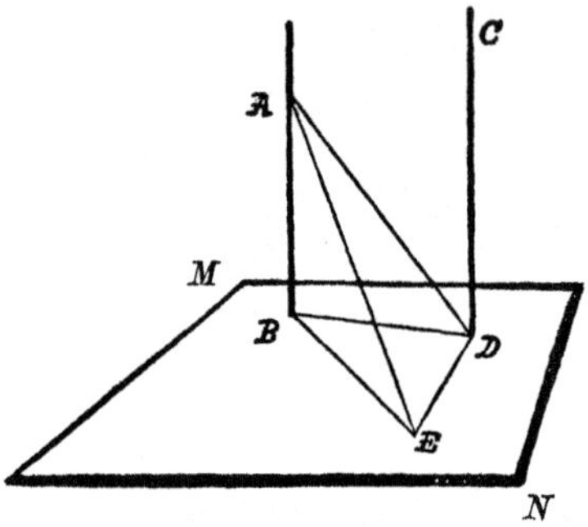

Hypothesis. AB, $CD \perp$ *plane* MN *at the points* B, D.
Conclusion. $AB \parallel CD$.
Proof. Join BD, and draw $DE \perp BD$ in the plane MN.
Make $DE = AB$, and join BE, AD, AE. Then

$$\triangle ABD \cong \triangle EDB,$$

(124. Triangles having two sides and the included angle equal are congruent.)

$$\therefore \quad AD = BE.$$

Then

$$\triangle ABE \cong \triangle EDA,$$

(129. Triangles having three sides in each respectively equal are congruent.)

$$\therefore \quad \measuredangle ABE = \measuredangle EDA.$$

But, by hypothesis, ABE is a rt. $\measuredangle$,

$$\therefore \quad EDA \text{ is a rt. } \measuredangle,$$

$$\therefore \quad AD, BD, CD, \text{ are all in one plane.}$$

(570. All lines perpendicular to another at the same point lie in the same plane.)

But AB is in this plane, since the points A and B are in it,

$$\therefore \quad AB \text{ and } CD \text{ lie in the same plane;}$$

and they are both $\perp BD$,

$$\therefore \quad AB \parallel CD.$$

(166. If two interior angles are supplemental, the lines are parallel.)

574. INVERSE OF 573. If one of two parallels is perpendicular to a plane, the other is also perpendicular to that plane.

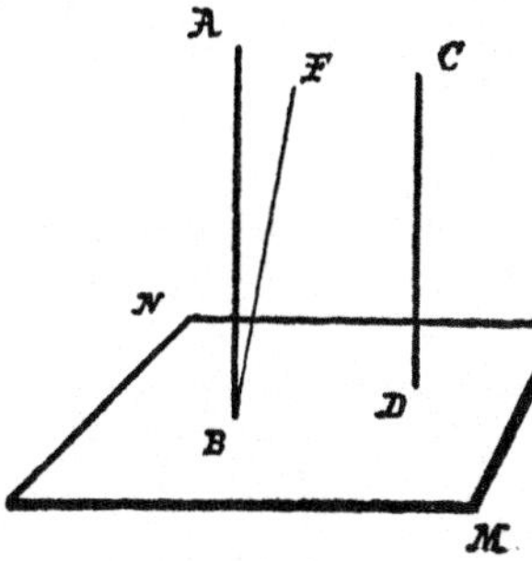

HYPOTHESIS. $AB \parallel CD$. $CD \perp$ plane MN.

CONCLUSION. $AB \perp$ plane MN.

PROOF. For if AB, meeting the plane MN at B, is not $\perp$ it, let BF be $\perp$ it;

therefore, by 573, $BF \parallel CD$, and, by hypothesis, $BA \parallel CD$.

But this is impossible,

(99. Two intersecting lines cannot both be parallel to the same line.)

$\therefore$ $AB \perp$ plane MN.

575. COROLLARY I. If one plane be perpendicular to one of two intersecting lines, and a second plane perpendicular to the second, their intersection is perpendicular to the plane of the two lines.

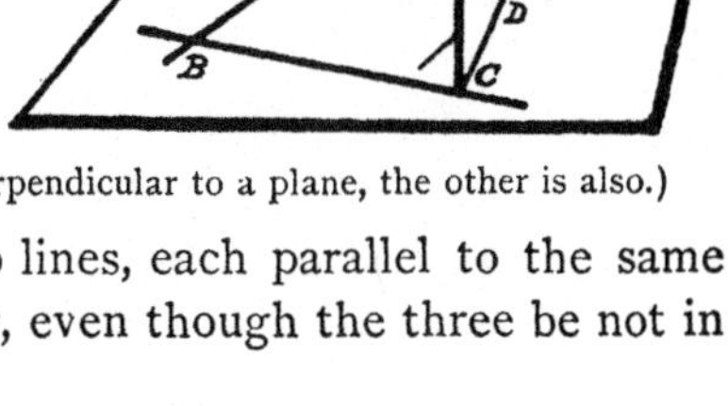

For their intersection DF is perpendicular to a line through D parallel to AB, and also perpendicular to a line through D parallel to BC.

(574. If, of two parallels, one be perpendicular to a plane, the other is also.)

576. COROLLARY II. Two lines, each parallel to the same line, are parallel to each other, even though the three be not in

one plane. For a plane perpendicular to the third line will, by 574, be perpendicular to each of the others; and therefore, by 573, they are parallel.

PROBLEM I.

577. *To draw a line perpendicular to a given plane from a given point without it.*

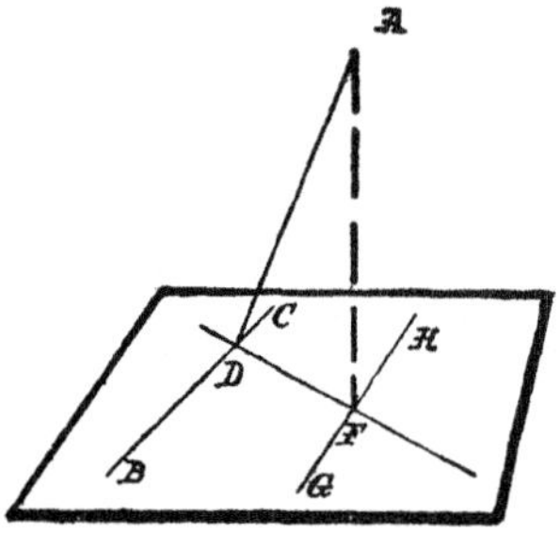

GIVEN, ***the plane** BH **and the point** A **without it.***

REQUIRED, *to draw from* A *a line* ⊥ *plane BH.*

CONSTRUCTION. In the plane draw any line *BC*, and, by 139, from *A* draw *AD* ⊥ *BC.*

In the plane, by 135, draw *DF* ⊥ *BC*; and from *A*, draw, by 139, *AF* ⊥ *DF*. *AF* will be the required perpendicular to the plane.

PROOF. Through *F* draw *GH* ∥ *BC*.

Then, since *BC* ⊥ the plane *ADF*,

(567. A line perpendicular to any two lines of a plane at their intersection is perpendicular to the plane.)

∴ *GH* ⊥ the plane *ADF*,

(574. If two lines are parallel, and one is perpendicular to a plane, the other is also.)

∴ *GH* ⊥ the line *AF* of the plane *ADF*,

∴ *AF* ⊥ *GH*.

But, by construction, *AF* ⊥ *DF*,

∴ *AF* ⊥ the plane passing through *GH*, *DF*.

PROBLEM II.

578. *To erect a perpendicular to a given plane from a given point in the plane.*

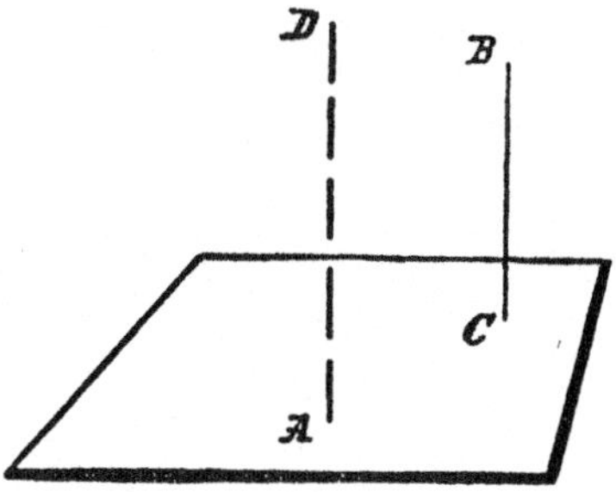

Let A be the given point.

From any point B, without the plane, draw, by 577, $BC \perp$ the plane, and from A, by 167, draw $AD \parallel BC$,

$\therefore \quad AD \perp$ the plane.

(574. If one of two parallels is perpendicular to a given plane, the other is also.)

579. The projection of a point upon a plane is the foot of the perpendicular drawn from the point to the plane.

580. The projection of a line upon a plane is the locus of the feet of the perpendiculars dropped from every point of the line upon the plane.

These perpendiculars are all in the same plane, since any two are parallel, and any third is parallel to either, and has a point in their common plane, and therefore lies wholly in that plane; therefore the projection of a line is a line.

THEOREM V.

581. *A line makes with its own projection upon a plane a less angle than with any other line in the plane.*

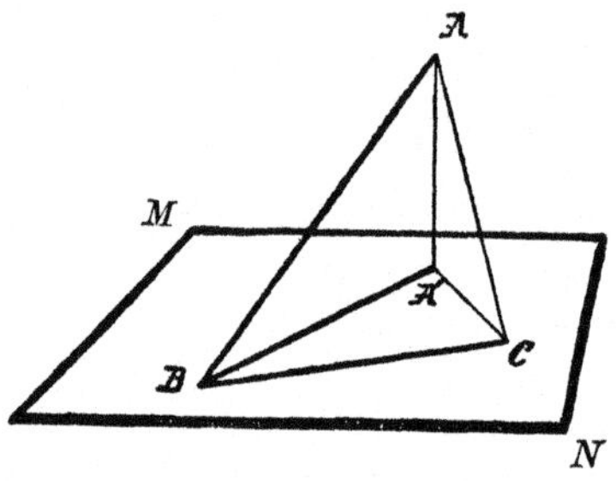

HYPOTHESIS. ***Let BA meet the plane MN at B, and let BA' be its projection upon the plane MN, and BC any other line drawn through B in the plane MN.***

CONCLUSION. $\measuredangle ABA' < \measuredangle ABC$.

PROOF. Take $BC = BA'$. Join AC and $A'C$.

Then, in $\triangle$s ABA' and ABC,

$$AB = AB,$$

$$BA' = BC,$$

$$AA' < AC,$$

(150. The perpendicular is the least sect from a point to a line.)

$$\therefore \measuredangle ABA' < \measuredangle ABC.$$

(161. In two triangles, if $a = a'$, $b = b'$, $c < c'$, therefore $C < C'$.)

582. The angle between a line and its projection on a plane is called the *Inclination* of the line to the plane.

583. Parallel planes are such as never meet, how far soever they may be produced.

584. A line is parallel to a plane when they never meet, how far soever they may be produced.

THEOREM VI.

585. *Planes to which the same line is perpendicular are parallel.*

For, if not, they intersect. Call any point of their intersection *X*. Draw from *X* a line in each plane to the foot of the common perpendicular. Then from this point *X* we would have two perpendiculars to the same line, which is impossible,

(145. There can be only one perpendicular from a point to a line.)

∴ the planes cannot intersect, and

∴ are parallel.

586. *If two lines are parallel, every plane through one of them, except the plane of the parallels, is parallel to the other.*

Let *AB* and *GH* be the parallels, and *DEF* any plane through *GH*; then the line *AB* and the plane *DEF* are parallel. For the plane of the parallels *ABHG* intersects the plane *DEF* in the line *GH*; and, if *AB* could meet the plane *DEF*, it could meet it only in some point of *GH*; but *AB* cannot meet *GH*, since they are parallel by hypothesis. Therefore *AB* cannot meet the plane *DEF*.

APPLICATIONS. (1) Through any given line a plane can be passed parallel to any other given line.

(2) Through any given point a plane can be passed parallel to any two given lines in space.

Theorem VII.

587. *If a pair of intersecting lines be parallel to another pair, but not in the same plane with them, the plane of the first pair is parallel to the plane of the second pair.*

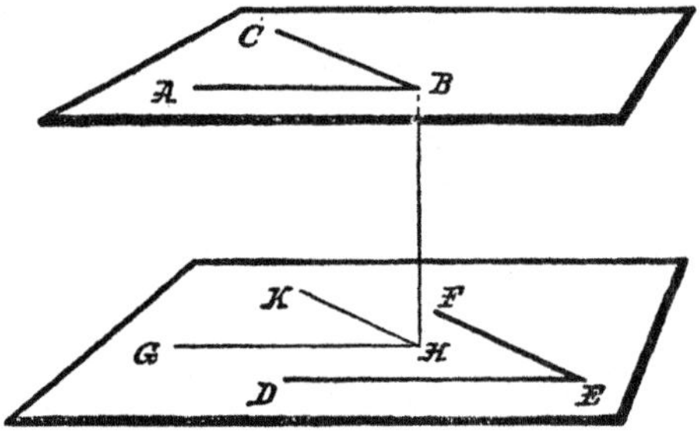

Hypothesis. $AB \parallel DE$, and $BC \parallel EF$.

Conclusion. Plane $ABC \parallel$ plane DEF.

Proof. From B draw $BH \perp$ plane DEF, meeting it in H.

Through H draw $GH \parallel DE$, and $HK \parallel EF$,

$$\therefore \quad GH \parallel AB, \quad \text{and} \quad HK \parallel BC,$$

(576. Two lines each parallel to the same line are parallel to each other.)

$$\therefore \quad \measuredangle ABH + \measuredangle BHG = \text{st.} \measuredangle.$$

But because BH was drawn $\perp$ plane DEF,

$$\therefore \quad \measuredangle BHG = \text{rt.} \measuredangle.$$

So

$$\therefore \quad \measuredangle ABH = \text{rt.} \measuredangle;$$

and, in same way,

$$\measuredangle CBH = \text{rt.} \measuredangle;$$

$$\therefore \quad HB \perp \text{plane } ABC,$$

$$\therefore \quad \text{plane } ABC \parallel \text{plane } DEF.$$

(585. Planes to which the same line is perpendicular are parallel.)

Theorem VIII.

588. *If a pair of intersecting lines be parallel to another pair, any angle made by the first pair is equal or supplemental to any angle made by the second pair.*

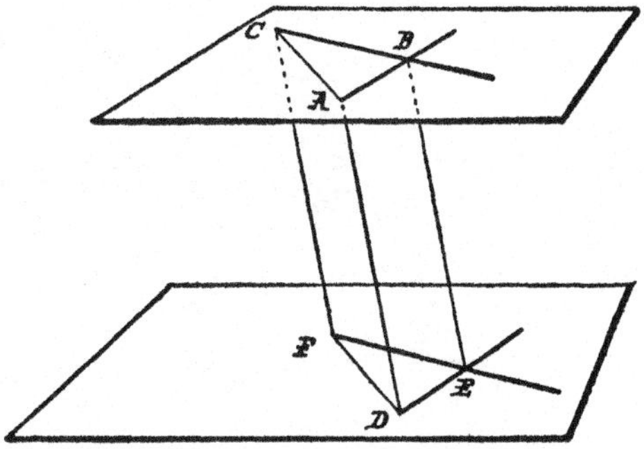

Hypothesis. $AB \parallel DE$, **and** $BC \parallel EF$.
Conclusion. $\measuredangle ABC =$ or supplemental to $\measuredangle DEF$.
Proof. Join BE.
Since $AB \parallel DE$,

$$\therefore \quad AB, BE, \text{ and } ED, \text{ are in one plane.}$$

In this plane, from A draw a line $\parallel BE$.
It must meet the line DE. Call the intersection point D.

$$\therefore \quad ABED \text{ is a } \square;$$

$$\therefore \quad AB = DE, \quad \text{and} \quad AD = BE.$$

In same way,

$$BC = EF, \quad \text{and} \quad CF = BE,$$

$$\therefore \quad AD = CF.$$

But since, by construction, $AD \parallel BE$, and $CF \parallel BE$, therefore, by 576, $AD \parallel CF$,

$$\therefore \quad ACFD \text{ is a } \square,$$

(216. If any two opposite sides of a quadrilateral are equal and parallel, it is a parallelogram.)

$$\therefore \quad AC = DF,$$

$$\therefore \quad \triangle ABC \cong \triangle DEF,$$

(129. Triangles with three sides of the one equal to three of the other are congruent.)

$$\therefore \quad \measuredangle ABC = \measuredangle DEF.$$

THEOREM IX.

589. *If two parallel planes be cut by another plane, their common sections with it are parallel.*

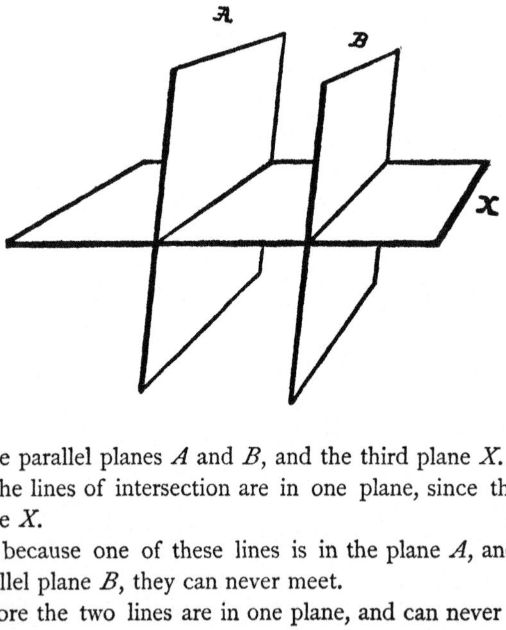

Call the parallel planes A and B, and the third plane X.

Then the lines of intersection are in one plane, since they both lie in the plane X.

Again, because one of these lines is in the plane A, and the other in the parallel plane B, they can never meet.

Therefore the two lines are in one plane, and can never meet; that is, they are parallel.

590. COROLLARY. Parallel sects included between two parallel planes are equal.

Theorem X.

591. *Parallel lines intersecting the same plane are equally inclined to it.*

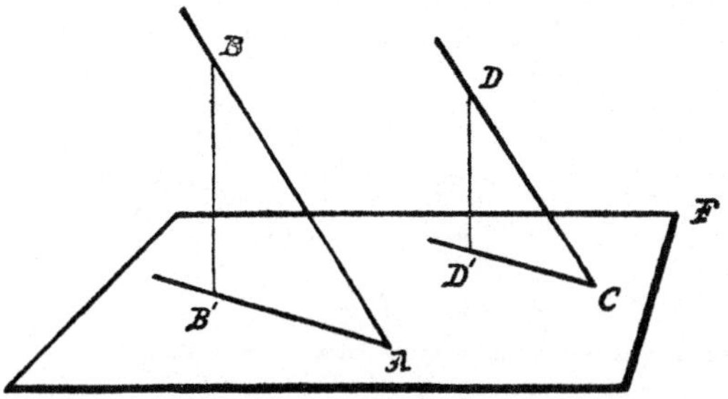

Hypothesis. $AB \parallel CD$,

$BB' \perp$ plane ACF,

$DD' \perp$ plane ACF.

Conclusion. $\measuredangle BAB' = \measuredangle DCD'$.

Proof. $\measuredangle ABB' =$ or supplemental to $\measuredangle CDD'$.

(588. If a pair of intersecting lines be parallel to another pair, any angle made by the first pair is equal or supplemental to any angle made by the second pair.)

But $\measuredangle ABB'$ cannot be supplemental to $\measuredangle CDD'$, since each is acute,

$$\therefore \quad \measuredangle ABB' = \measuredangle CDD';$$

and therefore their complements are equal.

592. Two lines not in the same plane are regarded as making with one another the angles included by two intersecting lines drawn parallel respectively to them.

We know, from 588, that these angles will always be the same, whatever the position of the point of intersection.

Theorem XI.

593. *If two lines be cut by three parallel planes, the corresponding sects are proportional.*

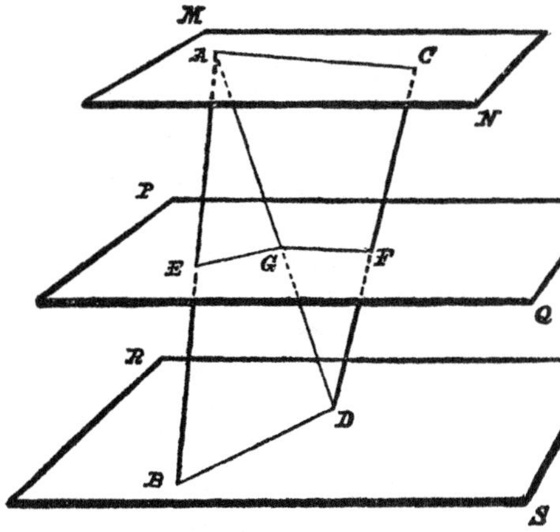

Let the lines AB, CD, be cut by the $\parallel$ planes MN, PQ, RS, in the points A, E, B, and C, F, D.

Conclusion. $AE : EB :: CF : FD$.

Join AD, cutting the plane PQ in G.

Join AC, BD, EG, FG. Then

$$EG \parallel BD.$$

(589. If parallel planes be cut by a third plane, their common sections with it are parallel.)

In the same way, $AC \parallel GF$;

$$\therefore\ AE : EB :: AG : GD,$$

and

$$AG : GD :: CF : FD;$$

(499. A line parallel to the base of a triangle divides the sides proportionally.)

$$\therefore\ AE : EB :: CF : FD.$$

THEOREM XII.

594. *Two lines not in the same plane have one, and only one, common perpendicular.*

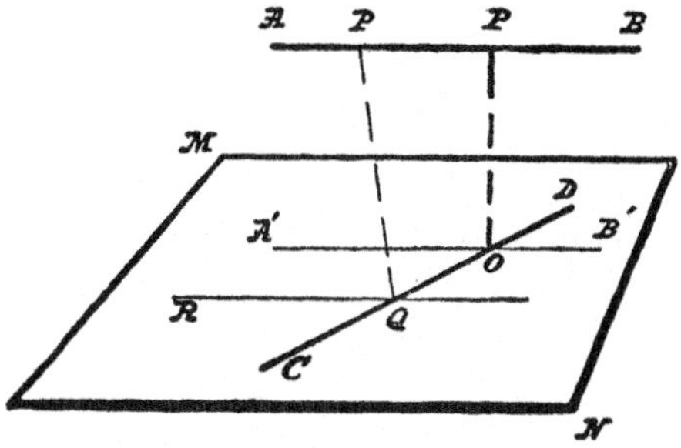

HYPOTHESIS. ***AB and CD, two lines not in the same plane, and therefore neither parallel nor intersecting each other.***

CONCLUSION. There is one line, and no more, perpendicular to both AB and CD.

PROOF. Through one of the lines, as CD, pass a plane, and let it revolve on CD as axis until it is parallel to AB. Call this plane MN. Let $A'B'$ be the projection of AB on the plane MN, and let O be the point in which this projection intersects CD. Then O, like every point in the projection, is the foot of a perpendicular to the plane from some point of AB. Call this point P. Then, since PO is perpendicular to the plane MN, it is perpendicular to CD and $A'B'$. But $A'B'$ is parallel to AB, because, being in a plane parallel to AB, it can never meet AB; and, being the projection of AB, it is, by 580, in the same plane with AB.

$\therefore$ since $PO \perp A'B'$, it is also $\perp AB$,

(170. A line perpendicular to one of two parallels is perpendicular to the other also.)

$\therefore$ $OP \perp$ both AB and CD.

If there could be any other common perpendicular, call it $P'Q$.

Through Q draw in the plane MN, $QR \parallel AB$.

Since $P'Q \perp AB$, $\therefore\ P'Q \perp QR$;

but, by hypothesis, $P'Q \perp CD$,

$\therefore\ P'Q \perp$ plane MN,

$\therefore\ Q$ is a point in $A'B'$, the projection of AB,

$\therefore\ Q$ is O, the only point common to $A'B'$ and CD,

$\therefore\ P'Q$ coincides with PO.

(569. At a given point in a given plane, only one perpendicular to the plane can be erected.)

THEOREM XIII.

595. *The smallest sect between two lines not in the same plane is their common perpendicular.*

For if any sect drawn from one to the other is not perpendicular to both, by dropping a perpendicular to the line it cuts obliquely from the point where it meets the other line, we get a smaller sect.

(150. The perpendicular is the smallest sect from a point to a line.)

596. REMARK. If two planes intersect, and two intersecting lines are drawn, one in each plane, these lines may, for the same

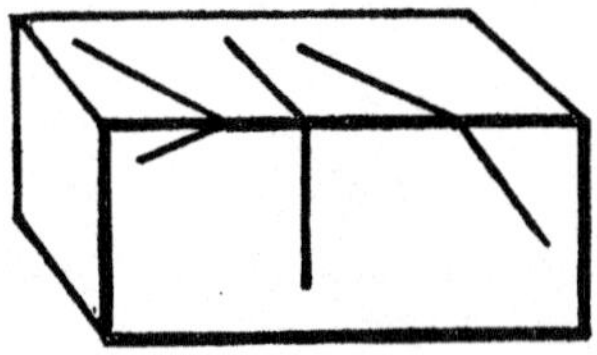

two planes, make an acute, right, or obtuse angle, according to their relation to the line of intersection of the planes.

THEOREM XIV.

597. *The smallest sect from a point to a plane is the perpendicular.*

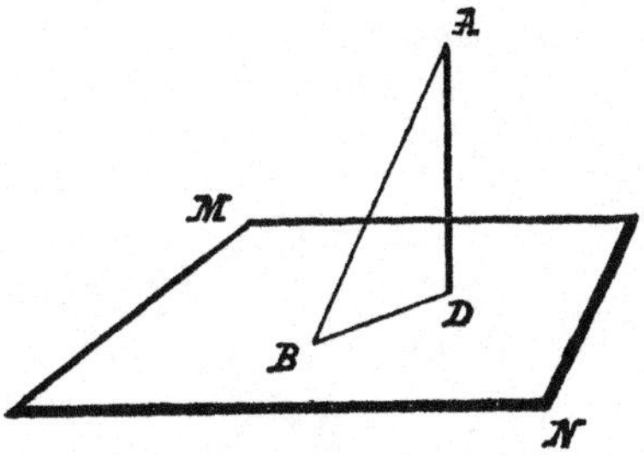

HYPOTHESIS. $AD \perp$ plane MN. B any other point of MN.
CONCLUSION. $AD < AB$.
PROOF. Join AB, BD. In $\triangle ABD$, $\measuredangle ADB$ is a rt. $\measuredangle$,

$$\therefore \quad AD < AB.$$

(150. The perpendicular is the least sect from a point to a line.)

THEOREM XV.

598. *Equal obliques from a point to a plane meet the plane in a circle whose center is the foot of the perpendicular from the point to the plane.*

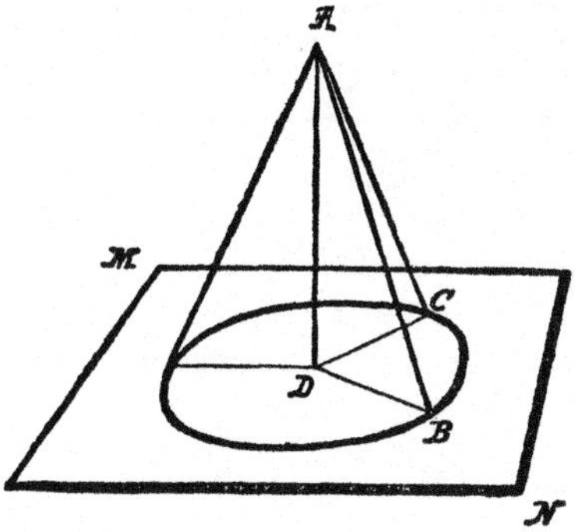

HYPOTHESIS. *AB, AC, equal sects from A to the plane MN. AD ⊥ plane MN.*

CONCLUSION. $DB = DC$.

PROOF. Rt. △ $ABD \cong$ rt. △ ACD.

(179. If two right triangles have the hypothenuse and one side respectively equal, they are congruent.)

599. COROLLARY I. To draw a perpendicular from a point to a plane, draw any oblique from the point to the plane; revolve this sect about the point, tracing a circle on the plane; find the center of this circle, and join it to the point.

600. COROLLARY II. If through the center of a circle a line be passed perpendicular to its plane, the sects from any point of this line to points on the circle are equal.

THEOREM XVI.

601. *The locus of all points from which the two sects drawn to two fixed points are equal, is the plane bisecting at right angles the sect joining the two given points.*

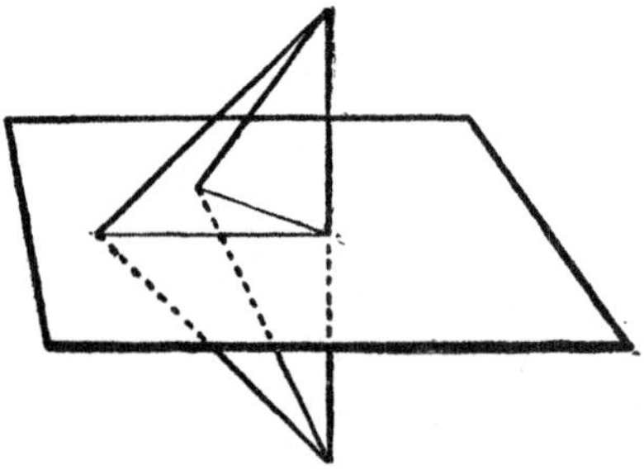

For the line from any such point to the mid point of the joining sect is perpendicular to that sect; therefore all such lines form a plane perpendicular to that sect at its mid point.

(570. All lines perpendicular to another line at the same point lie in the same plane.)

THEOREM XVII.

602. *The locus of all points from which the two perpendiculars onto the same sides of two fixed planes are equal, is a plane determined by one such point and the intersection line of the two given planes.*

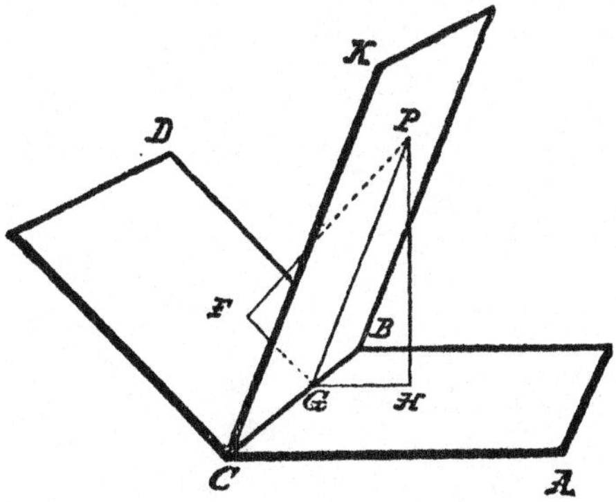

HYPOTHESIS. ***Let AB and CD be the two given planes, and K one such point.***

CONCLUSION. Perpendiculars dropped on to AB and CD from any point in the plane determined by K and the intersection line BC are equal.

PROOF. Take P any point in the plane KCB. Draw $PH \perp$ plane AB, and $PF \perp$ plane CD. Call G the point where the plane FPH cuts the line BC. Join FG, PG, GH.

Draw $KC \parallel PG$. Because the perpendiculars from K are given equal, therefore KC is equally inclined to the two planes. But PG has the same inclination to each as KC;

(591. Parallels intersecting the same plane are equally inclined to it.)

$$\therefore \quad \measuredangle PGF = \measuredangle PGH,$$

$$\therefore \quad \triangle PGF \cong \triangle PGH,$$

$$\therefore \quad PF = PH.$$

THEOREM XVIII.

603. *If three lines not in the same plane meet at one point, any two of the angles formed are together greater than the third.*

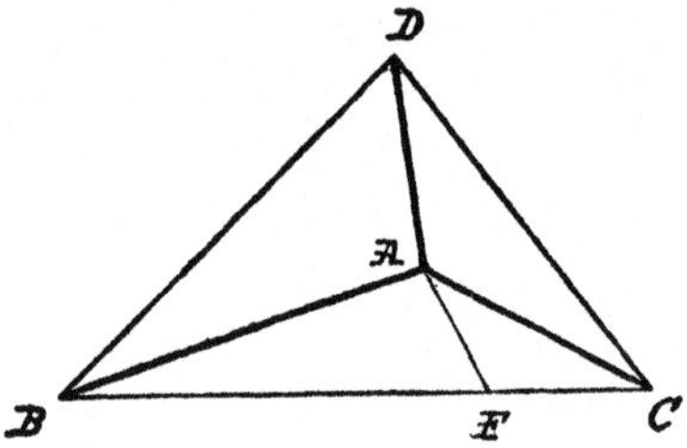

PROOF. The theorem requires proof only when the third angle considered is greater than each of the others. In the plane of the greatest ∡ BAC, make ∡ BAF = ∡ BAD. Make $AF = AD$. Through F draw a line BFC cutting AB in B, and AC in C. Join DB and DC.

$$\triangle BAD \cong \triangle BAF,$$

(124. Triangles having two sides and the included angle respectively equal are congruent.)

$$\therefore \quad BD = BF.$$

But from △ BCD we have $BD + DC > BC$.

(156. Any two sides of a triangle are together greater than the third.)

And, taking away the equals BD and BF,

$$DC > FC,$$

∴ in △s CAD and CAF, we have ∡ CAD > ∡ CAF.

(161. If two triangles have two sides respectively equal, but the third side greater in the first, its opposite angle is greater in the first.)

Adding the equal ∡s BAD and BAF gives

∡ BAD + ∡ CAD > ∡ BAC.

THEOREM XIX.

604. *If the vertices of a convex polygon be joined to a point not in its plane, the sum of the vertical angles of the triangles so made is less than a perigon.*

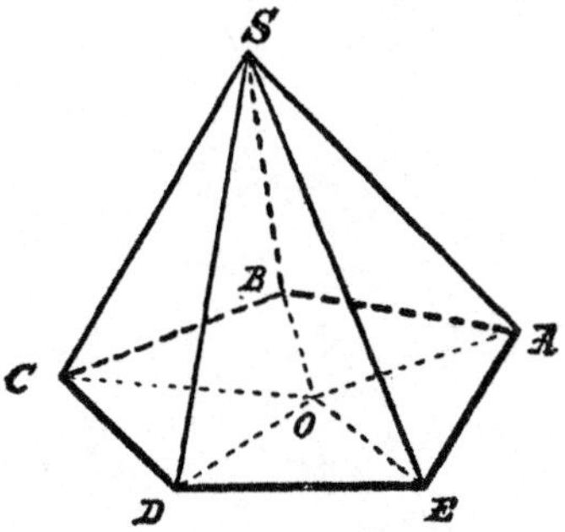

PROOF. The sum of the angles of the triangles which have the common vertex S is equal to the sum of the angles of the same number of triangles having their vertices at O in the plane of the polygon. But

$$\measuredangle SAB + \measuredangle SAE > \measuredangle BAE,$$

$$\measuredangle SBA + \measuredangle SBC > \measuredangle ABC, \text{ etc.}$$

(603. If three lines not in a plane meet at a point, any two of the angles formed are together greater than the third.)

Hence, summing all these inequalities, the sum of the angles at the bases of the triangles whose vertex is S, is greater than the sum of the angles at the bases of the triangles whose vertex is O; therefore the sum of the angles at S is less than the sum of the angles at O, that is, less than a perigon.

BOOK VIII.

TRI-DIMENSIONAL SPHERICS.

605. If one end point of a sect is fixed, the locus of the other end point is a *Sphere.*

606. The fixed end point is called the *Center* of the sphere.

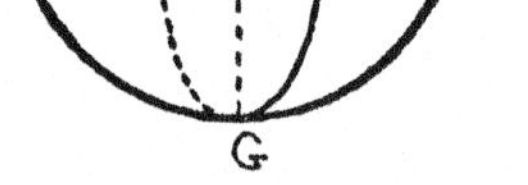

607. The moving sect in any position is called the *Radius* of the sphere.

608. As the motion of a sect does not change it, all radii are equal.

609. The sphere is a closed surface; for it has two points on every line passing through the center, and the center is midway between them.

610. Two such points are called *Opposite Points* of the sphere, and the sect between them is called a *Diameter.*

611. The sect from a point to the center is less than, equal to, or greater than, the radius, according as the point is within, on, or without, the sphere.

For, if a point is on the sphere, the sect drawn to it from the center is a radius; if the point is within the sphere, it lies on some radius; if without, it lies on the extension of some radius.

612. By 33, Rule of Inversion, a point is within, on, or without, the sphere, according as the sect to it from the center is less than, equal to, or greater than, the radius.

THEOREM I.

613. *The common section of a sphere and a plane is a circle.*

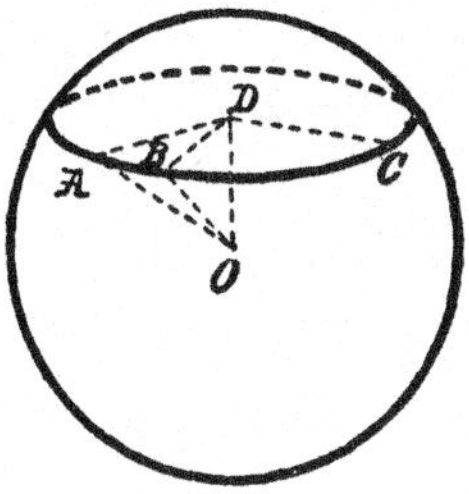

Take any sphere with center O.

Let A, B, C, etc., be points common to the sphere and a plane, and OD the perpendicular from O to the plane. Then

$$OA = OB = OC = \text{etc.},$$

being radii of the sphere,

$$\therefore \quad ABC, \text{ etc., is a circle with center } D.$$

(598. Equal obliques from a point to a plane meet the plane in a circle whose center is the foot of the perpendicular from the point to the plane.)

614. COROLLARY. The line through the center of any circle of a sphere, perpendicular to its plane, passes through the center of the sphere.

615. A *Great Circle* of a sphere is any section of the sphere made by a plane which passes through the center.

All other circles on the sphere are called *Small Circles.*

616. COROLLARY. All great circles of the sphere are equal, since each has for its radius the radius of the sphere.

617. The two points in which a perpendicular to its plane, through the center of a great or small circle of the sphere, intersects the sphere, are called the *Poles* of that circle.

618. COROLLARY. Since the perpendicular passes through the center of the sphere, the two poles of any circle are opposite points, and the diameter between them is called the *Axis* of that circle.

THEOREM II.

619. *Every great circle divides the sphere into two congruent hemispheres.*

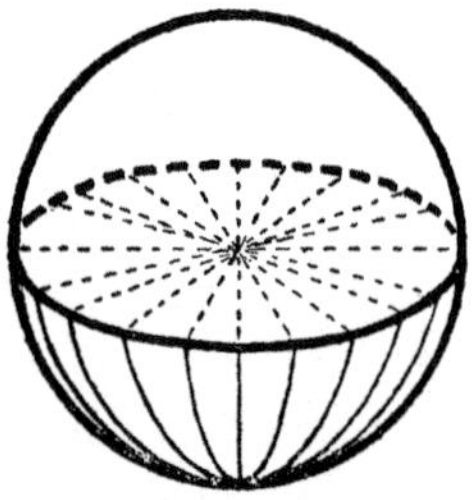

For if one hemisphere be turned about the fixed center of the sphere so that its plane returns to its former position, but inverted, the great circle will coincide with its own trace, and the two hemispheres will coincide.

620. Any two great circles of a sphere bisect each other.

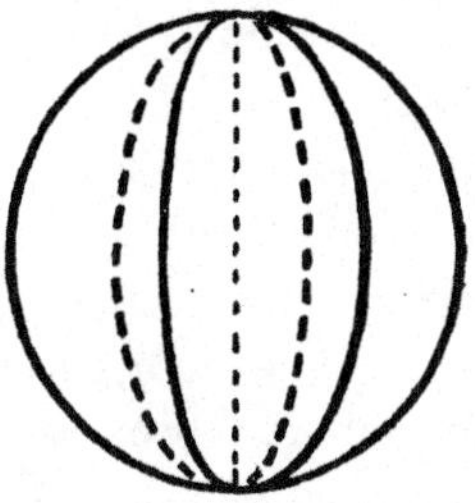

Since the planes of these circles both pass through the center of the sphere, their line of intersection is a diameter of the sphere, and therefore of each circle.

621. If any number of great circles pass through a point, they will also pass through the opposite point.

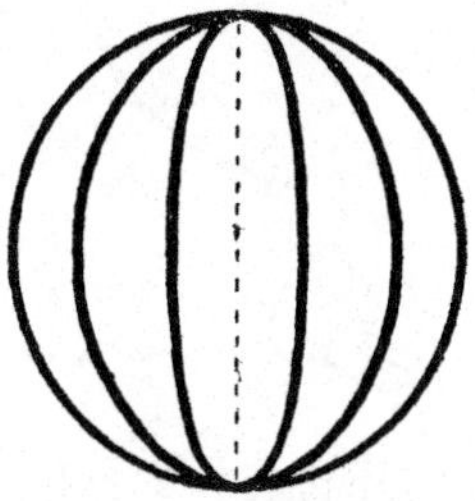

622. Through any two points in a sphere, not the extremi-

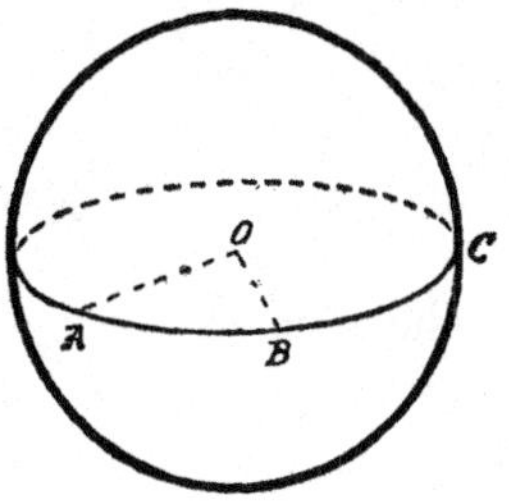

ties of a diameter, one, and only one, great circle can be passed; for the two given points and the center of the sphere determine its plane. Through opposite points, an indefinite number of great circles can be passed.

623. Through any three points in a sphere, a plane can be passed, and but one; therefore three points in a sphere determine a circle of the sphere.

624. A small circle is the less the greater the sect from its center to the center of the sphere. For, with the same hypothenuse, one side of a right-angled triangle decreases as the other increases.

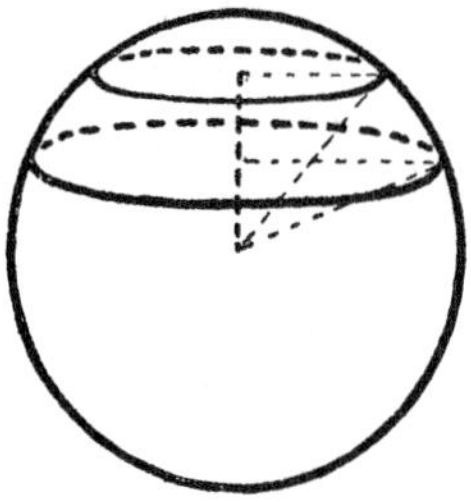

625. A *Zone* is a portion of a sphere included between two parallel planes. The circles made by the parallel planes are the *Bases* of the zone.

626. A line or plane is tangent to a sphere when it has one point, and only one, in common with the sphere.

627. Two spheres are tangent to each other when they have one point, and only one, in common.

EXERCISES. 105. If through a fixed point, within or without a sphere, three lines are drawn perpendicular to each other, intersecting the sphere, the sum of the squares of the three intercepted chords is constant. Also the sum of the squares of the six segments of these chords is constant.

106. If three radii of a sphere, perpendicular to each other, are projected upon any plane, the sum of the squares of the three projections is equal to twice the square of the radius of the sphere.

THEOREM III.

628. *A plane perpendicular to a radius of a sphere at its extremity is tangent to the sphere.*

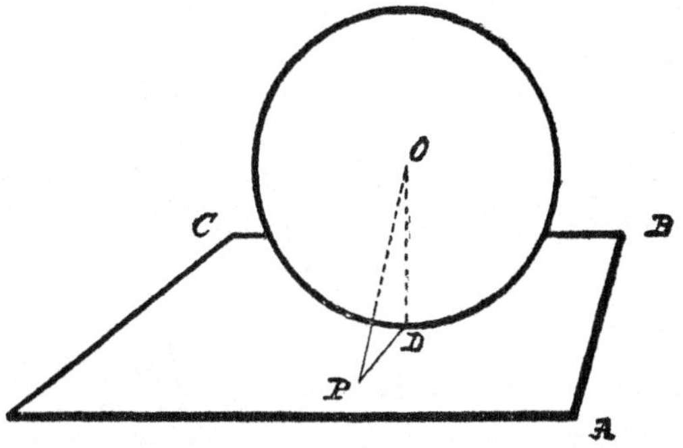

For, by 597, this radius, being perpendicular to the plane, is the smallest sect from the center to the plane; therefore every point of the plane is without the sphere except the foot of this radius.

629. COROLLARY. Every line perpendicular to a radius at its extremity is tangent to the sphere.

630. INVERSE OF 628. Every plane or line tangent to the sphere is perpendicular to the radius drawn to the point of contact. For since every point of the plane or line, except the point of contact, is without the sphere, the radius drawn to the point of contact is the smallest sect from the center of the sphere to the plane or line; therefore, by 597, it is perpendicular.

EXERCISES. 107. Cut a given sphere by a plane passing through a given line, so that the section shall have a given radius.

108. Find the locus of points whose sect from point *A* is *a*, and from point *B* is *b*.

THEOREM IV.

631. *If two spheres cut one another, their intersection is a circle whose plane is perpendicular to the line joining the centers of the spheres, and whose center is in that line.*

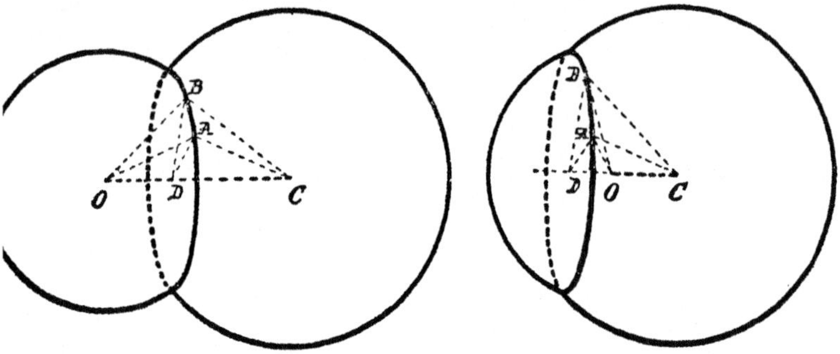

HYPOTHESIS. ***Let*** C ***and*** O ***be the centers of the spheres,*** A ***and*** B ***any two points in their intersection.***

CONCLUSION. A and B are on a circle having its center on the line OC, and its plane perpendicular to that line.

PROOF. Join CA, CB, OA, OB. Then

$$\triangle CAO \cong \triangle CBO,$$

because they have OC common, $CA = CB$, and $OA = OB$, radii of the same sphere.

Since these $\triangle$s are $\cong$, therefore perpendiculars from A and B upon OC are equal, and meet OC at the same point, D.

Then AD and DB are in a plane $\perp OC$; and, being equal sects, their extremities A and B are in a circle having its center at D.

632. COROLLARY. By moving the centers of the two intersecting spheres toward or away from each other, we can make their circle of intersection decrease indefinitely toward its center; therefore, if two spheres are tangent, either internally or externally, their centers and point of contact lie in the same line.

THEOREM V.

633. *Through any four points not in the same plane, one sphere, and only one, can be passed.*

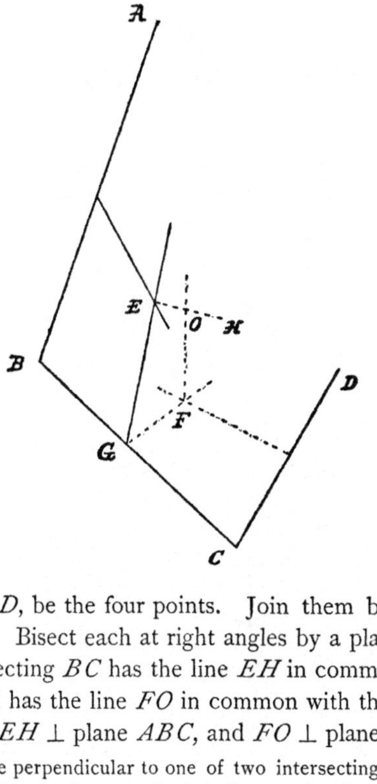

Let A, B, C, D, be the four points. Join them by any three sects, as AB, BC, CD. Bisect each at right angles by a plane.

The plane bisecting BC has the line EH in common with the plane bisecting AB, and has the line FO in common with the plane bisecting CD. Moreover, $EH \perp$ plane ABC, and $FO \perp$ plane BCD,

(575. If one plane be perpendicular to one of two intersecting lines, and a second plane perpendicular to the second, their intersection is perpendicular to the plane of the two lines.)

$$\therefore \quad EH \perp EG, \quad \text{and} \quad FO \perp FG,$$

$\therefore$ EH and FO meet, since, by hypothesis, $\measuredangle$ at $G >$ o and $<$ st. $\measuredangle$.

Call their point of meeting O. O shall be the center of the sphere containing A, B, C, and D.

For O is in three planes bisecting at right angles the sects AB, BC, CD.

(601. The locus of all points from which the two sects drawn to two fixed points are equal, is the plane bisecting at right angles the sect joining the two given points.)

634. A *Tetrahedron* is a solid bounded by four triangular plane surfaces called its faces. The sides of the triangles are called its edges.

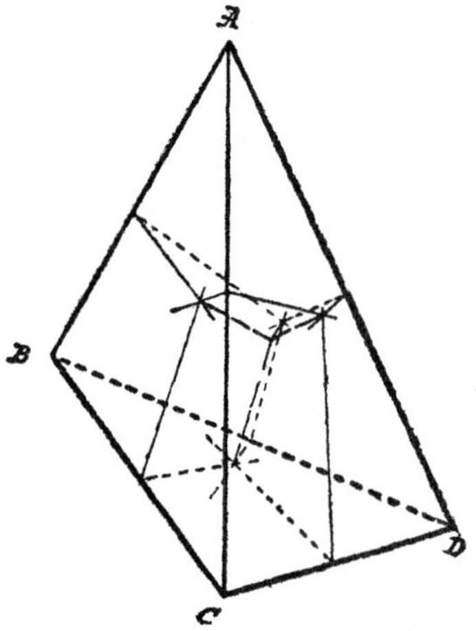

635. Corollary I. A sphere may be circumscribed about any tetrahedron.

636. Corollary II. The lines perpendicular to the faces of a tetrahedron through their circumcenters, intersect at a common point.

637. Corollary III. The six planes which bisect at right angles the six edges of a tetrahedron all pass through a common point.

PROBLEM I.

638. *To inscribe a sphere in a given tetrahedron.*

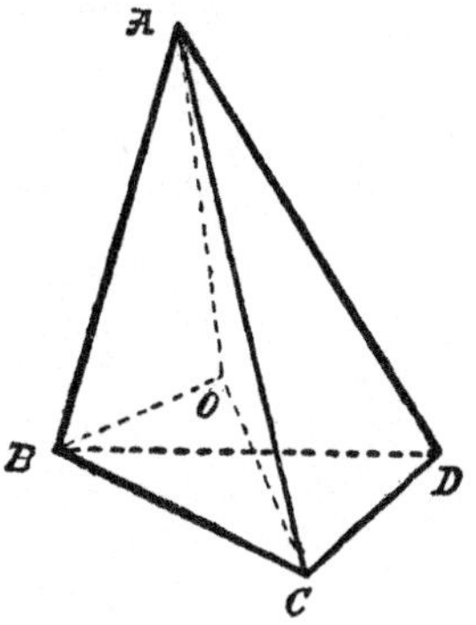

CONSTRUCTION. Through any edge and any point from which perpendiculars to its two faces are equal, pass a plane.

Do the same with two other edges in the same face as the first, and let the three planes so determined intersect at O. O shall be the center of the required sphere.

(602. The locus of all points from which the perpendiculars on the same sides of two planes are equal, is a plane determined by one such point and the intersection line of the given planes.)

639. COROLLARY. In the same way, four spheres may be escribed, each touching a face of the tetrahedron externally, and the other three faces produced.

640. The solid bounded by a sphere is called a *Globe.*

641. A globe-segment is a portion of a globe included between two parallel planes.

The sections of the globe made by the parallel planes are the bases of the segment.

Any sect perpendicular to, and terminated by, the bases, is the altitude of the segment.

642. A figure with two points fixed can still be moved by revolving it about the line determined by the two points.

This revolution can be performed in either of two senses, and continued until the figure returns to its original position. The fixed line is called the *Axis of Revolution.* If the axis of revolution is any line passing through the center, a sphere slides upon its trace. This is because every section of a sphere by a plane is a circle.

Any figure has central symmetry if it has a center which bisects all sects through it terminated by the surface. The sphere has central symmetry, and coincides with its trace throughout any motion during which the center remains fixed. Thus any figure drawn on a sphere may be moved about on the sphere without deformation. But, unlike planes, all spheres are not congruent. Only those with equal radii will coincide.

In general, a figure drawn upon one sphere will not fit upon another. So we cannot apply the test of superposition, except on the same sphere or spheres whose radii are equal. Again, if we wish the angles of a figure on a sphere to remain the same while the sides increase, we must magnify the whole sphere: on the same sphere similar figures cannot exist.

643. Two points are symmetrical with respect to a plane when this plane bisects at right angles the sect joining them. Two figures are symmetrical with respect to a plane when every point of one figure has its symmetrical point in the other. Any figure has planar symmetry if it can be divided by a plane into two figures symmetrical with respect to that plane. The sphere is symmetrical with respect to every plane through its center. Any two spheres are symmetrical with respect to every plane through their line of centers.

Every such plane cuts the spheres in two great circles; and the five relations between the center-sect, radii, and relative position of these circles given in 410, with their inverses, hold for the two spheres.

Any three spheres are symmetrical with respect to the plane determined by their centers.

THEOREM VI.

644. *If a sphere be tangent to the parallel planes containing opposite edges of a tetrahedron, and sections made in the globe and tetrahedron by one plane parallel to these are equivalent, sections made by any parallel plane are equivalent.*

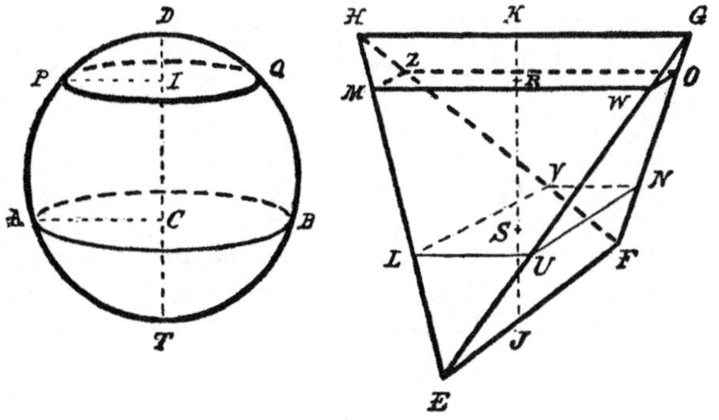

HYPOTHESIS. ***Let** KJ **be the sect** $\perp$ **the edges** EF **and** GH **in the** $\parallel$ **tangent planes.***

Then** $KJ = DT$**, the diameter.

***Let sections made by plane** $\perp$ KJ **at** R **and** $\perp$ DT **at** I**, where** $KR = DI$**, be equivalent; that is,** $\square\, MO = \odot\, PQ$.*

***Draw any parallel plane** $ACBLSN$.*

CONCLUSION. $\square\, LN = \odot\, AB$.

PROOF. Since $\triangle\, LEU \sim \triangle\, MEW$, and $\triangle\, LHV \sim \triangle\, MHZ$,

$$\therefore\quad MW : LU :: EM : EL :: JR : JS;$$

$$MZ : LV :: HM : HL :: KR : KS;$$

(593. If lines be cut by three parallel planes, the corresponding sects are proportional.)

$$\therefore\quad WM \,.\, MZ : UL \,.\, LV :: JR \,.\, RK : JS \,.\, SK;$$

$$\therefore\quad \square\, MO : \square\, LN :: JR \,.\, RK : JS \,.\, SK. \qquad (1)$$

(542. Mutually equiangular parallelograms have to one another the ratio which is compounded of the ratios of their sides.)

But

$\odot PQ : \odot AB :: PI^2 : AC^2 :: TI \, . \, ID : TC \, . \, CD.$ (2)

(546. Similar figures are to each other as the squares on their corresponding sects.)

(522. If from any point in a circle a perpendicular be dropped upon a diameter, it will be a mean proportional between the segments of the diameter.)

By hypothesis and construction, in proportions (1) and (2), the first, third, and fourth terms are respectively equal,

$$\therefore \ \square LN = \odot AB.$$

Theorem VII.

645. *The sects joining its pole to points on any circle of a sphere are equal.*

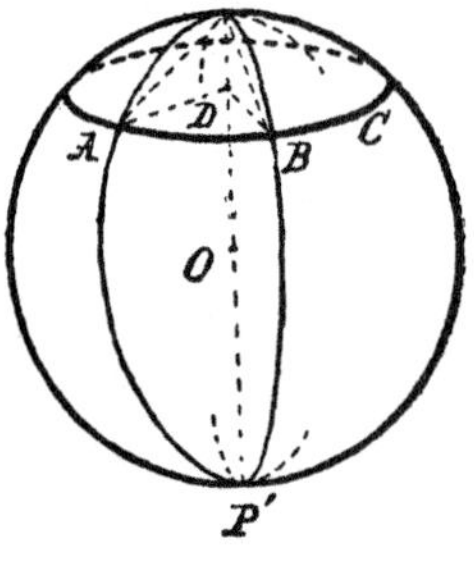

Proof. (600. If through the center of a circle a line be passed perpendicular to its plane, the sects from any point of this line to points on the circle are equal.)

646. Corollary. Since chord PA equals chord PB, therefore the arc subtended by chord PA in the great circle PA equals the arc subtended by chord PB in the great circle PB.

Hence the great-circle-arcs joining a pole to points on its circle are equal.

So, if an arc of a great circle be revolved in a sphere about one of its extremities, its other extremity will describe a circle of the sphere.

647. One-fourth a great circle is called a *Quadrant.*

648. The great-circle-arc joining any point in a great circle with its pole is a quadrant.

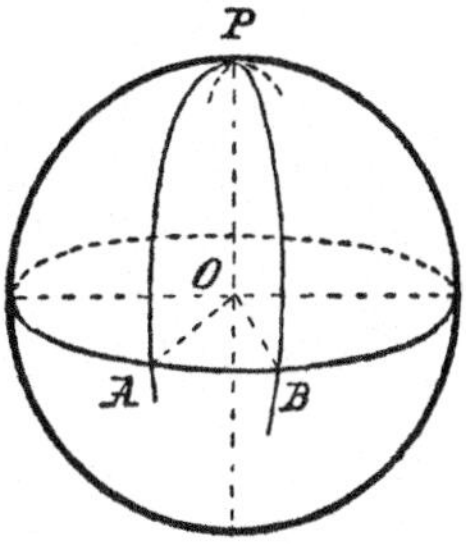

649. If a point P be a quadrant from two points, A, B, which are not opposite, it is the pole of the great circle through A, B; for each of the angles POA, POB, is right, and therefore PO is perpendicular to the plane OAB.

650. The angle between two intersecting curves is the angle between their tangents, at the point of intersection.

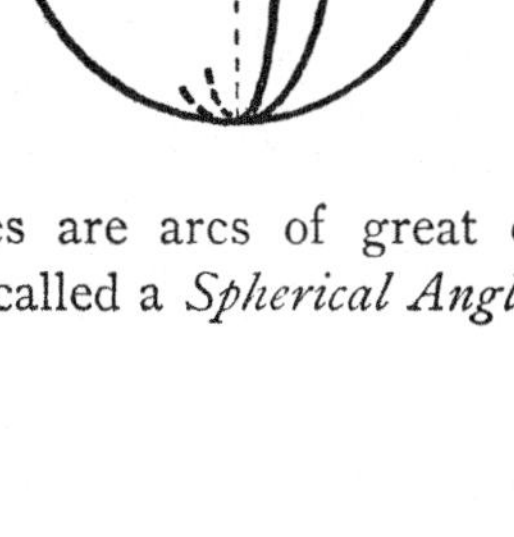

When the curves are arcs of great circles of the same sphere, the angle is called a *Spherical Angle.*

651. If from the vertices, *A* and *F*, of any two angles in a sphere, as poles, great circles, *BC* and *GH*, be described, the angles will be to one another in the ratio of the arcs of these circles intercepted between their sides (produced if necessary).

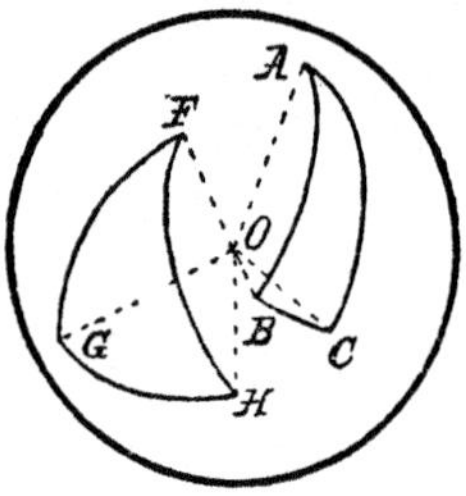

For the angles *A* and *F* are equal respectively to the angles *BOC* and *GOH*.

(591. Parallels intersecting the same plane are equally inclined to it.)

But *BOC* and *GOH* are angles at the centers of equal circles, and therefore, by 506, are to one another in the ratio of the arcs *BC* and *GH*.

652. COROLLARY. Any great-circle-arc drawn through the pole of a given great circle is perpendicular to that circle.

FG is ⊥ *GH*. For, by hypothesis, *FG* is a quadrant; therefore the great circle described with *G* as pole passes through *F*, and so the arc intercepted on it between *GF* and *GH* is, by 648, also a quadrant. But, by 651, the angle at *G* is to this quadrant as a straight angle is to a half-line.

Inversely, any great-circle-arc perpendicular to a great circle will pass through its pole.

For if we use *G* as pole when the angle at *G* is a right angle, then *FH*, its corresponding arc, is a quadrant when *GF* is a quadrant; therefore, by 649, *F* is the pole of *GH*.

THEOREM VIII.

653. *The smallest line in a sphere, between two points, is the great-circle-arc not greater than a semicircle, which joins them.*

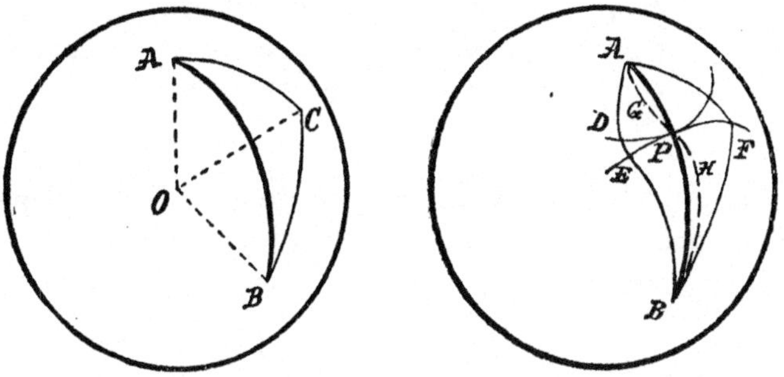

HYPOTHESIS. ***AB is a great-circle-arc, not greater than a semicircle, joining any two points A and B on a sphere.***

FIRST, let the points A and B be joined by the broken line ACB, which consists of the two great-circle-arcs AC and CB.

CONCLUSION. $AC + CB > AB$.

PROOF. Join O, the center of the sphere, with A, B, and C.

$$\measuredangle AOC + \measuredangle COB > \measuredangle AOB.$$

(603. If three lines not in the same plane meet at one point, any two of the angles formed are together greater than the third.)

But the corresponding arcs are in the same ratio as these angles,

$$\therefore \quad AC + CB > AB.$$

SECOND, let P be any point whatever on the great-circle-arc AB. The smallest line on the sphere from A to B must pass through P.

For by revolving the great-circle-arcs AP and BP about A and B as poles, describe circles.

These circles touch at P, and lie wholly without each other; for let F be any other point in the circle whose pole is B, and join FA, FB by great-circle-arcs, then, by our *First*,

$$FA + FB > AB,$$

$$\therefore \quad FA > PA, \text{ and } F \text{ lies without the circle whose pole is } A.$$

Now let *ADEB* be any line on the sphere from *A* to *B* not passing through *P*, and therefore cutting the two circles in different points, one in *D*, the other in *E*. A portion of the line *ADEB*, namely, *DE*, lies between the two circles. Hence if the portion *AD* be revolved about *A* until it takes the position *AGP*, and the portion *BE* be revolved about *B* into the position *BHP*, the line *AGPHB* will be less than *ADEB*. Hence the smallest line from *A* to *B* passes through *P*, that is, through any or every point in *AB*; consequently it must be the arc *AB* itself.

654. COROLLARY. A sect is the smallest line in a plane between two points.

BOOK IX.

TWO-DIMENSIONAL SPHERICS.

INTRODUCTION.

655. Book IX. will develop the Geometry of the Sphere, from theorems and problems almost identical with those whose assumption gave us Plane Geometry. In Book VIII., these have been demonstrated by considering the sphere as contained in ordinary tri-dimensional space. But, if we really confine ourselves to the sphere itself, they do not admit of demonstration, except by making some more difficult assumption: and so they are the most fundamental properties of this surface and its characteristic line, the great circle; just as the assumptions in our first book were the most fundamental properties of the plane and its characteristic line.

So now we will call a great circle simply the spherical *line;* and, whenever in this book the word line is used, it means spherical line. Sect now means a part of a line less than a half-line.

Two-Dimensional Definition of the Sphere.

656. Suppose a closed line, such that any portion of it may be moved about through every portion of it without any other change. Suppose a portion of this line is such, that, when moved on the line until its first end point comes to the trace of its second end point, that second end point will have moved to the trace of its first end point. Call such a portion a half-line, and any lesser portion a sect. Suppose, that, while the extremities of a half-line are kept fixed, the whole line can be so moved that the slightest motion takes it completely out of its trace, except in the two fixed points. Such motion would generate a surface which we will call the Sphere.

Fundamental Properties of the Sphere.

ASSUMPTIONS.

657. A figure may be moved about in a sphere without any other change; that is, figures are independent of their place on the sphere.

658. Through any two points in the sphere can be passed a line congruent with the generating line of the sphere. In Book IX., the word line will always mean such a line, and sect will mean a portion of it less than half.

659. Two sects cannot meet twice on the sphere; that is, if two sects have two points in common, the two sects coincide between those two points.

660. If two lines have a common sect, they coincide throughout. Therefore through two points, not end points of a half-line, only one distinct line can pass.

661. A sect is the smallest path between its end points in the sphere.

662. A piece of the sphere from along one side of a line will fit either side of any other portion of the line.

DEFINITIONS.

663. If one end point of a sect is kept fixed, the other end point moving in the sphere describes what is called an arc, while the sect describes at the fixed point what is called a spherical angle. The angle and arc are greater as the amount of turning in the sphere is greater.

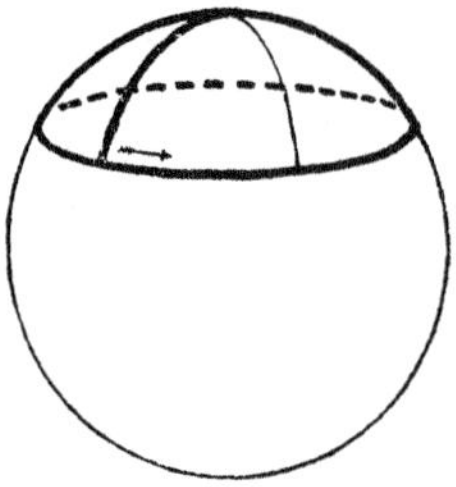

664. When a sect has turned sufficiently to fall again into the same line, but on the other side of the fixed point or vertex, the angle described is called a straight angle, and the arc described is called a semicircle.

665. Half a straight angle is called a right angle.

666. The whole angle about a point in a sphere is called a perigon: the whole arc is called a circle.

The point is called the pole of the circle, and the equal sects are called its spherical radii.

ASSUMPTIONS.

667. A circle can be described from any pole, with any sect as spherical radius.

668. All straight angles are equal.

669. Corollary I. All perigons are equal.

670. Corollary II. If one extremity of a sect is in a line, the two angles on the same side of the line as the sect are together a straight angle.

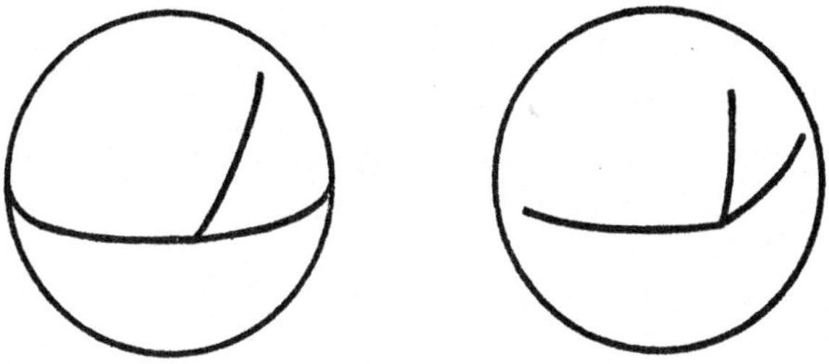

671. Corollary III. Defining adjacent angles as two angles having a common vertex, a common arm, and not overlapping, it follows, that, if two adjacent angles together equal a straight angle, their two exterior arms fall into the same line.

672. Corollary IV. If two sects cut one another, the vertical angles are equal.

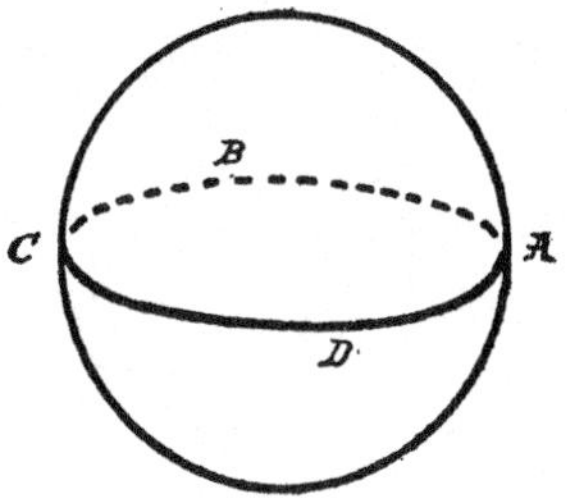

673. Any line turning in the sphere about one of its points, through a straight angle, comes to coincidence with its trace, and has described the sphere.

674. Corollary I. The sphere is a closed surface.

675. Corollary II. Every line bisects the sphere, — cuts it into hemispheres.

Reduced Properties of the Sphere.

Theorem I.

676. *All lines in a sphere intersect, and bisect each other at their points of intersection.*

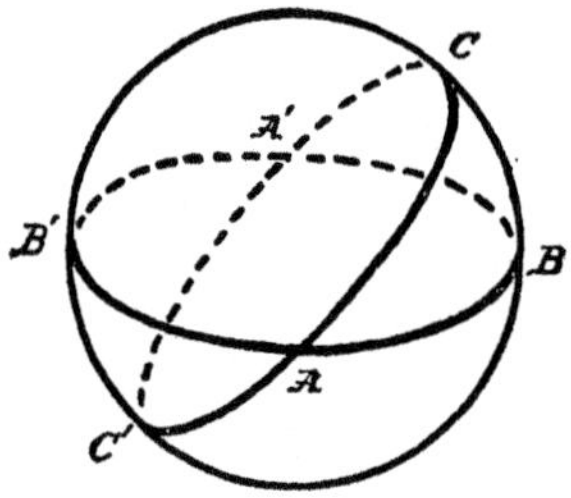

Let BB' and CC' be any two lines.

Since, by 675, each of them bisects the sphere, therefore the second cannot lie wholly in one of the hemispheres made by the first, therefore they intersect at two points; let them intersect at A and A'.

If BAC be revolved in the sphere about A until some point in the sect AB coincides with some point in the sect AB', then ABA' will lie along $AB'A'$.

(660. Through two points, not end points of a half-line, only one distinct line can pass.)

Then, also, since the angles BAC and $B'AC'$, being vertical, are, by 672, equal, ACA' will lie along $AC'A'$. Therefore the second point of intersection of AB and AC must coincide with the second point of intersection of AB' and AC', and ABA' be a half-line, as also ACA'.

677. A spherical figure such as *ABA'CA*, which is contained by two half-lines, is called a *Lune*.

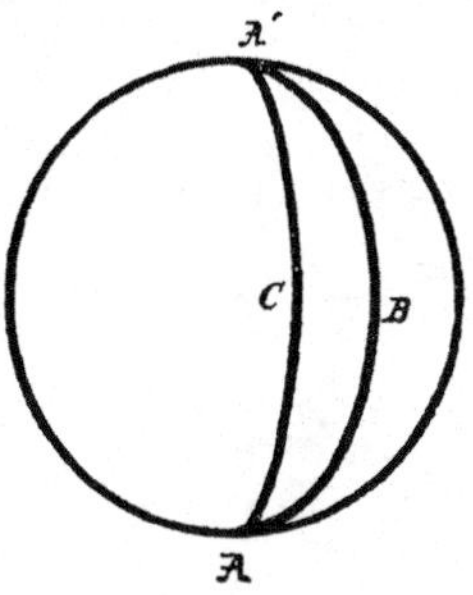

THEOREM II.

678. *The angle contained by the sides of a lune at one of their points of intersection equals the angle contained at the other.*

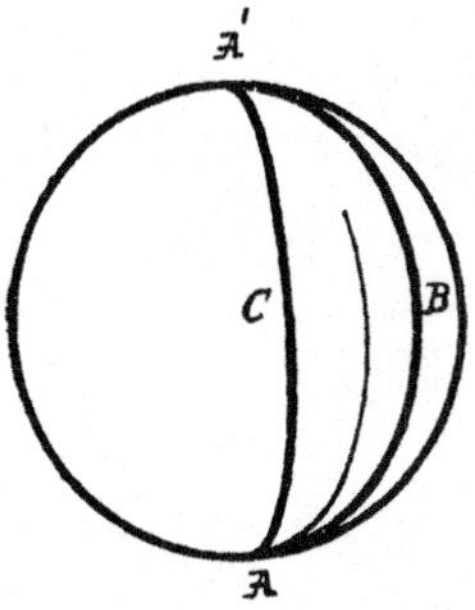

For, if the two angles are not equal, one must be the greater. Suppose $\measuredangle A > \measuredangle A'$.

Move the lune in the sphere until point A' coincides with the trace of point A, and half-line $A'BA$ coincides with the trace of half-line ACA'. Then, since we have supposed $\measuredangle A > \measuredangle A'$, the half-line $A'CA$ would start from the trace of A between the trace of ACA' and the trace of ABA'. Starting between them, it could meet neither again until it reached the trace of A'; and so we would have the surface of the lune less than its trace, which is contrary to our first assumption (657).

∴ the two angles cannot be unequal.

DEFINITIONS.

679. The *Supplement of a Sect* is the sect by which it differs from a half-line.

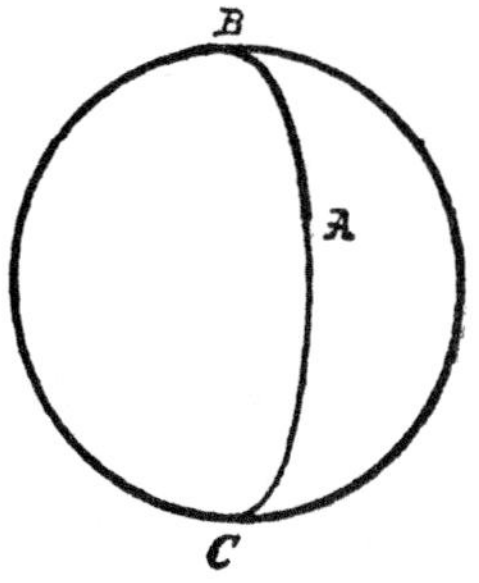

680. One-quarter of a line is called a *Quadrant.*

618. A *Spherical Polygon* is a closed figure in the sphere formed by sects.

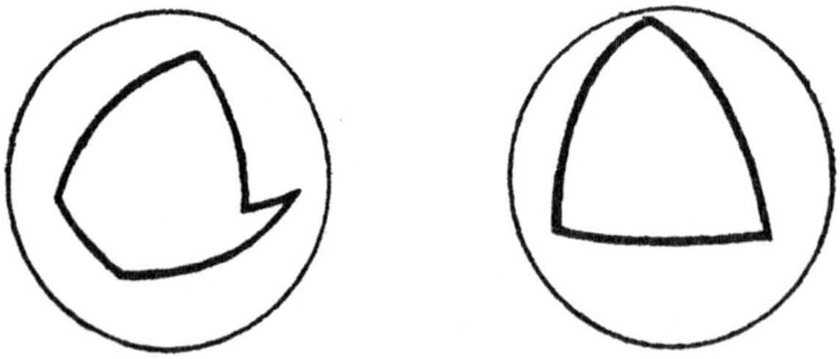

682. A *Spherical Triangle* is a convex spherical polygon of three sides.

683. *Symmetrical Spherical Polygons* are those in which the sides and angles of the one are respectively equal to those of

the other, but arranged in the reverse order. If one end of a sect were pivoted within one polygon, and one end of another sect pivoted within the symmetrical polygon, and the two sects revolved so as to pass over the equal parts at the same time, one sect would move clockwise, while the other moved counter-clockwise.

684. Two points are symmetrical with respect to a fixed line, called the axis of symmetry, when this axis bisects at right angles the sect joining the two points.

Any two figures are symmetrical with respect to an axis when every point of one has its symmetrical point on the other.

Theorem III.

685. *The perimeter of a convex spherical polygon wholly contained within a second spherical polygon is less than the perimeter of the second.*

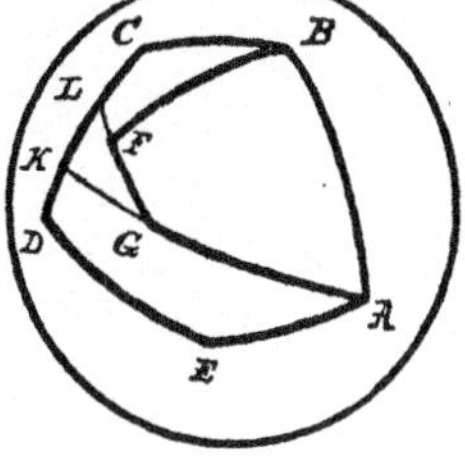

Let *ABFGA* be a convex spherical polygon wholly contained in *ABCDEA*.

Produce the sides AG and GF to meet the containing perimeter at K and L respectively.

By 661, a sect is the smallest path between its end points,

$$\therefore \quad BF < BCLF, \quad \text{and} \quad LG < LKG, \quad \text{and} \quad KA < KDEA;$$

$$\therefore \quad BFGA < BCLGA < BCKA < BCDEA.$$

THEOREM IV.

686. *The sum of the sides of a convex spherical polygon is less than a line.*

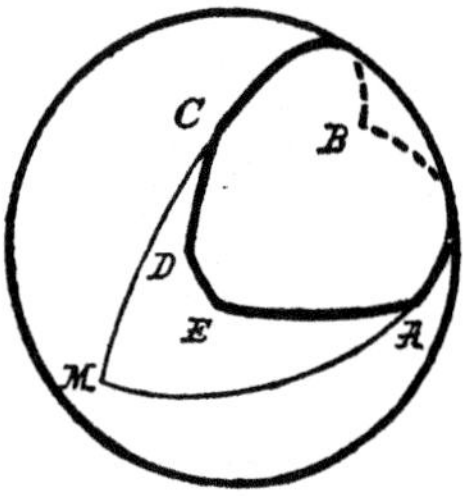

Let $ABCDEA$ be the polygon, and let its sides BA and BC be produced to meet again at M. Since the polygon is convex, the ∡ BCD < st. ∡, and also ∡ BAE < st. ∡; therefore $ABCDEA$ lies wholly within $ABCMA$; therefore, by 685, the perimeter of $ABCDEA$ < the perimeter of $ABCMA$, that is, less than a line.

SYMMETRY WITHOUT CONGRUENCE.

687. If two figures have central symmetry in a plane, either can be made to coincide with the other by turning it in the plane through a straight angle. This holds good when for "plane" we substitute "sphere."

If two figures have axial symmetry in a plane, they can be made to coincide by folding the plane over along the axis, but not by any sliding in their plane. That is, we must use the third dimension of space, and then their congruence depends on the property of the plane that its two sides are indistinguishable, so that any piece will fit its trace after being turned over. This procedure, folding along a line, can have no place in a strictly two-dimensional geometry; and, were we in tri-dimensional spherics, we could say, that from the outside a sphere is convex, while from the inside it is concave, and that a piece of it, after being turned over, will not fit its trace, but only touch it at one point. So figures with axial symmetry, according to 684, on a sphere cannot be made to coincide; and the word symmetrical is henceforth devoted entirely to such.

Theorem V.

688. *Two spherical triangles having two sides and the included angle of one equal respectively to two sides and the included angle of the other, are either congruent or symmetrical.*

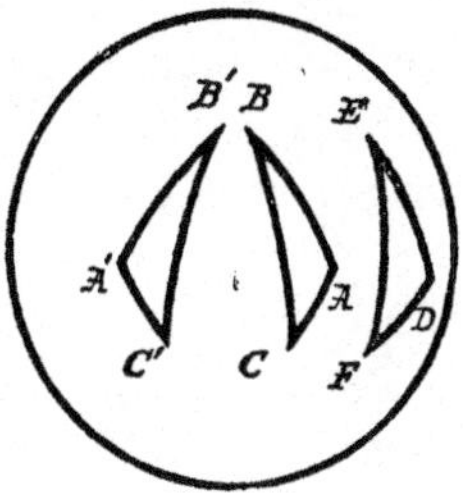

(1) If the parts given equal are arranged in the same order, as in *DEF* and *ABC*, then the triangle *DEF* can be moved in the sphere until it coincides with *ABC*.

(2) If the parts given equal are arranged in reverse order, as in *DEF* and *A'B'C'*, by making a triangle symmetrical to *A'B'C'*, as *ABC*, we get the equal parts arranged in the same order as in *DEF*, which proves *DEF* congruent to any triangle symmetrical to *A'B'C'*.

Theorem VI.

689. *If two spherical triangles have two sides of the one equal respectively to two sides of the other, but the included angle of the first greater than the included angle of the second, then the third side of the first will be greater than the third side of the second.*

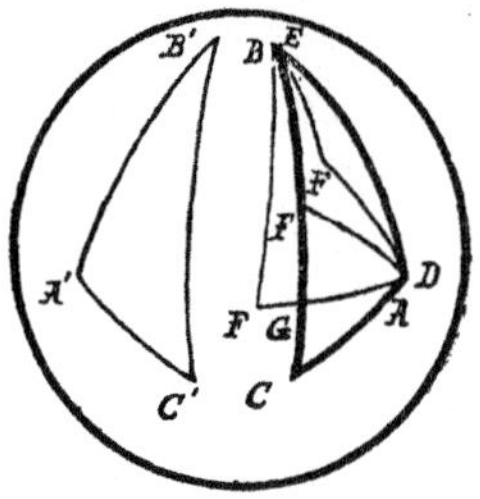

Hypothesis. $AB = DE$, and $AC = EF$, but $\measuredangle BAC > \measuredangle EDF$.

Conclusion. $BC > EF$.

Proof. When AB coincides with DE, if the points C and F are on the same side of the line AB, then, since $\measuredangle BAC > \measuredangle EDF$, the side DF will lie between BA and AC, and DF must stop either within the triangle ABC, on BC, or after cutting BC.

When F is within the triangle, by 685, $BC + CA > EF + FD$; but $AC = DF$,

$$\therefore \quad BC > EF.$$

When F lies on BC, then EF is but a part of BC. In case DF cuts BC, call their intersection point G. Then $BG + GF > EF$, and $GC + GD > AC$,

$$\therefore \quad BC + FD > AC + EF;$$

but $FD = AC$,

$$\therefore \quad BC > EF.$$

If, when one pair of equal sides coincide, the other pair lie on opposite sides of the line of coincidence, the above proof will show the third side of a triangle symmetrical to the first to be greater than the third side of the second triangle, and therefore the third side of the first greater than the third side of the second.

690. From 688 and 689, by 33, Rule of Inversion, if two spherical triangles have two sides of the one equal respectively to two sides of the other, the included angle of the first is greater than, equal to, or less than, the included angle of the second, according as the third side of the first is greater than, equal to, or less than, the third side of the second.

691. COROLLARY. Therefore, by 688, two spherical triangles having three sides of the one equal respectively to three sides of the other are either congruent or symmetrical.

PROBLEM I.

692. *To bisect a given spherical angle.*

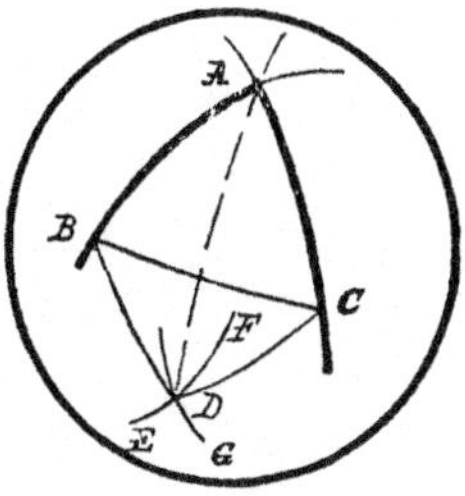

Let BAC be the given $\measuredangle$.

In its arms take equal sects AB and AC each less than a quadrant.

Join BC. With B as pole, and any sect greater than half BC, but not greater than a quadrant, as a spherical radius, describe the circle EDF. With an equal spherical radius from C as pole describe an arc intersecting the circle EDF at D within the lune BAC. Join DB, DC, DA. DA shall bisect the angle BAC.

For the $\widehat{\triangle}$s ABD, ACD, having AD common and $AB = AC$, and $BD = DC$ (spherical radii of equal circles), are, by 691, symmetrical,

$$\therefore \measuredangle BAD = \measuredangle CAD.$$

693. Corollary I. To bisect the re-entrant angle BAC, produce the bisector of the angle BAC.

694. Corollary II. To erect a perpendicular to a given line from a given point in the line, bisect the straight angle at that point.

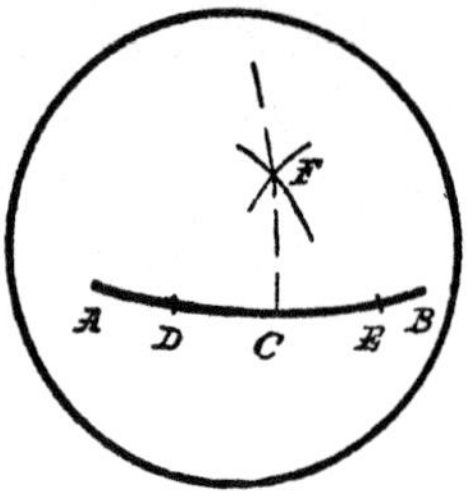

PROBLEM II.

695. *To bisect a given sect.*

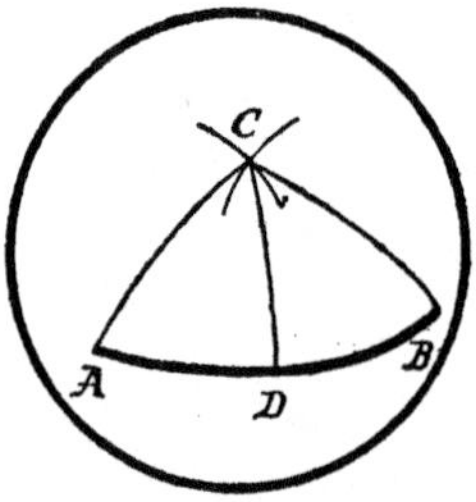

With A and B, the extremities of the given sect, as poles, and equal spherical radii greater than half AB, but less than a quadrant, describe arcs intersecting at C.

Join AC and BC; and, by 692, bisect the angle ACB, and produce the bisector to meet AB at D.

D is the mid point of the sect AB.

For, by 688, $\widehat{\triangle}\, ACD$ is symmetrical to $\widehat{\triangle}\, BCD$.

696. Corollary. The angles at the base of an isosceles spherical triangle are equal.

PROBLEM III.

697. *To draw a perpendicular to a given line from a given point in the sphere not in the line.*

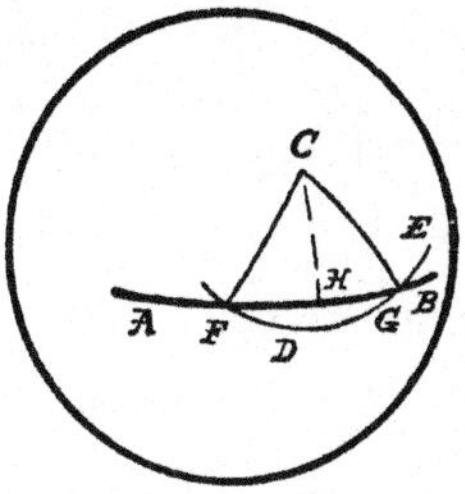

GIVEN, *the line AB, and point C.*

Take a point *D* on the other side of the line *AB*, and with *C* as pole, and *CD* as spherical radius, describe an arc cutting *AB* in *F* and *G*. Bisect the sect *FG* at *H*, and join *CH*. *CH* shall be ⊥ *AB*.

For, by 691, $\hat{\triangle}$s *GCH* and *FCH* are symmetrical.

THEOREM VII.

698. *If two lines be drawn in a sphere, at right angles to a given line, they will intersect at a point from which all sects drawn to the given line are equal.*

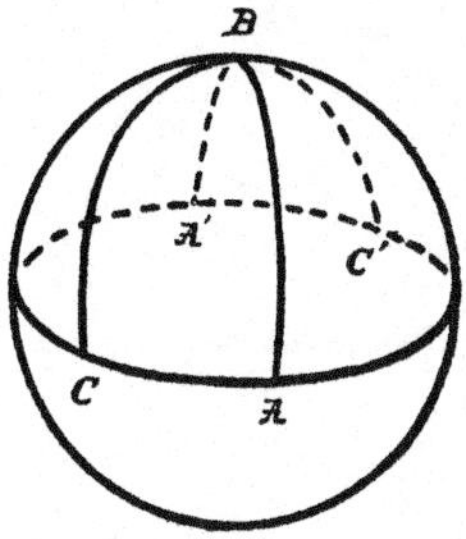

Let *AB* and *CB*, drawn at right angles to *AC*, intersect at *B*, and meet *AC* again at *A'* and *C'* respectively.

Then $\measuredangle BA'C' = \measuredangle BAC'$, and $\measuredangle BC'A' = \measuredangle BCA'$;

(678. The angles contained by the sides of a lune, at their two points of intersection, are equal.)

moreover,

$$AC = A'C',$$

for they have the common supplement AC'. Hence, keeping A and C on the line AC, slide ABC until AC comes into coincidence with $A'C'$. Then, the angles at A, C, A', C', being all right, AB will lie along $A'B$, and CB along $C'B$, and hence the figures ABC and $A'BC'$ coincide.

$\therefore$ each of the half-lines ABA' and CBC' is bisected at B.

In like manner, any other line drawn at right angles to AC passes through B, the mid point of ABA'.

Hence every sect from AC to B is a quadrant.

699. Corollary I. A line is a circle whose spherical radius is a quadrant.

700. Corollary II. A point which is a quadrant from two points in a line, and not in the line, is its pole.

701. Corollary III. Any sect from the pole of a line to the line is perpendicular to it.

702. Corollary IV. Equal angles at the poles of lines intercept equal sects on those lines.

703. Corollary V. If K be the pole, and FG a sect of

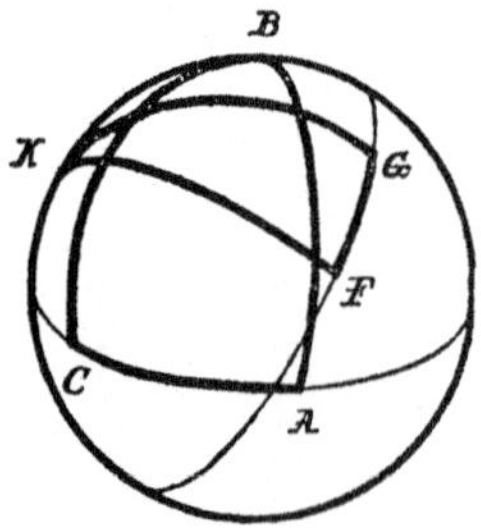

any other line, the angles and semilunes ABC and FKG are to one another as AC to FG.

704. We see, from 698, that a spherical triangle may have two or even three right angles.

If a spherical triangle ABC has two right angles, B and C, it is called a *bi-rectangular* triangle. By 698, the vertex A is the pole of BC, and therefore AB and AC are quadrants.

705. The *Polar* of a given spherical triangle is a spherical triangle, the poles of whose sides are respectively the vertices of the given triangle, and its vertices each on the same side of a side of the given triangle as a given vertex.

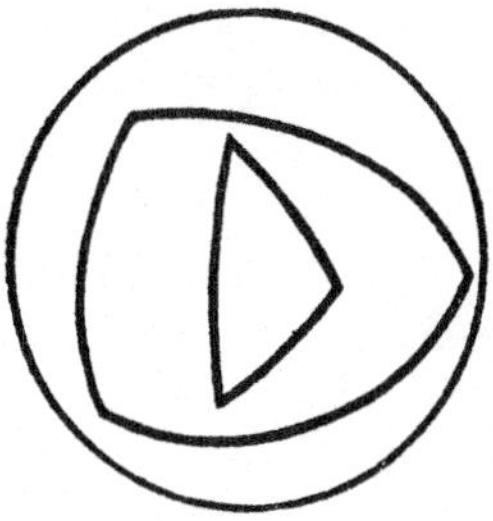

THEOREM VIII.

706. *If, of two spherical triangles, the first is the polar of the second, then the second is the polar of the first.*

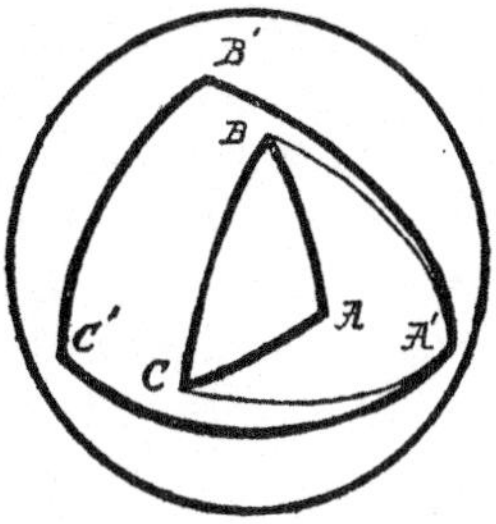

HYPOTHESIS. **Let $A'B'C'$ be the polar of ABC.**
CONCLUSION. Then ABC is the polar of $A'B'C'$.

PROOF. Join $A'B$ and $A'C$.

Since B is the pole of $A'C'$, therefore BA' is a quadrant; and since C is the pole of $A'B'$, therefore CA' is a quadrant;

$\therefore$ by 700, A' is the pole of BC.

In like manner, B' is the pole of AC, and C' of AB.

Moreover, A and A' are on the same side of $B'C'$, B and B' on the same side of $A'C'$, and C and C' on the same side of $A'B'$.

$\therefore$ ABC is the polar of $A'B'C'$.

THEOREM IX.

707. *In a pair of polar triangles, any angle of either intercepts, on the side of the other which lies opposite to it, a sect which is the supplement of that side.*

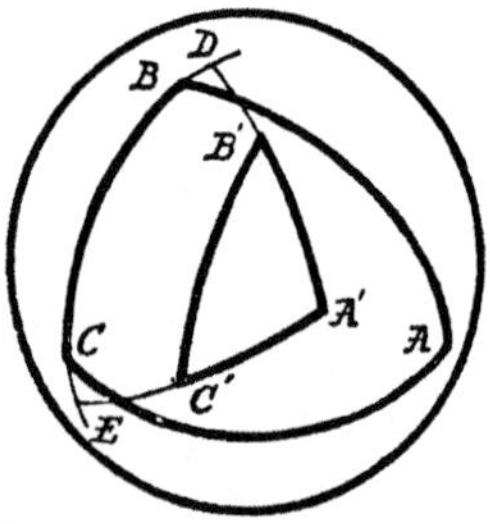

Let ABC and $A'B'C'$ be two polar triangles.

Produce $A'B'$ and $B'C'$ to meet BC at D and E respectively. Since B is the pole of $B'C'$, therefore BE is a quadrant; and since C is the pole of $A'B'$, therefore CD is a quadrant; therefore $BE + CD$ = half-line, but $BE + CD = BC + DE$. Therefore DE, the sect of BC which A' intercepts, is the supplement of BC.

EXERCISES. 109. Any lune is to a tri-rectangular triangle as its angle is to half a right angle.

Theorem X.

708. *If two angles of a spherical triangle be equal, the sides which subtend them are equal.*

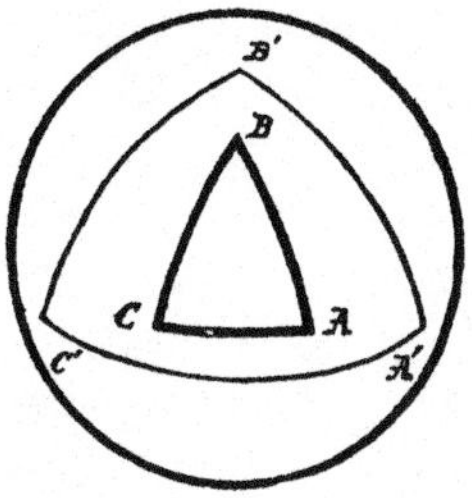

Hypothesis. *In* $\widehat{\triangle}\, ABC$ *let* $\measuredangle A = \measuredangle C$.

Conclusion. $BC = AB$.

Proof. For draw $A'B'C'$, the polar of ABC.

Now, on $B'C'$ and $A'B'$ the equal $\measuredangle$s A and C intercept equal sects.

Therefore $B'C'$ and $A'B'$, being, by 707, the supplements of these equal sects, are equal, $\therefore \measuredangle A' = \measuredangle C'$,

(696. The angles at the base of an isosceles spherical triangle are equal.)

$\therefore$ the supplements of BC and AB are equal,

$\therefore BC = AB$.

Theorem XI.

709. *If one angle of a spherical triangle be greater than a second, the side opposite the first must be greater than the side opposite the second.*

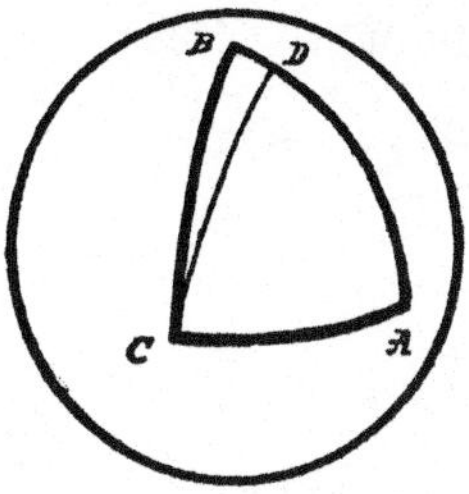

In $\widehat{\triangle}\, ABC$, $\measuredangle C > \measuredangle A$. Make $\measuredangle ACD = \measuredangle A$;

$\therefore$ by 708, $AD = CD$.

But, by 661, $CD + DB > BC$,

$$\therefore \quad AD + DB > BC.$$

710. From 708 and 709, by 33, Rule of Inversion,

If one side of a spherical triangle be greater than a second, the angle opposite the first must be greater than the angle opposite the second.

Theorem XII.

711. *Two spherical triangles having two angles and the included side of the one equal respectively to two angles and the included side of the other, are either congruent or symmetrical.*

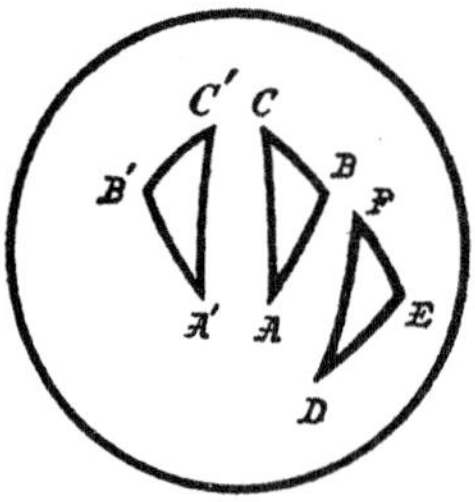

For the first triangle can be moved in the sphere into coincidence with the second, or with a triangle made symmetrical to the second.

Theorem XIII.

712. *Two spherical triangles having three angles of the one equal respectively to three angles of the other, are either congruent or symmetrical.*

Since the given triangles are respectively equiangular, their polars are respectively equilateral.

(702. Equal angles at the poles of lines intercept equal sects on those lines; and, by 707, these equal sects are the supplements of corresponding sides.)

Hence these polars, being, by 691, congruent or symmetrical, are respectively equiangular, and therefore the original spherical triangles are respectively equilateral.

THEOREM XIV.

713. *An exterior angle of a spherical triangle is greater than, equal to, or less than, either of the interior opposite angles, according as the medial from the other interior opposite angle is less than, equal to, or greater than, a quadrant.*

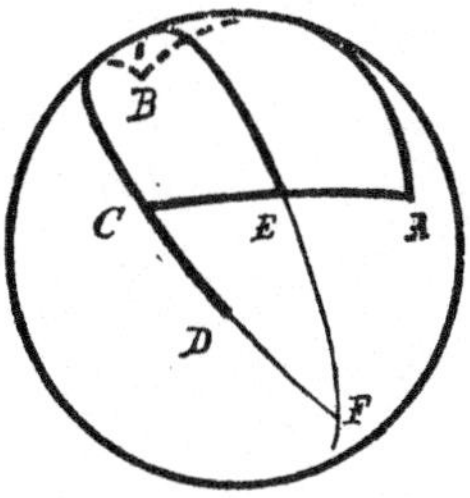

Let ACD be an exterior angle of the $\widehat{\triangle}\, ABC$. By 695, bisect AC at E. Join BE, and produce to F, making $EF = BE$. Join FC.

$$\widehat{\triangle}\, ABE \cong \widehat{\triangle}\, CFE,$$

(688. Spherical triangles having two sides and the included angle equal are congruent or symmetrical.)

$$\therefore \measuredangle BAE = \measuredangle FCE.$$

If, now, the medial BE be a quadrant, BEF is a half-line, and, by 676, F lies on BD;

$$\therefore \measuredangle DCE \text{ coincides with } \measuredangle FCE,$$

$$\therefore \measuredangle DCE = \measuredangle BAE.$$

If the medial BE be less than a quadrant, BEF is less than a half-line, and F lies between AC and CD;

$$\therefore \measuredangle DCA > \measuredangle FCE,$$

$$\therefore \measuredangle DCA > \measuredangle BAC.$$

And if BE be greater than a quadrant, BEF is greater than a half-line, and F lies between CD and AC produced;

$$\therefore \measuredangle DCA < \measuredangle FCE,$$

$$\therefore \measuredangle DCA < \measuredangle BAC.$$

Thus, according as *BE* is greater than, equal to, or less than, a quadrant, the exterior ∡ *ACD* is less than, equal to, or greater than, the interior opposite ∡ *BAC*.

714. From 713, by 33, Rule of Inversion,

According as the exterior angle *ACD* is greater than, equal to, or less than, the interior opposite angle *BAC*, the medial *BE* is less than, equal to, or greater than, a quadrant.

THEOREM XV.

715. *Two spherical triangles having the angles at the base of the one equal to the angles at the base of the other, and the sides opposite one pair of equal angles equal, are either congruent or symmetrical, provided that in neither triangle is the third angular point a quadrant from any point in that half of its base not adjacent to one of the sides equal by hypothesis.*

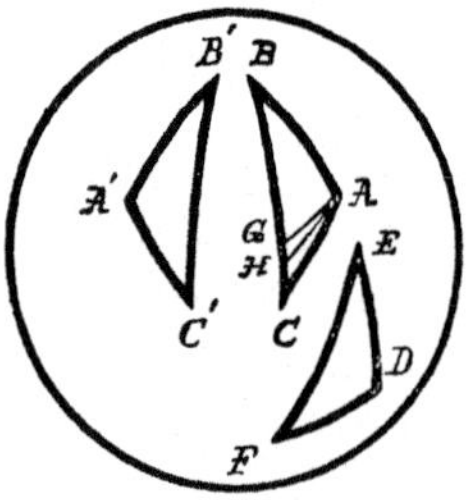

FIRST, if the parts given equal lie clockwise in the two spherical triangles, as in *ABC* and *DEF*, where we suppose ∡ *B* = ∡ *E*, ∡ *C* = ∡ *F*, and *AB* = *DE*, make *DE* coincide with *AB*; then *EF* will lie along *BC*, and *DF* must coincide with *AC*. For if it could take any other position, as *AG*, it would make a $\widehat{\triangle}$ *AGC* with exterior ∡ *AGB* = interior opposite ∡ *C*, and therefore, by 714, with medial *AH* a quadrant, which is contrary to our hypothesis.

SECOND, if the equal parts lie in one spherical triangle clockwise, in the other counter-clockwise, as in $\widehat{\triangle}$ *A'B'C'* and $\widehat{\triangle}$ *DEF*, then $\widehat{\triangle}$ *DEF* ≅ $\widehat{\triangle}$ *ABC*, which is symmetrical to $\widehat{\triangle}$ *A'B'C'*.

THEOREM XVI.

716. *If a sect drawn in the sphere from a point perpendicular to a line be less than a quadrant, it is the smallest which can be drawn in the sphere from the point to the line; of others, that which is nearer the perpendicular is less than that which is more remote; also to every sect drawn on one side of the perpendicular there can be drawn one, and only one, sect equal on the other side.*

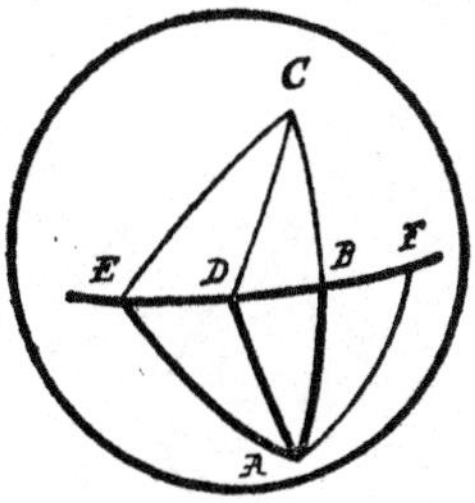

Let A be the point, BD the line, $AB \perp BD$, and AD nearer than AE to AB.

Produce AB to C, making $BC = AB$. Join CD, CE.

$$\text{Then } AD = CD.$$

(688. Spherical triangles having two sides and the included angle equal are congruent or symmetrical.)

But, by 661,

$$AD + CD \text{ or } 2AD > AC \text{ or } 2AB,$$

$$\therefore\ AD > AB.$$

Also, by 685,

$$AE + CE \text{ or } 2AE > AD + CD \text{ or } 2AD.$$

Again, make $BF = BD$. Join AF.

As before, $AF = AD$, and we have already shown that no two sects on the same side of the perpendicular can be equal.

717. Corollary. The greatest sect that can be drawn from A to BD is the supplement of AB.

Theorem XVII.

718. *If two spherical triangles have two sides of the one equal to two sides of the other, and the angles opposite one pair of equal sides equal, the angles opposite the other pair are either equal or supplemental.*

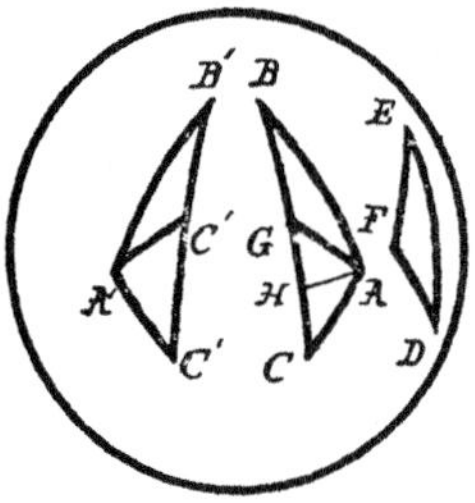

First, given the equal parts $AB = DE$, $AC = DF$, and $\measuredangle B = \measuredangle E$ arranged clockwise in the two spherical triangles.

If $\measuredangle A = \measuredangle D$, the spherical triangles are congruent.

If $\measuredangle A$ is not $= \measuredangle D$, one must be the greater.

Suppose $\measuredangle A > \measuredangle D$. Make DE coincide with AB. Then, since $\measuredangle B = \measuredangle E$, side EF will lie along BC; since $\measuredangle A > \measuredangle D$, side DF will lie between AB and AC, as at AG.

Now, the two angles at G are supplemental; but one is $\measuredangle F$ and the other $= \measuredangle C$, because $\widehat{\triangle}\ CFG$ is isosceles.

Second, if the parts given equal lie clockwise in one spherical triangle and counter-clockwise in the other, as in $\widehat{\triangle}\ A'B'C'$ and $\widehat{\triangle}\ DEF$, then, taking the $\widehat{\triangle}\ ABC$ symmetrical to $A'B'C'$, the above proof shows $\measuredangle F$ equal or supplemental to $\measuredangle C$, that is, to $\measuredangle C'$.

Theorem XVIII.

719. *Symmetrical isosceles spherical triangles are congruent.*

For, since four sides and four angles are equal, the distinction between clockwise and counter-clockwise is obliterated.

Theorem XIX.

720. *The locus of a point from which the two sects drawn to two given points are equal is the line bisecting at right angles the sect joining the two given points.*

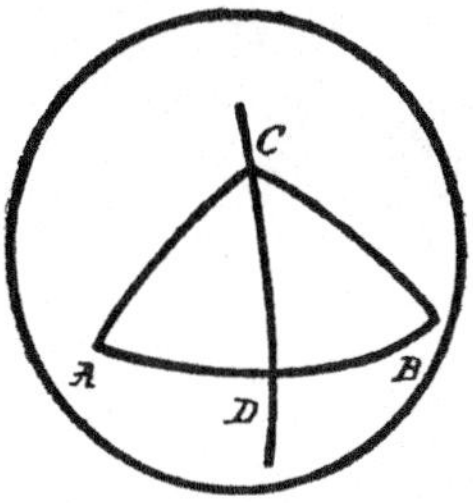

For every point in this perpendicular bisector, and no point out of it, possesses the property.

PROBLEM IV.

721. *To pass a circle through auy three points, or to find the circumcenter of any spherical triangle.*

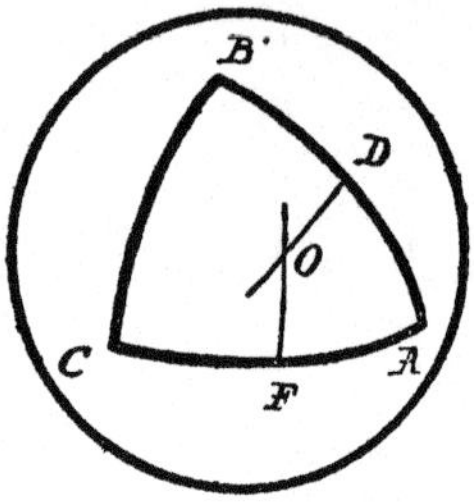

Find the intersection point of perpendiculars erected at the mid-points of two sides.

THEOREM XX.

722. *Any angle made with a side of a spherical triangle by joining its extremity to the circumcenter, equals half the angle-sum less the opposite angle of the triangle.*

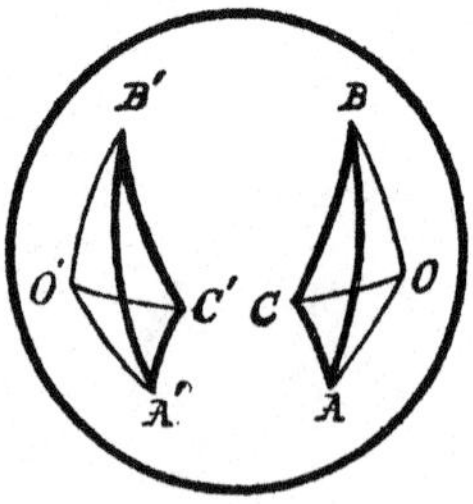

For $\measuredangle A + \measuredangle B + \measuredangle C = 2 \measuredangle OCA + 2 \measuredangle OCB \pm 2 \measuredangle OAB$,

$$\therefore \quad \measuredangle OCA = \frac{\measuredangle A + \measuredangle B + \measuredangle C}{2} - (\measuredangle OCB \pm \measuredangle OAB)$$

$$= \frac{\measuredangle A + \measuredangle B + \measuredangle C}{2} - \measuredangle B.$$

723. COROLLARY. *Symmetrical spherical triangles are equivalent.*

For the three pairs of isosceles triangles formed by joining the vertices to the circumcenters having respectively a side and two adjacent angles equal, are congruent.

THEOREM XXI.

724. *When three lines mutually intersect, the two triangles on opposite sides of any vertex are together equivalent to the lune with that vertical angle.*

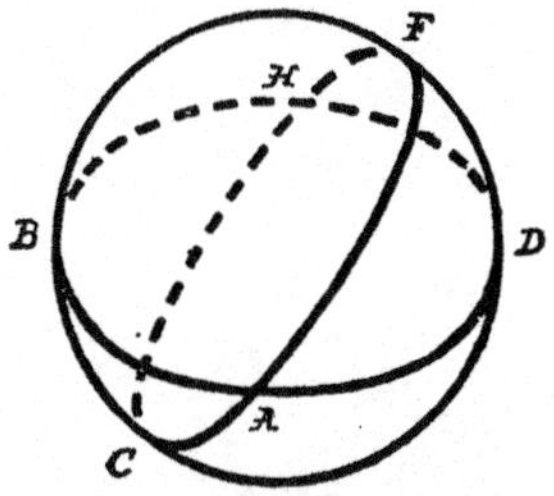

$$\hat{\triangle} ABC + \hat{\triangle} ADF = \text{lune } ABHCA.$$

For $DF = BC$, having the common supplement CD; and $FA = CH$, having the common supplement HF; and $AD = BH$, having the common supplement HD;

$$\therefore \quad \hat{\triangle} ADF = \hat{\triangle} BCH,$$

$$\therefore \quad \hat{\triangle} ABC + \hat{\triangle} ADF = \hat{\triangle} ABC + \hat{\triangle} BCH = \text{lune } ABHCA.$$

725. The *Spherical Excess* of a spherical triangle is the excess of the sum of its angles over a straight angle. In general, the spherical excess of a spherical polygon is the excess of the sum of its angles over as many straight angles as it has sides, less two.

THEOREM XXII.

726. *A spherical triangle is equivalent to a lune whose angle is half the triangle's spherical excess.*

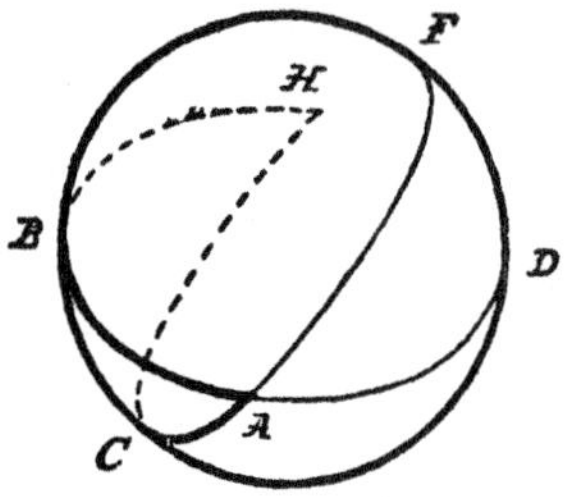

PROOF. Produce the sides of the $\widehat{\triangle}\ ABC$ until they meet again, two and two, at D, F, and H. The $\widehat{\triangle}\ ABC$ now forms part of three lunes, whose angles are A, B, and C respectively.

But, by 724, lune with $\measuredangle\ A = \widehat{\triangle}\ ABC + \widehat{\triangle}\ ADF$.

Therefore the lunes whose angles are A, B, and C, are together equal to a hemisphere plus twice $\widehat{\triangle}\ ABC$.

But a hemisphere is a lune whose angle is a straight angle,

$$\therefore\ 2\,\widehat{\triangle}\ ABC = \text{lune whose } \measuredangle \text{ is } (A + B + C - \text{st. } \measuredangle)$$
$$= \text{lune whose } \measuredangle \text{ is } e.$$

727. COROLLARY I. The sum of the angles of a spherical triangle is greater than a straight angle and less than 3 straight angles.

728. COROLLARY II. Every angle of a spherical triangle is greater than $\frac{1}{2}e$.

729. COROLLARY III. A spherical polygon is equivalent to a lune whose angle is half the polygon's spherical excess.

730. COROLLARY IV. Spherical polygons are to each other as their spherical excesses, since, by 703, lunes are as their angles.

731. COROLLARY V. To construct a lune equivalent to any spherical polygon, add its angles, subtract ($n - 2$) straight angles, halve the remainder, and produce the arms of a half until they meet again.

THEOREM XXIII.

732. *If the line be completed of which the base of a given spherical triangle is a sect, and the other two sides of the triangle be produced to meet this line, and a circle be passed through these two points of intersection and the vertex of the triangle, the locus of the vertices of all triangles equivalent to the given triangle, and on the same base with it, and on the same side of that base, is the arc of this circle terminated by the intersection points and containing the vertex.*

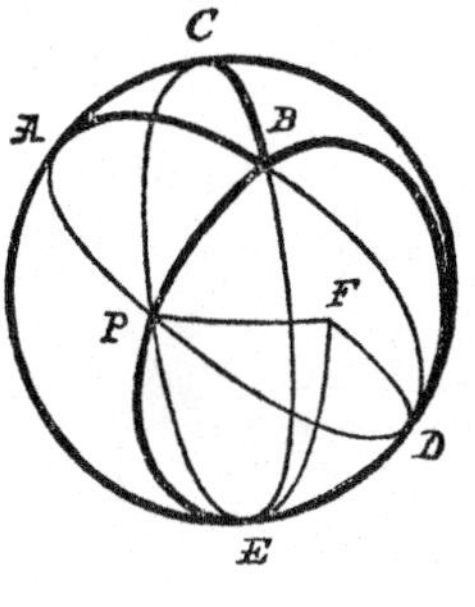

Let ABC be the given spherical triangle, AC its base. Produce AB and CB to meet AC produced in D and E respectively. By 719, pass a circle through B,D,E. Let P be any point in the arc EBD. Join AP and CP. AP produced passes through the opposite point D, and CP through E, forming with DE the $\widehat{\triangle}$ PDE. Join F, its circumcenter, with P, D, and E. Since, by 678, the two angles of a lune are equal,

$$\therefore \angle PAC = \text{st.} \angle - \angle PDE,$$
$$\angle PCA = \text{st.} \angle - \angle PED,$$
$$\angle APC = \angle DPE;$$

$\therefore\ \measuredangle(PAC + PCA + APC) = 2\text{ st.}\measuredangle\text{s} - \measuredangle(PDE + PED - DPE)$
$= 2\text{ st.}\measuredangle\text{s} - 2\measuredangle FDE =$ a constant.

(722. Any angle made with a side of a spherical triangle by joining its extremity to the circumcenter, equals half the angle-sum less the opposite angle of the spherical triangle.)

733. In a sphere a line is said to touch a circle when it meets the circle, but will not cut it.

THEOREM XXIV.

734. *The line drawn at right angles to the spherical radius of a circle at its extremity touches the circle.*

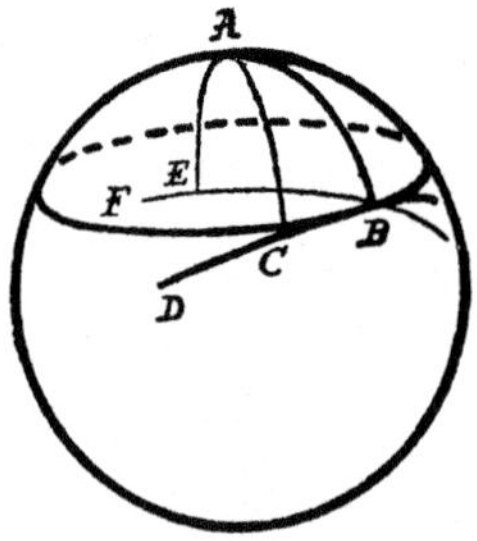

Let BD be perpendicular to the spherical radius AB. Join A with any point C in BD.

By 716, $AC > AB$, therefore C is without the circle.

And no other line through B, as BF, can be tangent.

For draw $AE \perp BF$. By 716, $AB > AE$,

$\therefore$ E is within the circle.

THEOREM XXV.

735. *In a sphere the sum of one pair of opposite angles of a quadrilateral inscribed in a circle equals the sum of the other pair.*

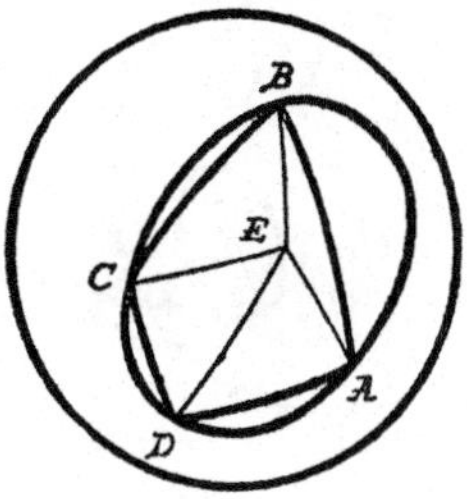

Join E, the circumcenter, with A, B, C, D, the vertices of the inscribed quadrilateral.

By 696, $\measuredangle ABC = \measuredangle BAE + \measuredangle BCE$, and $\measuredangle ADC = \measuredangle DAE + \measuredangle DCE$,

$$\therefore \quad \measuredangle ABC + \measuredangle ADC = \measuredangle BAD + \measuredangle BCD.$$

THEOREM XXVI.

736. *In equal circles, equal angles at the pole stand on equal arcs.*

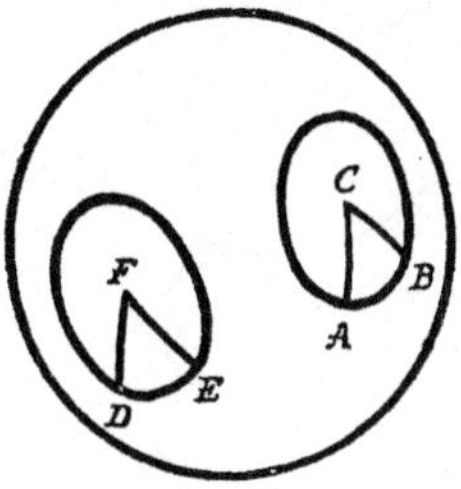

For the figures may be slidden into coincidence.

Theorem XXVII.

737. *Equal spherical chords cut equal circles into the same two arcs.*

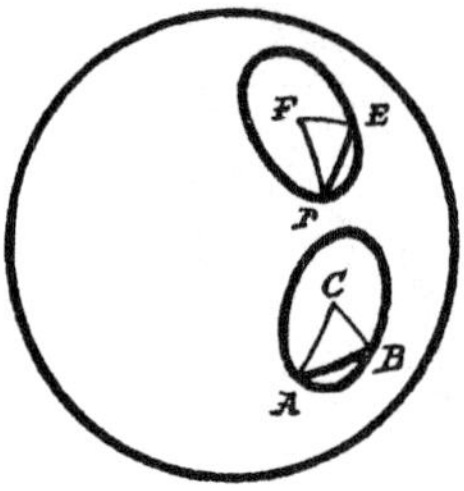

Theorem XXVIII.

738. *In equal circles, angles at the corresponding poles have the same ratio as their arcs.*

Theorem XXIX.

739. *Four pairs of equal circles can be drawn to touch three non-concurrent lines in a sphere.*

BOOK X.

POLYHEDRONS.

740. A *Polyhedron* is a solid bounded by planes.

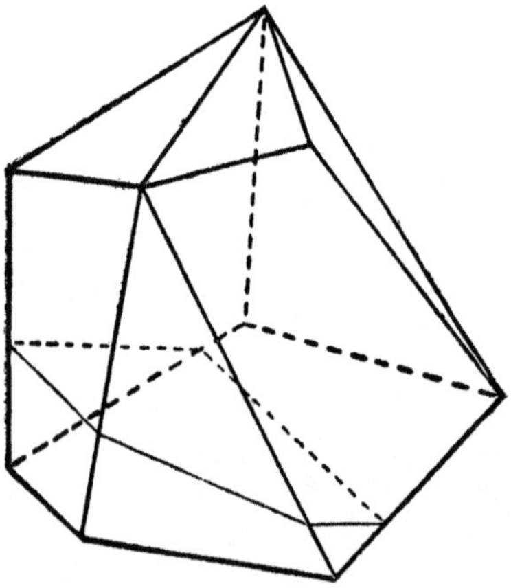

741. The bounding planes, by their intersections, determine the *Faces* of the polyhedron, which are polygons.

742. The *Edges* of a polyhedron are the sects in which its faces meet.

743. The *Summits* of a polyhedron are the points in which its edges meet.

744. A *Plane Section* of a polyhedron is the polygon in which a plane passing through it cuts its faces.

745. A *Pyramid* is a polyhedron of which all the faces, except one, meet in a point.

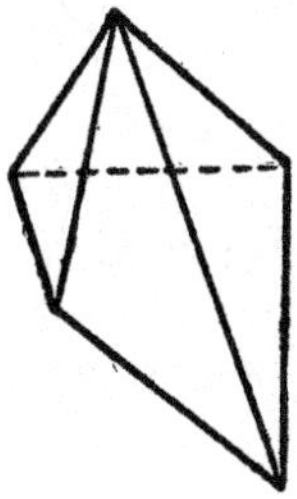

746. The point of meeting is called the *Apex*, and the face not passing through the apex is taken as the *Base.*

747. The faces and edges which meet at the apex are called *Lateral Faces and Edges.*

748. Two polygons are said to be parallel when each side of the one is parallel to a corresponding side of the other.

749. A *Prism* is a polyhedron two of whose faces are congruent parallel polygons, and the other faces are parallelograms.

750. The *Bases* of a prism are the congruent parallel polygons.

751. The *Lateral Faces* of a prism are all except its bases.

752. The *Lateral Edges* are the intersections of the lateral faces.

753. A *Right Section* of a prism is a section by a plane perpendicular to its lateral edges.

754. The *Altitude of a Prism* is any sect perpendicular to both bases.

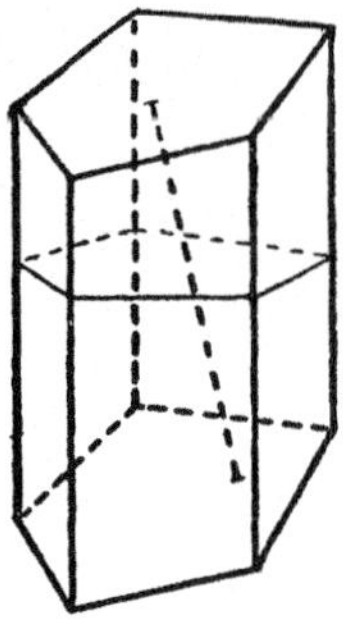

755. The *Altitude of a Pyramid* is the perpendicular from its vertex to the plane of its base.

756. A *Right Prism* is one whose lateral edges are perpendicular to its bases.

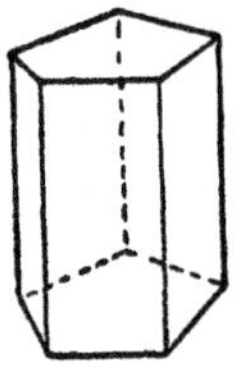

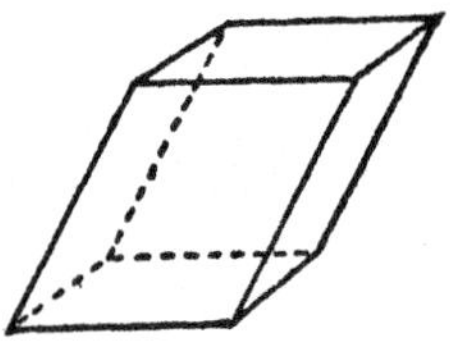

757. Prisms not right are *oblique.*

758. A *Parallelopiped* is a prism whose bases are parallelograms.

759. A *Quader* is a parallelopiped whose six faces are rectangles.

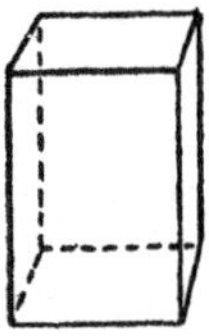

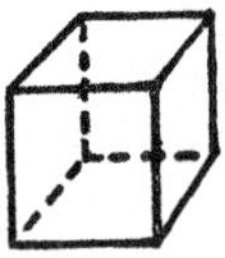

760. A *Cube* is a quader whose six faces are squares.

THEOREM I.

761. *All the summits of any polyhedron may be joined by one closed line breaking only in them, and lying wholly on the surface.*

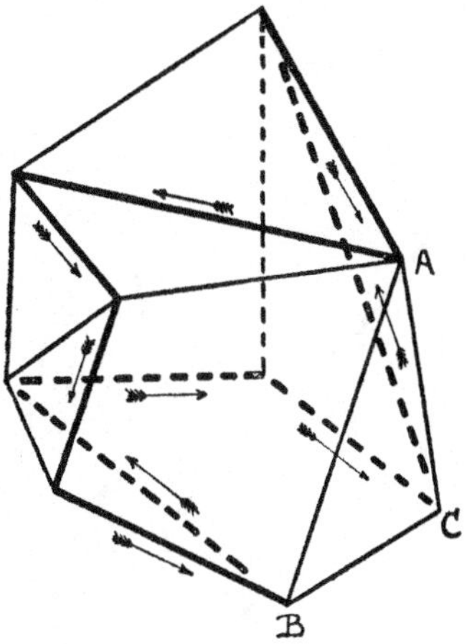

For, starting from one face, *ABC* . . . , each side belongs also to a neighboring polygon.

Therefore, to join *A* and *B*, we may omit *AB*, and use the remainder of the perimeter of the neighboring polygon *a*. In the same way, to join *B* and *C*, we may omit *BC*, and use the remainder of the perimeter of the neighboring polygon *b*, unless the polygons *a* and *b* have in common an edge from *B*. In such a case, draw from *B* in *b* the diagonal nearest the edge common to *a* and *b*; take this diagonal and the perimeter of *b* beyond it around to *C*, as continuing the broken line; and proceed in the same way from *C* around the neighboring polygon *c*.

When this procedure has taken in all summits in faces having an edge in common with *ABC* . . . , we may, by proceeding from the closed broken line so obtained, in the same way take in the summits on the next series of contiguous faces, etc.

So continue until the single closed broken line goes once, and only once, through every summit.

Theorem II.

762. *Cutting by diagonals the faces not triangles into triangles, the whole surface of any polyhedron contains four less triangles than double its number of summits.*

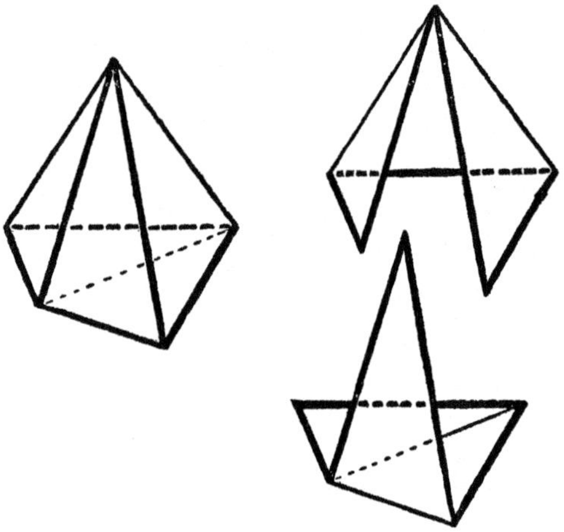

$$T = 2(S - 2).$$

For, joining all the summits by a single closed broken line, this cuts the surface into two bent polygons, each of which contains $S - 2$ triangles, where S is the number of summits.

763. Corollary. The sum of all the angles in the faces of any polyhedron is as many perigons as the polyhedron has summits, less two.

764. Remark. Theorem II. is called Descartes' Theorem, and is really the fundamental theorem on polyhedrons, though this place has long been held by Theorem III., called Euler's Theorem, which follows from it with remarkable elegance.

Theorem III.

765. *The number of faces and summits in any polyhedron, taken together, exceeds by two the number of its edges.*

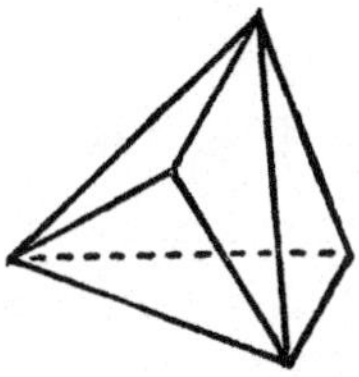

Case I. If all the faces are triangles. Then, by 762,

$$F = 2(S - 2).$$

But also

$$2E = 3F,$$

for each edge belongs to two faces, and so we get a triangle for every time 3 is contained in $2E$.

By adding, we have $2E = 2F + 2(S - 2)$; that is,

$$F + S = E + 2.$$

Case II. If not all the faces are triangles.

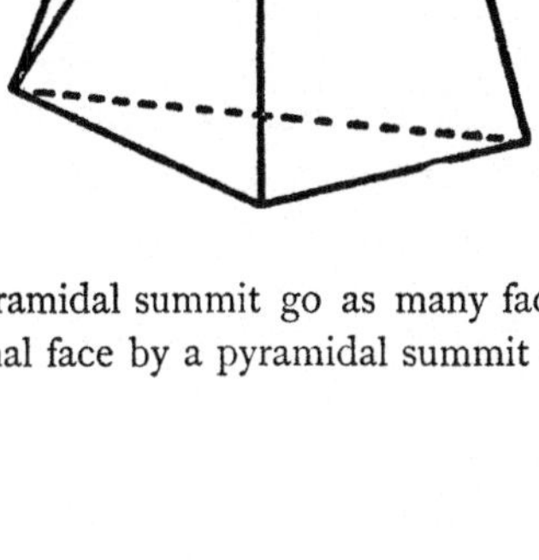

Since to any pyramidal summit go as many faces as edges, we may replace any polygonal face by a pyramidal summit without changing the equality or inequality relation of $F + S$ to $E + 2$; for such replacement only adds the same number to F as to E, and changes one face to a summit. But, after all polygonal faces have been so replaced, $F + S = E + 2$, by Case I. Therefore always the relation was equality.

THEOREM IV.

766. *Quaders having congruent bases are to each other as their altitudes.*

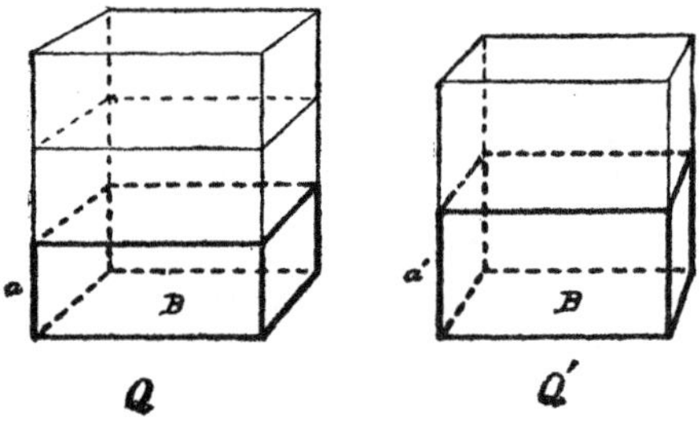

HYPOTHESIS. ***Let*** a ***and*** a' ***be the altitudes of two quaders,*** Q ***and*** Q'***, having congruent bases*** B***.***

CONCLUSION. $Q : Q' :: a : a'$.

PROOF. Of a take any multiple, ma; then the quader on base B with altitude ma is mQ.

In the same way, take equimultiples na', nQ'.

According as mQ is greater than, equal to, or less than, nQ', we have ma greater than, equal to, or less than na'; therefore, by definition,

$$Q : Q' :: a : a'.$$

EXERCISES. 110. In no polyhedron can triangles and three-faced summits both be absent; together are present at least eight. Not all the faces nor all the summits have more than five sides.

111. There is no seven-edged polyhedron.

THEOREM V.

767. *Quaders having equal altitudes are to each other as their bases.*

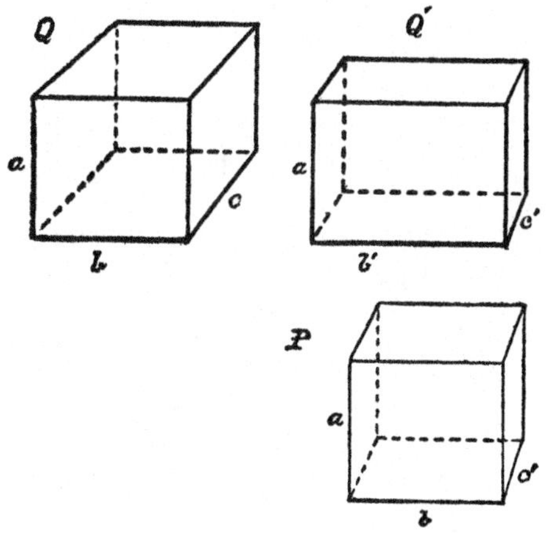

HYPOTHESIS. ***Let the rectangles*** bc ***and*** $b'c'$ ***be the bases of two quaders,*** Q ***and*** Q'***, of altitude*** a.

CONCLUSION. $Q : Q' :: bc : b'c'$.

PROOF. Make P a third quader of altitude a and base bc'.

Now, considering the rectangles ab as the bases of Q and P, by 766,

$$Q : P :: c : c';$$

considering the rectangles ac' as bases of P and Q', by 766,

$$P : Q' :: b : b'.$$

Therefore, compounding the ratios,

$$Q : Q' :: bc : b'c'.$$

Theorem VI.

768. *Two quaders are to each other in the ratio compounded of the ratios of their bases and altitudes.*

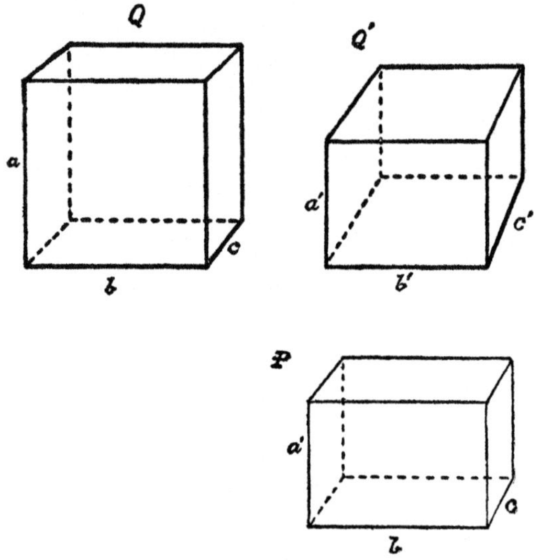

Hypothesis. ***Let Q and Q' be two quaders, of altitude a and a', and base bc and $b'c'$, respectively.***

Conclusion. $Q : Q' :: abc : a'b'c'$.

Proof. Make P a third quader of altitude a' and base bc.

Then, by 766,

$$Q : P :: a : a';$$

and, by 767,

$$P : Q' :: bc : b'c'.$$

Therefore, compounding,

$$Q : Q' :: abc : a'b'c'.$$

Theorem VII.

769. *Any parallelopiped is equivalent to a quader of equivalent base and equal altitude.*

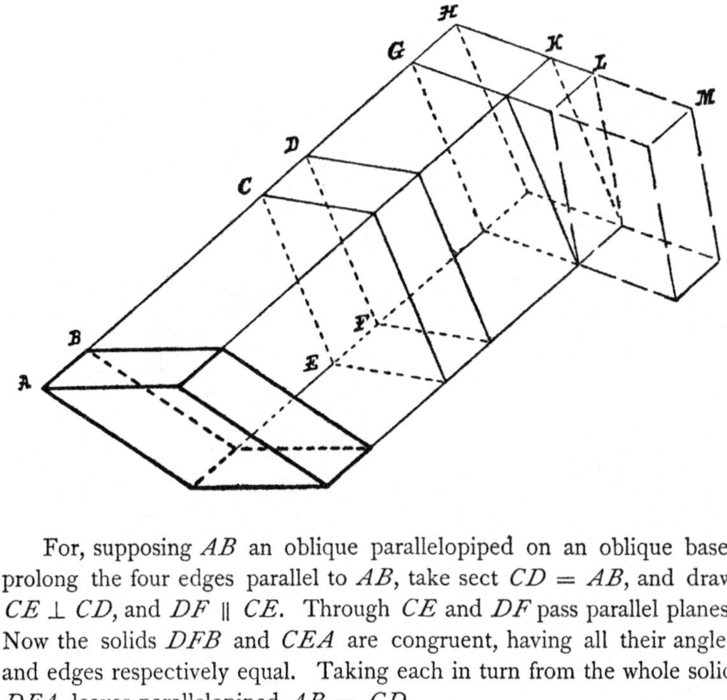

For, supposing AB an oblique parallelopiped on an oblique base, prolong the four edges parallel to AB, take sect $CD = AB$, and draw $CE \perp CD$, and $DF \parallel CE$. Through CE and DF pass parallel planes. Now the solids DFB and CEA are congruent, having all their angles and edges respectively equal. Taking each in turn from the whole solid DFA, leaves parallelopiped $AB = CD$.

In the line CD take $GH = CD$, and through G and H pass planes perpendicular to GH.

The solids DFH and CEG are congruent, therefore parallelopiped $CD = GH$.

Now prolong the four edges not parallel to GH, and take $LM = HK$, and through L and M pass planes perpendicular to LM.

As before, parallelopiped $GH = LM$; but LM is a quader of equivalent base and equal altitude to parallelopiped AB.

THEOREM VIII.

770. *A plane passed through two diagonally opposite edges of a parallelopiped divides it into two equivalent triangular prisms.*

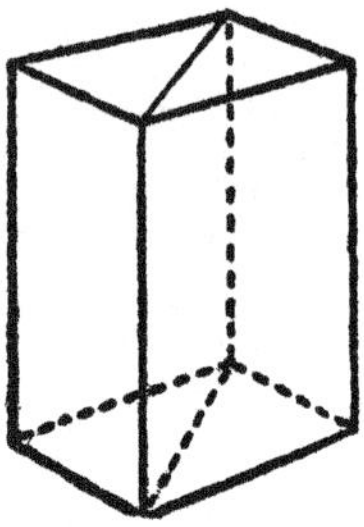

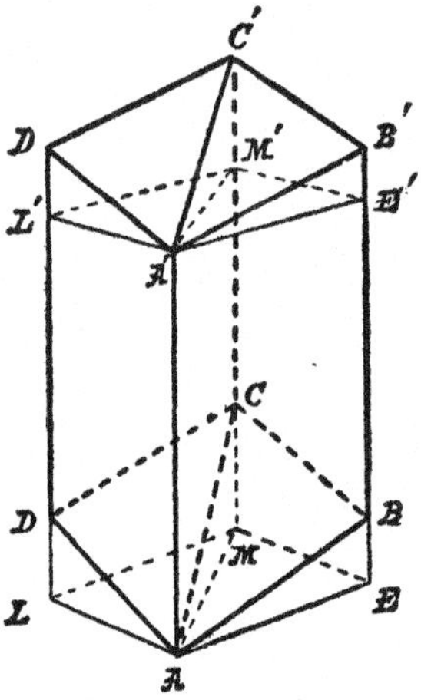

If the lateral faces are all rectangles, the two prisms are congruent; if not, the prisms are still equivalent.

For draw planes perpendicular to AA' at the points A and A'. Then the prism $ABC\ A'B'C'$ is equivalent to the right prism $AEM\ A'E'M'$, because the pyramid $AEBCM$ is congruent to the pyramid $A'E'B'C'M'$. In the same way, $ADC\ A'D'C'$ is equivalent to $ALM\ A'L'M'$. But $AEM\ A'E'M'$ and $ALM\ A'L'M'$ are congruent,

$$\therefore\ ABC\ A'B'C' \text{ and } ADC\ A'D'C' \text{ are equivalent.}$$

771. COROLLARY. Any triangular prism is half a parallelopiped of twice its base but equal altitude.

THEOREM IX.

772. *If a pyramid be cut by a plane parallel to its base, the section is to the base as the square of the perpendicular on it, from the vertex, is to the square of the altitude of the pyramid.*

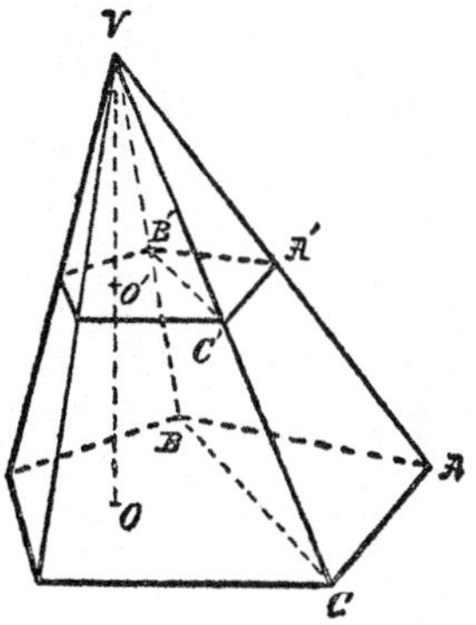

The section and base are similar, since corresponding diagonals cut them into triangles similar in pairs because having all their sides respectively proportional.

For

$$A'C' : AC :: VC' : VC :: VO' : VO,$$

and

$$B'C' : BC :: VC' : VC,$$

$$\therefore \quad AC' : AC :: B'C' : BC;$$

and in the same way,

$$B'C' : BC :: A'B' : AB,$$

$$\therefore \quad \triangle A'B'C' \sim \triangle ABC, \text{ etc.},$$

$$\therefore \quad \text{section} : \text{base} :: \overline{A'C'}^2 : \overline{AC}^2 :: \overline{VO'}^2 : \overline{VO}^2.$$

773. COROLLARY. In pyramids of equivalent bases and equal altitudes, two sections having equal perpendiculars from the vertices are equivalent.

Theorem X.

774. *Tetrahedra (triangular pyramids) having equivalent bases and equal altitudes are equivalent.*

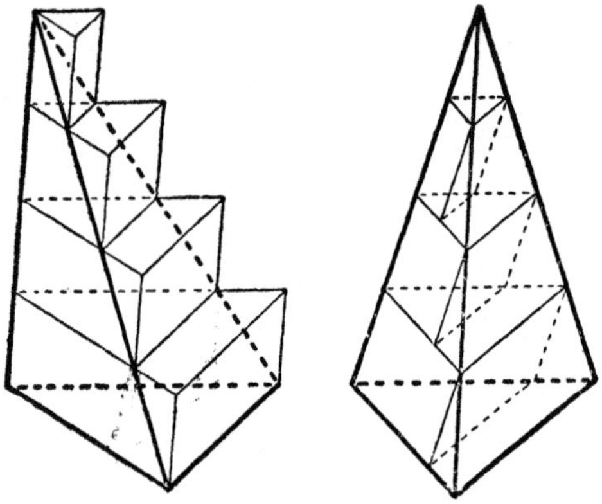

Divide the equal altitudes a into n equal parts, and through each point of division pass a plane parallel to the base.

By 773, all sections in the first tetrahedron are triangles equivalent to the corresponding sections in the second.

Beginning with the base of the first tetrahedron, construct on each section, as lower base, a prism $\frac{a}{n}$ high, with lateral edges parallel to one of the edges of the tetrahedron.

In the second, similarly construct prisms on each section, as upper base.

Since the first prism-sum is greater than the first tetrahedron, and the second prism-sum is less than the second tetrahedron, therefore the difference of the tetrahedra is less than the difference of the prism-sums.

But, by 771, each prism in the second tetrahedron is equivalent to the prism next above it on the first tetrahedron.

So the difference of the prism-sums is simply the lowest prism of the first series. As n increases, this decreases, and can be made as small as we please by taking n sufficiently great; but it is always greater than the constant difference between the tetrahedra, and so that constant difference must be nought.

THEOREM XI.

775. *A triangular pyramid is one-third of a triangular prism of the same base and altitude.*

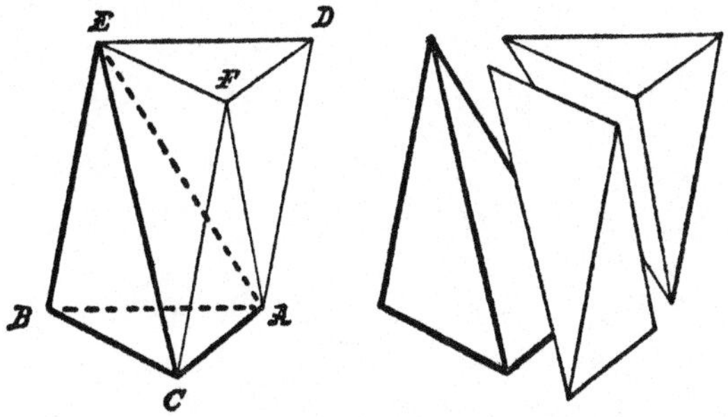

Let $E\,ABC$ be a triangular pyramid. Through one edge of the base, as AC, pass a plane parallel to the opposite lateral edge, EB, and through the vertex E pass a plane parallel to the base. The prism $ABC\,DEF$ has the same base and altitude as the given pyramid. The plane AFE cuts the part added, into two triangular pyramids, each equivalent to the given pyramid; for $E\,ABC$ and $A\,DEF$ have the same altitude as the prism, and its bottom and top respectively as bases; while $E\,AFC$ and $E\,AFD$ have the same altitude and equal bases.

BOOK XI.

MENSURATION, OR METRICAL GEOMETRY.

CHAPTER I.

THE METRIC SYSTEM. — LENGTH, AREA.

776. In practical science, every quantity is expressed by a phrase consisting of two components, — one a number, the other the name of a thing of the same kind as the quantity to be expressed, but agreed on among men as a standard or *Unit*.

777. The *Measurement* of a magnitude consists in finding this number.

778. Measurement, then, is the process of ascertaining approximately the ratio a magnitude bears to another chosen as the standard; and *the measure* of a magnitude is this ratio expressed approximately in numbers.

779. For the continuous-quantity space, the fundamental unit actually adopted is the METER, which is a bar of platinum preserved at Paris, the bar supposed to be taken at the temperature of melting ice.

780. This material meter is the ultimate standard universally chosen, because of the advantages of the metric system

of subsidiary units connected with it, which uses only decimal multiples and sub-multiples, being thus in harmony with the decimal nature of the notation of arithmetic.

781. The metric system designates multiples by prefixes derived from the Greek numerals, and sub-multiples by prefixes from the Latin numerals.

Prefix.	Meaning as used.	Derivation.	Abbreviation.
myria-	ten thousand	μύριοι	
kilo-	one thousand	χίλιοι	k-
hecto-	one hundred	ἑκατόν	h-
deka-	ten	δέκα	da-
deci-	one-tenth	decem	d-
centi-	one-hundredth	centum	c-
milli-	one-thousandth	mille	m-

The abbreviation for meter is m.; hence km. for kilometer, mm. for millimeter.

782. The adoption of the meter gives the world one standard sect as fundamental unit.

783. The *Length* of any sect is its ratio to the meter expressed approximately in numbers.

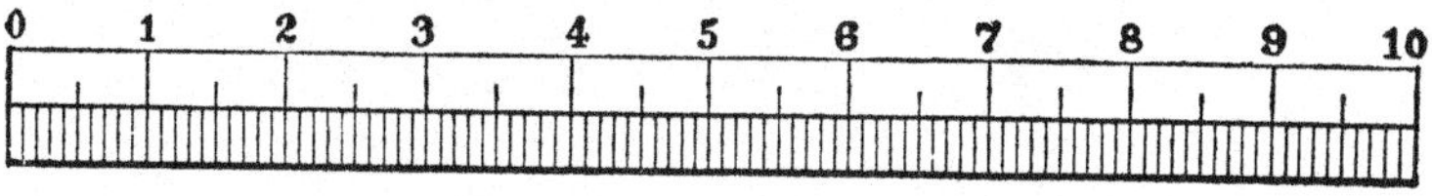

1 decimeter = 10 centimeters = 100 millimeters.

Men of science often express their measurements in terms of a subsidiary unit, the *Centimeter.*

The length of a sect referred to the centimeter as unit is one hundred times as great as referred to the meter.

784. An accessible sect may be practically measured by the direct application of a known sect, such as the edge of a ruler suitably divided.

But because of incommensurability, any description of a sect in terms of the standard sect must be usually imperfect and merely approximate.

Moreover, in few physical measurements of any kind are more than six figures of such approximations accurate; and that degree of exactness is very seldom obtainable, even by the most delicate instruments.

785. For the measurement of surfaces the standard is the square on the linear unit.

786. The *Area* of any surface is its ratio to this square.

787. If the unit for length be a meter, the unit for area, a square on the meter, is called a square meter ($m.^2$; better, $^{m.2}$).

In science the *Square Centimeter* ($^{cm.2}$) is adopted as the primary unit of surface.

788. ***To find the area of a rectangle.***

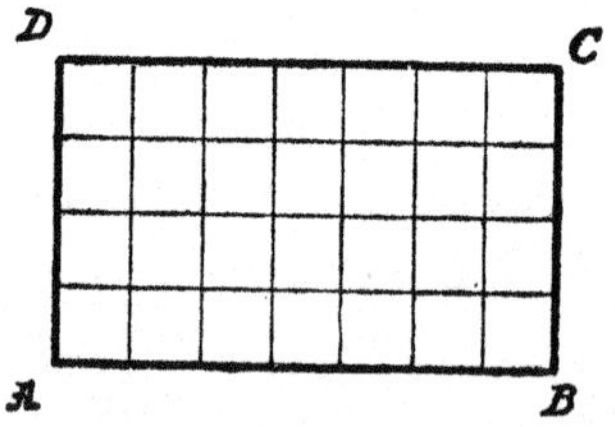

Rule. *Multiply the base by the altitude.*

Formula. $R = ab$.

Proof. — Special Case: When the base and altitude of the rectangle are commensurable. In this case, there is always a sect which will divide both base and altitude exactly. If this sect be assumed as linear unit, the lengths a and b are integral numbers. In the rectangle $ABCD$, divide AD into a, and AB into b, equal parts. Through the points of division draw lines parallel to the sides of the rectangle. These lines divide the rectangle into a number of squares, each of which equals the assumed unit of surface. In the bottom row, there are b such squares; and, since there are a rows, we have b squares repeated a times, which gives, in all, ab squares.

NOTE. The composition of ratios includes numerical multiplication as a particular case.

But ordinary multiplication is also an independent growth from addition.

In this latter point of view, the multiplier indicates the number of additions or repetitions, while the multiplicand indicates the thing added or repeated. This is not a mutual operation, and the product is always in terms of the unit of the multiplicand. The multiplicand may be any aggregate; the multiplier is an aggregate of repetitions. To repeat a thing does not change it in kind, so the result is an aggregate of the same sort exactly as the multiplicand.

But if the multiplicand itself is also an aggregate of repetitions, the two factors are the same in kind, and the multiplication is commutative.

This is the only sort of multiplication needed in mensuration; for all ratios are supposed to be expressed exactly or approximately in numbers, and in our rules it is only of these numbers that we speak. Thus, when the rule says, "Multiply the base by the altitude," it means, Multiply the number taken as the length of the base, by the number which is the measure of the altitude in terms of the same linear unit. The product is a *number*, which we prove to be the *area* of the rectangle; that is, its numerical measure in terms of the superficial unit. This is the meaning to be assigned whenever in mensuration we speak of the product of one sect by another.

GENERAL PROOF. If L represent the unit for length, and S the unit of surface, and ▱ ab the rectangle, the length of whose base is b and the length of whose altitude is a, then, by 542,

$$\frac{\square ab}{S} = \frac{aL}{L} \cdot \frac{bL}{L}.$$

But the first member of this equation is the *area* of the rectangle, which number we may represent by R; and the second member is equal to the product of the numbers a and b;

$$\therefore \quad R = ab.$$

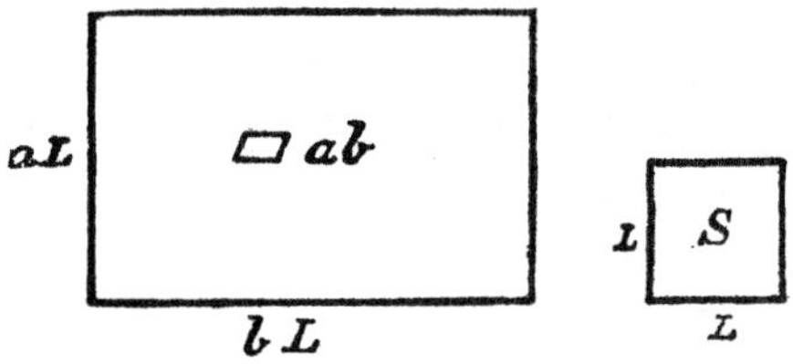

EXAMPLE 1. Find the area of a ribbon 1m. long and 1cm. wide.

Answer. $\frac{1}{100}$m.² = 100cm.².

Since a square is a rectangle having its length and breadth equal, therefore

789. *To find the area of a square.*

RULE. *Take the second power of the number denoting the length of its side.*

NOTE. This is why the product of a number into itself is called the square of that number.

790. *Given, the area of a square, to find the length of a side.*

RULE. *Extract the square root of the number denoting the area.*

791. METRIC UNITS OF SURFACE.

1 hectar (ha.)	=	1 sq. hectometer	=	10000 sq. meters.
1 dekar	=		=	1000 sq. meters.
1 ar (a.)	=	1 sq. dekameter	=	100 sq. meters.
1 deciar	=		=	10 sq. meters.
1 centiar	=	1 sq. meter	=	1 sq. meter.
1 milliar	=		=	.1 sq. meter.
		1 sq. decimeter	=	.01 sq. meter.
		1 sq. centimeter	=	.0001 sq. meter.
		1 sq. millimeter	=	.000001 sq. meter.

EXAMPLE 2. How many square centimeters in 10 millimeters square?

Answer. $(10^{\text{mm.}})^2 = 100^{\text{mm.}^2} = 1^{\text{cm.}^2}$.

792. REMARK. Distinguish carefully between *square meters* and *meters square.*

We say 10 square kilometers ($10^{\text{km.}^2}$), meaning a surface which would contain 10 others, each a square kilometer; while the expression "5 kilometers square" $(5^{\text{km.}})^2$ means a square whose sides are each 5 kilometers long, so that the figure contains $25^{\text{km.}^2}$.

EXAMPLE 3. A square is $1000^{\text{m.}^2}$. Find its side.

Answer. $\sqrt{1000^{\text{m.}}} = 31.623^{\text{m.}}$

793. Because the sum of the squares on the two sides of a right-angled triangle is the square of the hypothenuse, therefore, also,

Given, the hypothenuse and one side, to find the other side.

RULE. *Multiply their sum by their difference, and extract the square root.*

FORMULA. $c^2 - a^2 = (c + a)(c - a) = b^2$.

From this it follows, that, in an acute-angled triangle, if we are given two sides and the projection of one on the other, or two sides and an altitude, we can find the third side.

EXERCISES. 112. What must be given in order to find the medials of a triangle?

113. If on the three sides of any triangle squares are described outward, the sects joining their outer corners are twice the medials of the triangle, and perpendicular to them.

CHAPTER II.

RATIO OF ANY CIRCLE TO ITS DIAMETER.

Problem I.

794. *Given, the perimeters of a regular inscribed and a similar circumscribed polygon, to compute the perimeters of the regular inscribed and circumscribed polygons of double the number of sides.*

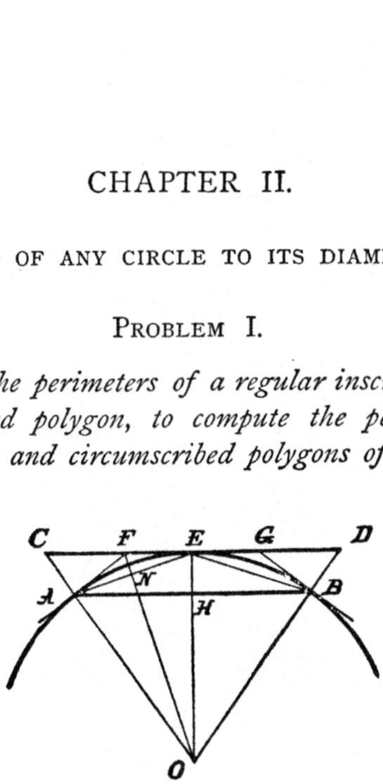

Take AB a side of the given inscribed polygon, and CD a side of the similar circumscribed polygon, tangent to the arc AB at its mid-point E.

Join AE, and at A and B draw the tangents AF and BG; then AE is a side of the regular inscribed polygon of double the number of sides, and FG is a side of the circumscribed polygon of double the number of sides.

Denote the perimeters of the given inscribed and circumscribed polygons by p and q respectively, and the required perimeters of the inscribed and circumscribed polygons of double the number of sides by p' and q' respectively.

Since OC is the radius of the circle circumscribed about the polygon whose perimeter is q,

$$\therefore \quad q : p :: OC : OE.$$

(548. The perimeters of two similar regular polygons are as the radii of their circumscribed circles.)

But, since OF bisects the $\measuredangle\, COE$,

$$\therefore \quad OC : OE :: CF : FE,$$

(523. The bisector of an angle of a triangle divides the opposite side in the ratio of the other two sides of the triangle.)

$$\therefore \quad q : p :: CF : FE,$$

whence, by composition,

$$p + q : 2p :: CF + FE : 2FE = CE : FG :: q : q',$$

since FG is a side of the polygon whose perimeter is q', and is contained as many times in q' as CE is contained in q.

$$\therefore \quad p + q : 2p :: q : q'.$$

If, now, the letters be taken to represent lengths in terms of the unit sect L, this proportion is

$$pL + qL : 2pL :: qL : q'L,$$

which gives the number

$$q' = \frac{2pq}{p + q}. \qquad (1)$$

Again, right $\triangle\, AEH \sim EFN$, since acute $\measuredangle\, EAH = FEN$,

$$\therefore \quad AH : AE :: EN : EF,$$

$$\therefore \quad p : p' :: p' : q',$$

since AH and AE are contained the same number of times in p and p' respectively, and EN and EF are contained twice that number of times in p' and q' respectively. If, now, the letters be taken to represent length in terms of the same unit sect L, this proportion is

$$pL : p'L :: p'L : q'L,$$

which gives the number

$$p' = \sqrt{pq'}. \qquad (2)$$

Therefore from the given lengths p and q we compute q' by equation (1), and then with p and q' we compute p' by equation (2).

THEOREM I.

795. *The length of a circle whose diameter is unity is* 3.141592+.

The length of the perimeter of the circumscribed square is 4. A side of the inscribed square is $\frac{1}{2}\sqrt{2}$, therefore its perimeter is $2\sqrt{2}$.

Now, putting $p = 2\sqrt{2}$ and $q = 4$ in 794, we find, for the perimeters of the circumscribed and inscribed octagons,

$$q' = \frac{2pq}{p + q} = 3.3137085,$$

$$p' = \sqrt{pq'} = 3.0614675.$$

Then, taking these as given quantities, we put $p = 3.0614675$, and $q = 3.3137085$, and find by the same formulæ, for the polygons of sixteen sides, $q' = 3.1825979$, and $p' = 3.1214452$.

Continuing the process, the results will be found as in the following table: —

Number of Sides.	Perimeter of Circumscribed Polygon.	Perimeter of Inscribed Polygon.
4	4.0000000	2.8284271
8	3.3137085	3.0614675
16	3.1825979	3.1214452
32	3.1517249	3.1365485
64	3.1441184	3.1403312
128	3.1422236	3.1412773
256	3.1417504	3.1415138
512	3.1416321	3.1415729
1024	3.1416025	3.1415877
2048	3.1415951	3.1415914
4096	3.1415933	3.1415923
8192	3.1415928	3.1415926

But since, by 654, a chord is shorter than its arc, therefore the circle is longer than any inscribed perimeter; and, assuming that two tangents from an external point cannot be together shorter than the included arc, it follows that the circle is not longer than a circumscribed perimeter; therefore the circle whose diameter is unity is longer than 3.1415926 and not longer than 3.1415928.

796. A *Variable* is a quantity which may have successively an indefinite number of different values.

797. *If a variable* which changes its value according to some law *can be made to approach some fixed value as nearly as we please, but can never become equal to it,* the constant is called the *Limit* of the variable.

EXAMPLE 4. The limit of the fraction $\frac{1}{x}$, *as* x *increases indefinitely,* is zero; for, by taking x sufficiently great, we can make $\frac{1}{x}$ less than any assigned quantity, but we can never make it zero.

PRINCIPLE OF LIMITS.

798. *If,* while tending toward their respective limits, *two variables are always in the same ratio, their limits will have that ratio.*

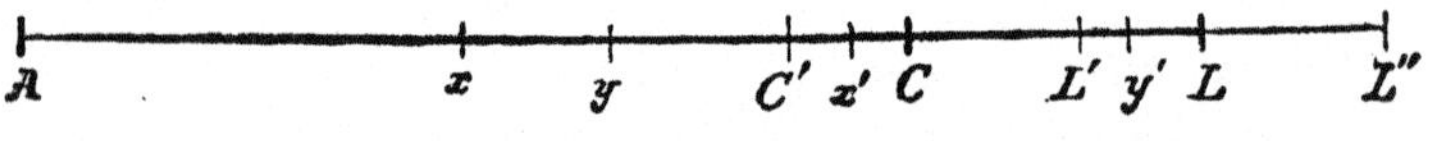

Let the sects AC and AL represent the limits of any two variable magnitudes which are always in the same ratio, and let Ax, Ay, represent two corresponding values of the variables themselves; then

$$Ax : Ay :: AC : AL.$$

For if not, then

$$Ax : Ay :: AC : \text{to some sect} > \text{or} < AL.$$

Suppose, in the first place, that

$$Ax : Ay :: AC : AL',$$

where AL' is less than AL.

By hypothesis, the variable Ay continually approaches AL, and may be made to differ from it by less than any given quantity.

Let Ax and Ay, then, continue to increase, always remaining in the same ratio, until Ay differs from AL by less than the quantity $L'L$; or, in other words, until the point y passes the point L', and reaches some point, as y', between L' and L, and x reaches the corresponding point x' on the sect AC. Then, since the ratio of the two variables is always the same,

$$Ax : Ay :: Ax' : Ay'.$$

But, by hypothesis,

$$Ax : Ay :: AC : AL',$$

$$\therefore Ax' : Ay' :: AC : AL'.$$

But

$$Ax' < AC, \qquad \therefore Ay' < AL',$$

which is absurd. Hence the supposition that $Ax : Ay :: AC : AL'$, or to any quantity less than AL, is absurd.

Suppose, then, in the second place, that

$$Ax : Ay :: AC : AL'',$$

where $AL'' > AL$. Now, there is some sect, as AC', less than AC, which is to AL as AC is to AL''.

Substituting this ratio for that of AC to AL'', we have

$$Ax : Ay :: AC' : AL,$$

which, by a process of reasoning similar to the above, may be shown to be absurd. Hence, since the fourth term of the proportion can be neither greater nor less than AL, it must equal AL; that is,

$$Ax : Ay :: AC : AL.$$

799. COROLLARY. If two variables are always equal, their limits are equal.

THEOREM II.

800. *Any two circles are to each other as their radii.*

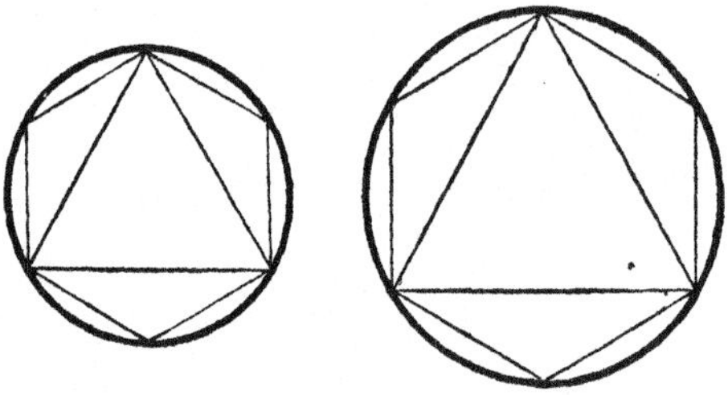

PROOF. By 548, the perimeters of any two regular polygons of the same number of sides have the same ratio as the radii of their circumscribed circles.

The inscribed regular polygons remaining similar to each other when the number of sides is doubled, their perimeters continue to have the same ratio. Assuming the circle to be the limit toward which the perimeter of the inscribed polygon increases, by 798, Principle of Limits, the circles have the same ratio as their radii.

801. Since $c : c' :: r : r' :: 2r : 2r'$,

$$\therefore \frac{c}{2r} = \frac{c'}{2r'};$$

that is, the ratio of any one circle to its diameter is the same as the ratio of any other circle to its diameter.

This constant ratio is denoted by the Greek letter π. But, by 795, this ratio for the circle with unit diameter, and therefore for every circle, is

$$\pi = 3.141592+.$$

802. For any circle.

FORMULA. $c = 2r\pi$.

CIRCULAR MEASURE OF AN ANGLE.

803. When its vertex is at the center of the circle, by 506,

$$\frac{\text{any } \measuredangle}{\text{st. } \measuredangle} = \frac{\text{its intercepted arc}}{\text{semicircle}} = \frac{\text{arc}}{r\pi},$$

$$\therefore \quad \frac{\text{any } \measuredangle}{\frac{1}{\pi}\text{st. } \measuredangle} = \frac{\text{arc}}{r}.$$

So, if we adopt as unit angle the *radian*, or that part of a perigon denoted by $\frac{\text{st. } \measuredangle}{\pi}$, that is, the angle subtended at the center

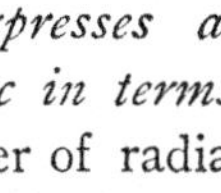

of every circle by an arc equal to its radius, and hence named a radian, then

The number which expresses any angle in radians, also expresses its intercepted arc in terms of the radius.

If u denote the number of radians in any angle, and l the length of its intercepted arc, then

$$u = \frac{l}{r}.$$

The fraction *arc divided by radius*, or *u*, is called the circular measure of an angle.

804. Arcs are said to measure the angles at the center which include them, because these arcs contain their radius as often as the including angle contains the radian. In this sense an angle at the center is measured by the arc intercepted between its sides.

CHAPTER III.

MEASUREMENT OF SURFACES.

805. By 248, any parallelogram is equivalent to the rectangle of its base and altitude; therefore,

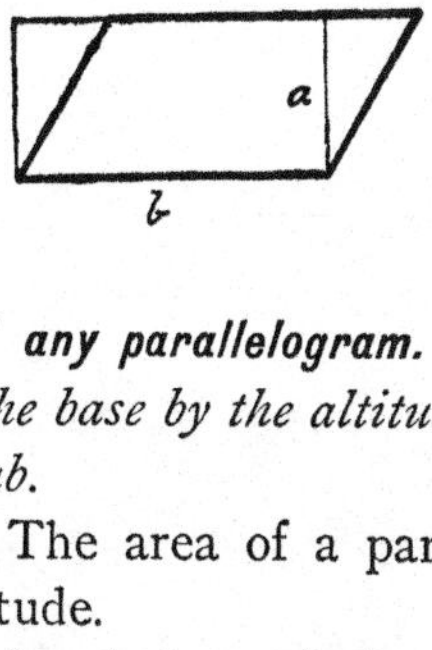

To find the area of any parallelogram.

RULE. *Multiply the base by the altitude.*

FORMULA. $\square = ab.$

806. COROLLARY. The area of a parallelogram divided by the base gives the altitude.

807. By 252, any triangle is equivalent to one-half the rectangle of its base and altitude; therefore,

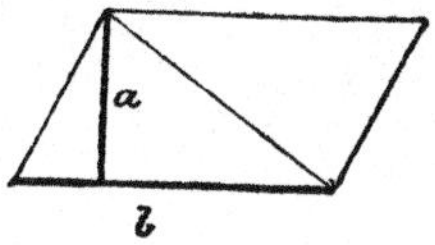

Given, one side and the perpendicular upon it from the opposite vertex, to find the area of a triangle.

RULE. *Take half the product of the base into the altitude.*

FORMULA. $\triangle = \frac{1}{2}ab.$

808. *Given, the three sides, to find the area of a triangle.*

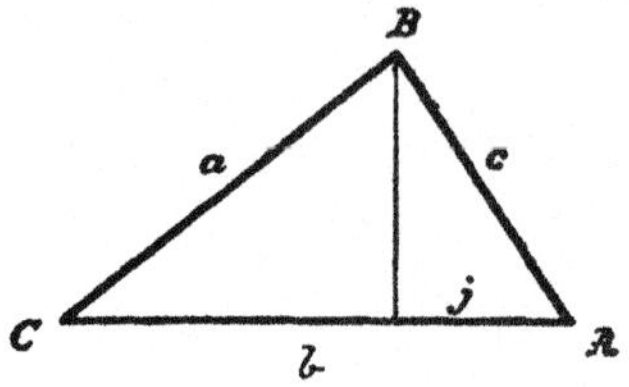

RULE. *From half the sum of the three sides subtract each side separately; multiply together the half-sum and the three remainders. The square root of this product is the area.*

FORMULA. $\triangle = \sqrt{s(s-a)(s-b)(s-c)}$.

PROOF. Calling j the projection of c on b, by 306,

$$a^2 = b^2 + c^2 - 2bj,$$

$$\therefore\ j = \frac{b^2 + c^2 - a^2}{2b}.$$

Calling the altitude h, this gives

$$h^2 = c^2 - j^2 = c^2 - \frac{(b^2 + c^2 - a^2)^2}{4b^2},$$

$$\therefore\ 4h^2b^2 = 4b^2c^2 - (b^2 + c^2 - a^2)^2,$$

$$\therefore\ 2hb = \sqrt{4b^2c^2 - (b^2 + c^2 - a^2)^2},$$

$$\therefore\ 2hb = \sqrt{(2bc + b^2 + c^2 - a^2)(2bc - b^2 - c^2 + a^2)},$$

$$\therefore\ 2hb = \sqrt{(a + b + c)(b + c - a)(a + b - c)(a - b + c)},$$

$$\therefore\ \frac{1}{2}hb = \sqrt{\frac{(a + b + c)}{2}\,\frac{(b + c - a)}{2}\,\frac{(a + b - c)}{2}\,\frac{(a - b + c)}{2}}.$$

Writing $s = \dfrac{a + b + c}{2}$ gives $\dfrac{b + c - a}{2} = s - a$,

$$\therefore\ \tfrac{1}{2}hb = \sqrt{s.(s - a)(s - b)(s - c)}.$$

But, by 807, $\frac{1}{2}hb = \triangle$.

809. *To find the area of a regular polygon.*

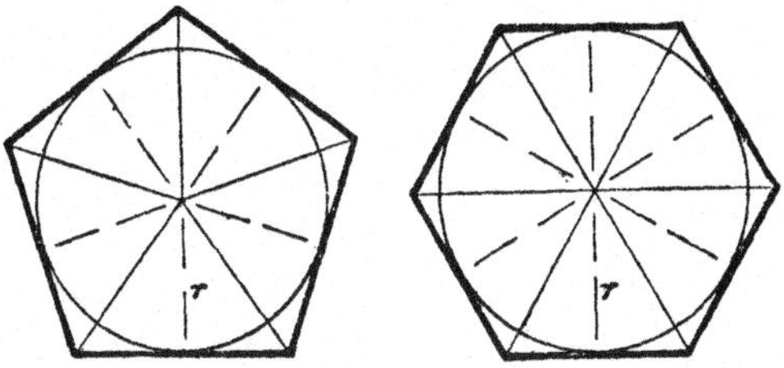

RULE. *Take half the product of its perimeter by the radius of the inscribed circle.*

FORMULA. $N = \frac{rp}{2}$.

PROOF. Sects from the center to the vertices divide the polygon into congruent isosceles triangles whose altitude is the radius of the inscribed circle, and the sum of whose bases is the perimeter of the polygon.

810. *To find the area of a circle.*

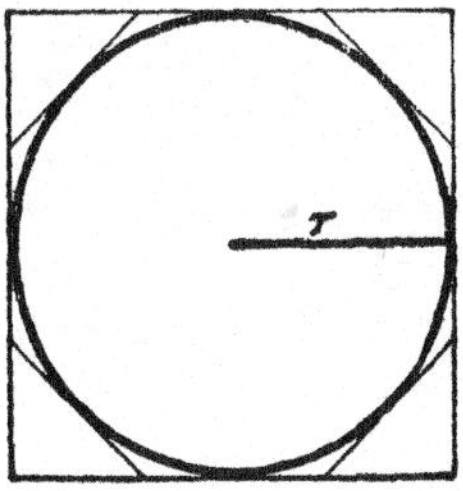

RULE. *Multiply its squared radius by* π.

FORMULA. $\odot = r^2\pi$.

If a regular polygon be circumscribed about the circle, its area N, by 809, is $\frac{1}{2}rp$.

If, now, the number of sides of the regular polygon be continually doubled, the perimeter p decreases toward c as limit, and N toward circle. But the variables N and p are always in the constant ratio $\frac{1}{2}r$; therefore, by 798, Principle of Limits, their limits are in the same ratio,

$$\therefore \quad \odot = \tfrac{1}{2}rc.$$

But, by 802, $c = 2r\pi$,

$$\therefore \quad \odot = r^2\pi.$$

811. ***To find the area of a sector.***

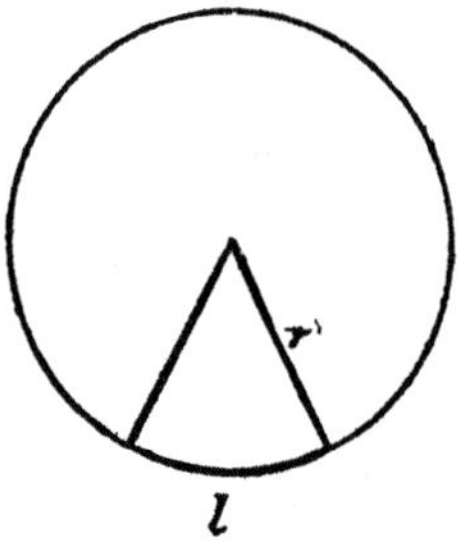

RULE. *Multiply the length of the arc by half the radius.*

FORMULA. $S = \frac{1}{2}lr = \frac{1}{2}ur^2$.

PROOF. By 506, $S : \odot :: l : c :: u : 2\pi$,

$$\therefore \quad S = \frac{\odot l}{c} = \frac{\odot u}{2\pi},$$

$$\therefore \quad S = \frac{\frac{1}{2}rcl}{c} = \frac{r^2\pi u}{2\pi},$$

$$\therefore \quad S = \tfrac{1}{2}lr = \tfrac{1}{2}ur^2.$$

812. An *Annulus* is the figure included between two concentric circles. Its height is the difference between the radii.

813. ***To find the area of a sector of an annulus.***

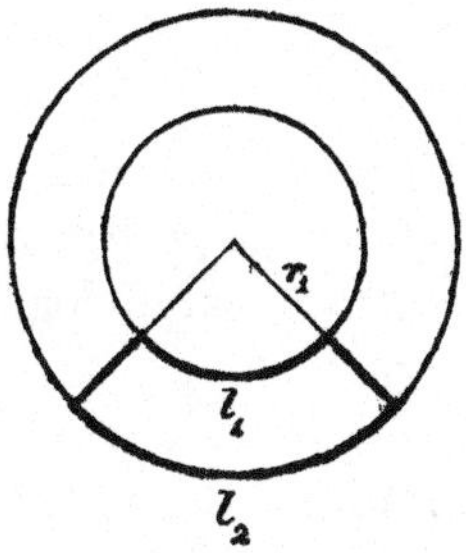

RULE. *Multiply the sum of the bounding arcs by half the difference of their radii.*

FORMULA. $S.A. = \frac{1}{2}(r_2 - r_1)(l_1 + l_2) = \frac{1}{2}h(l_1 + l_2)$.

PROOF. The annular sector is the difference between the two sectors $\frac{1}{2}r_2l_2$ and $\frac{1}{2}r_1l_1$.

But l_1 and l_2 are arcs subtending the same angle; therefore, by 804,

$$\frac{l_1}{r_1} = \frac{l_2}{r_2},$$

$$\therefore \quad l_1r_2 = l_2r_1;$$

$$\therefore \quad \tfrac{1}{2}r_2l_2 - \tfrac{1}{2}r_1l_1 = \tfrac{1}{2}r_2l_1 + \tfrac{1}{2}r_2l_2 - \tfrac{1}{2}r_1l_1 - \tfrac{1}{2}r_1l_2$$
$$= \tfrac{1}{2}(r_2 - r_1)(l_1 + l_2).$$

814. ***To find the lateral area of a prism.***

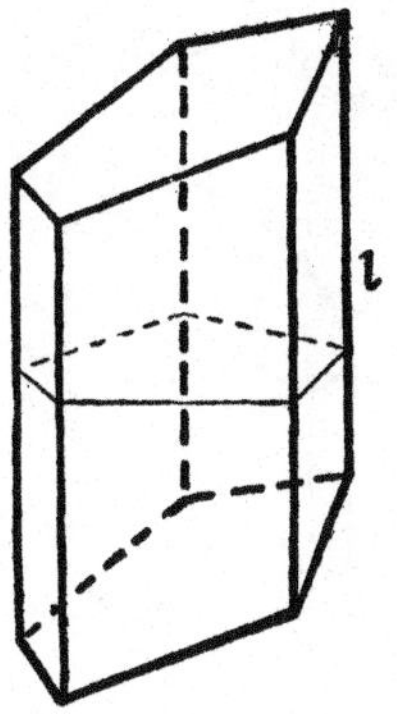

RULE. *Multiply a lateral edge by the perimeter of a right section.*

FORMULA. $P = lp$.

PROOF. The lateral edges of a prism are all equal. The sides of a right section, being perpendicular to the lateral edges, are the altitudes of the parallelograms which form the lateral surface of the prism.

815. A *Cylindric Surface* is generated by a line so moving that every two of its positions are parallel.

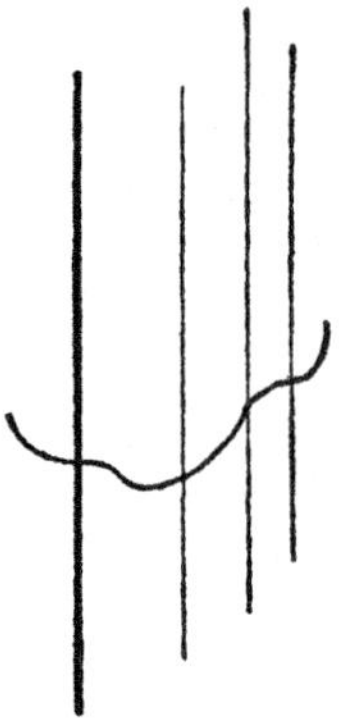

816. The generatrix in any position is called an *Element* of the surface.

817. A *Cylinder* is a solid bounded by a cylindric surface and two parallel planes.

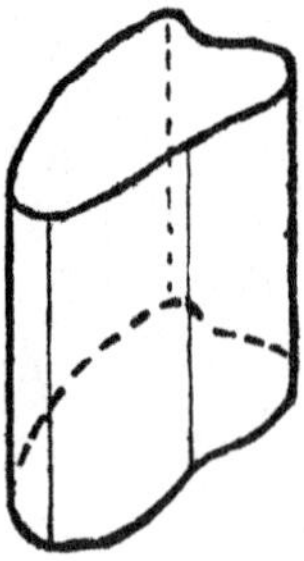

818. The *Axis* of a circular cylinder is the sect joining the centers of its bases.

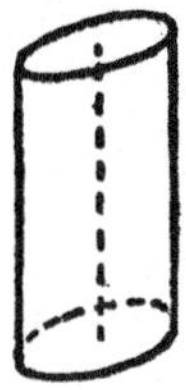

819. A *Truncated Cylinder* is the portion between the base and a non-parallel section.

820. ***To find the lateral area of a right circular cylinder.***

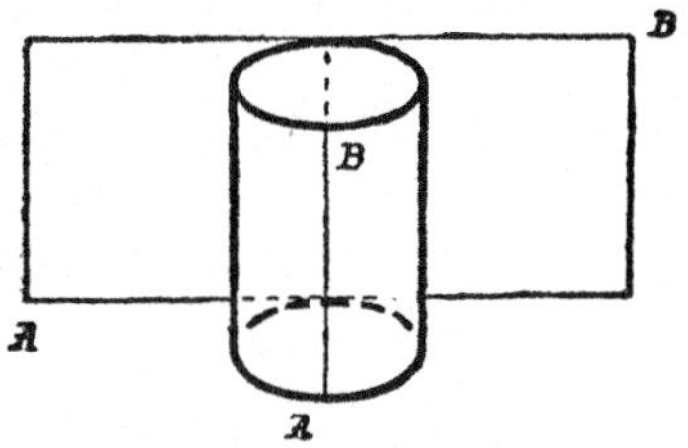

RULE. *Multiply its length by its circle.*
FORMULA. $C = cl.$

PROOF. Imagine the curved surface slit along an element and then spread out flat. It thus becomes a rectangle, having for one side the circle, and for the adjacent side an element.

821. COROLLARY I. The curved surface of a truncated circular cylinder is the product of the circle of the cylinder by the intercepted axis.

For, by symmetry, substituting an oblique for the right section through the same point of the axis, alters neither the curved surface nor the volume, since the solid between the two sections will be the same above and below the right section.

822. COROLLARY II. The lateral area of any cylinder on any base equals the length of the cylinder multiplied by the perimeter of a right section.

823. The lateral surface of a prism or cylinder is called its *Mantel.*

EXERCISES. 114. The mantel of a right circular cylinder is equivalent to a circle whose radius is a mean proportional between the altitude of the cylinder and the diameter of its base.

115. A plane through two elements of a right circular cylinder cuts its base in a chord which subtends at the center an angle whose circular measure is u; find the ratio of the curved surfaces of the two parts of the cylinder.

824. A *Conical Surface* is generated by a straight line moving so as always to pass through a fixed point called the apex.

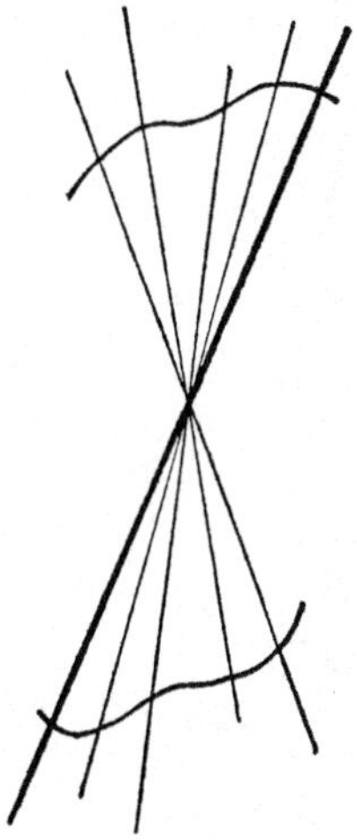

825. A *Cone* is a solid bounded by a conical surface and a plane.

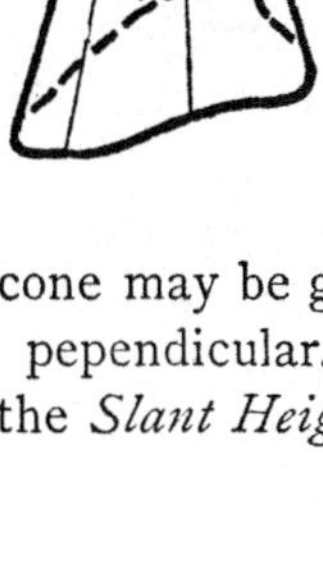

826. A right circular cone may be generated by revolving a right triangle about one pependicular. All the elements are equal, and each is called the *Slant Height* of the cone.

827. The *Frustum* of a pyramid or cone is the portion included between the base and a parallel section.

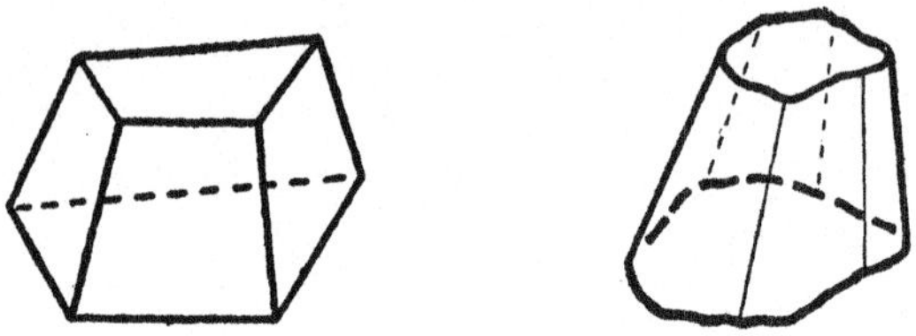

828. *To find the lateral area of a right circular cone.*

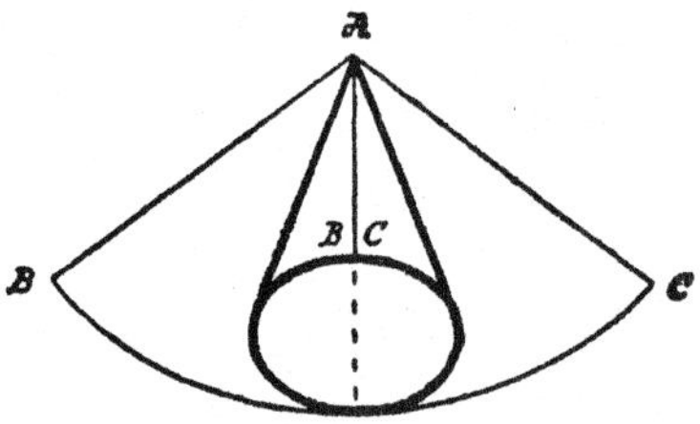

RULE. *Multiply the circle of its base by half the slant height.*

FORMULA. $K = \frac{1}{2}ch = \pi rh.$

PROOF. If the curved surface be slit along a slant height, and spread out flat, it becomes a sector of a circle, with slant height as radius, and the circle of cone's base as arc; therefore, by 811, its area is $\frac{1}{2}ch$.

829. COROLLARY. Since, by 810, the base $\odot = \frac{1}{2}cr$, therefore the curved surface is to the base as the slant height is to the radius of base, as the length of circle with radius h is to circle with radius r, as the perigon is to the sector angle.

830. ***To find the lateral area of the frustum of a right circular cone.***

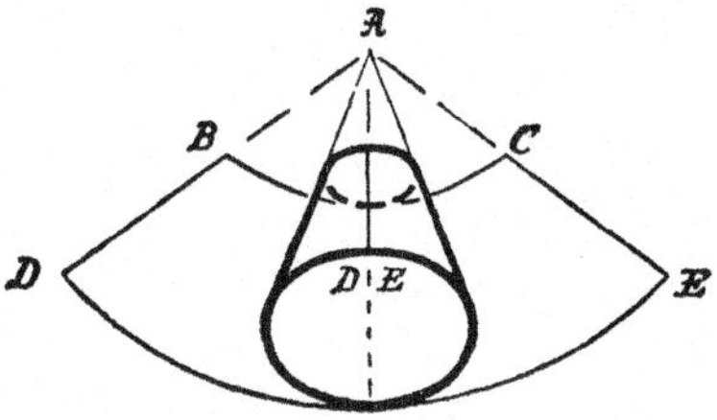

RULE. *Multiply the slant height of the frustum by half the sum of the circles of its bases.*

FORMULA. $F = \frac{1}{2}h(c_1 + c_2) = \pi h(r_1 + r_2)$.

PROOF. Completing the cone, and slitting it along a slant height, the curved surface of the frustum develops into the difference of two similar sectors having a common angle, the arcs of the sectors being the circles of the bases of the frustum. By 813, the area of this annular sector,

$$F = \frac{1}{2}h(c_1 + c_2).$$

831. The axis of a right circular cone, or cone of revolution, is the line through its vertex and the center of its base.

EXERCISES. 116. Given the two sides of a right-angled triangle. Find the area of the surface described when the triangle revolves about its hypothenuse.

HINT. Calling a and b the given altitude and base, and x the perpendicular from the right angle to the hypothenuse, by 828, the area of the surface described by a is πxa, and by b is πxb. Also $a : x :: \sqrt{a^2 + b^2} : b$.

117. In the frustum of a right circular cone, on each base stands a cone with its apex in the center of the other base; from the basal radii r_1 and r_2 find the radius of the circle in which the two cones cut.

THEOREM III.

832. *The lateral area of a frustum of a cone of revolution is the product of the projection of the frustum's slant height on the axis by twice π times a perpendicular erected at the mid-point of this slant height, and terminated by the axis.*

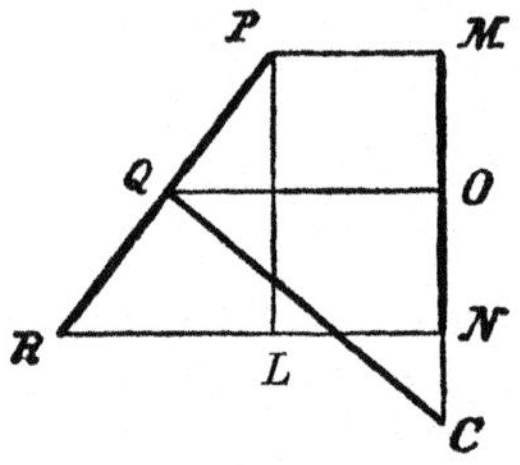

PROOF. By 830, the lateral area of the frustum whose slant height is PR and axis MN is

$$F = \pi(PM + RN)PR.$$

But if Q is the mid-point of PR, then $PM + RN = 2QO$,

$$\therefore \quad F = 2\pi \times PR \times QO.$$

But $\triangle PRL \sim \triangle QCO$, since the three sides of one are drawn perpendicular to the sides of the other;

$$\therefore \quad PR \times QO = PL \times QC,$$

$$\therefore \quad F = 2\pi(LP \times QC) = 2\pi(MN \times CQ).$$

EXERCISES. 118. Reckon the mantel from the two radii when the inclination of a slant height to one base is half a right angle.

119. If in the frustum of a right cone the diameter of the upper base equals the slant height, reckon the mantel from the altitude a and perimeter p of an axial section.

833. ***To find the area of a sphere.***

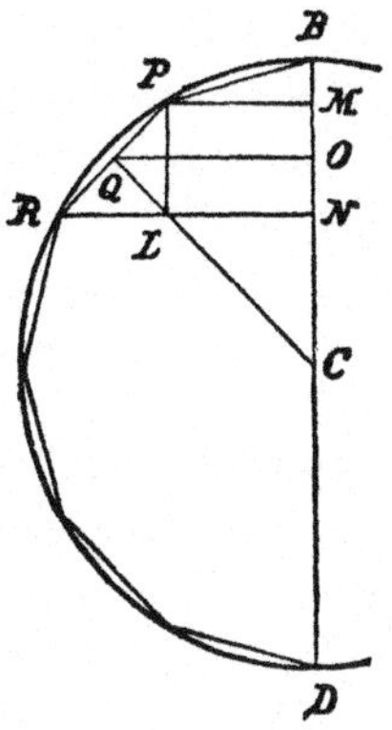

RULE. *Multiply four times its squared radius by π.*

FORMULA. $H = 4r^2\pi$.

PROOF. In a circle inscribe a regular polygon of an even number of sides. Then a diameter through one vertex passes through the opposite vertex, halving the polygon symmetrically.

Let PR be one of its sides; draw PM, RN, perpendicular to the diameter BD. From the center C the perpendicular CQ bisects PR. Drop the perpendiculars PL, QO.

Now, if the whole figure revolve about BD as axis, the semicircle will generate a sphere, while each side of the inscribed polygon, as PR, will generate the curved surface of the frustum of a cone. By 832, this

$$F = 2\pi(MN \times CQ);$$

and the sum of all the frustums, that is, the surface of the solid generated by the revolving semi-polygon, equals $2\pi CQ$ into the sum of the projections, BM, MN, NC, etc., which sum is BD.

$\therefore$ Sum of surfaces of frustums $= 2\pi CQ \times BD$.

As we continually double the number of sides of the inscribed polygon, its semi-perimeter approaches the semicircle as limit, and its surface of revolution approaches the sphere as limit, while CQ, its apothem, approaches r, the radius of the sphere, as limit. Representing the sum of the surfaces of the frustums by ΣF, and BD by $2r$, we have

$$\frac{\Sigma F}{CQ} = 4r\pi.$$

That is, the variable sum is to the variable CQ in the constant ratio $4r\pi$; therefore, by 798, Principle of Limits, their limits have the same ratio,

$$\therefore \quad \frac{H}{r} = 4r\pi,$$

$$\therefore \quad H = 4r^2\pi.$$

834. A *Calot* is a zone of only one base.

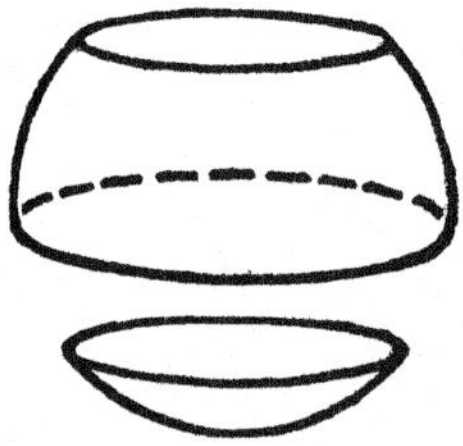

835. The last proof gives also the following rule:—

To find the area of a zone.

RULE. *Multiply the altitude of the segment by twice π times the radius of the sphere.*

FORMULA. $Z = 2ar\pi$.

CHAPTER IV.

SPACE ANGLES.

836. A *Plane Angle* is the divergence of two straight lines which meet in a point.

837. A *Space Angle* is the spread of two or more planes which meet in a point.

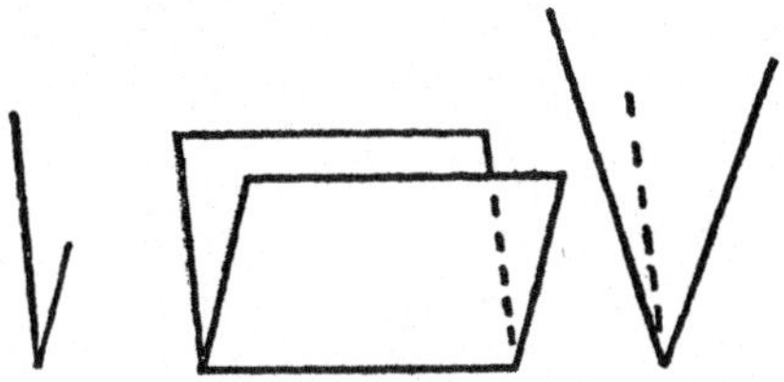

838. *Symmetrical Space Angles* are those which cut out symmetrical spherical polygons on a sphere, when their vertices are placed at its center.

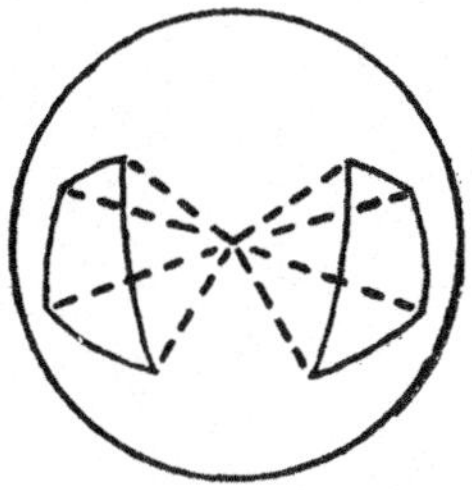

839. A *Steregon* is the whole amount of space angle round about a point in space.

840. A *Steradian* is the angle subtended at the center by that part of every sphere equal to the square of its radius.

841. The space angle made by only two planes corresponds to the lune intercepted on any sphere whose center is in the common section of the two planes.

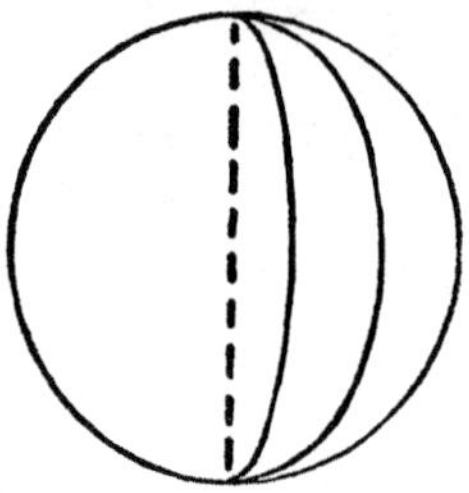

842. A *Spherical Pyramid* is a portion of a globe bounded by a spherical polygon and the planes of the sides of the polygon. The center of the sphere is the apex of the pyramid; the spherical polygon is its base.

843. Just as plane angles at the center of a circle are proportional to their intercepted arcs, and also sectors, so *space*

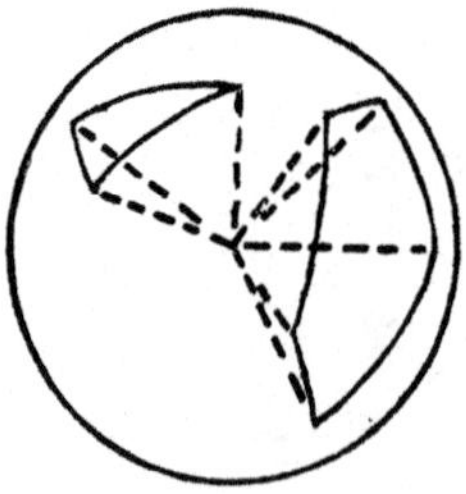

angles at the center of a sphere are proportional to their intercepted spherical polygons, and also spherical pyramids.

EXAMPLE. Find the ratio of the space angles of two right cones of altitude a_1 and a_2, but having the same slant height, h.

These space angles are as the corresponding calots (or zones of one base) on the sphere of radius h. Therefore, by 835, the required ratio is

$$\frac{2\pi h(h - a_1)}{2\pi h(h - a_2)} = \frac{h - a_1}{h - a_2},$$

the ratio of the calot altitudes.

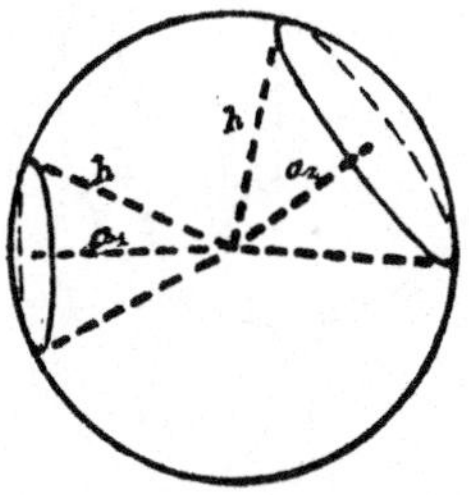

For the equilateral and right-angled cones this becomes

$$\frac{2 - \sqrt{3}}{2 - \sqrt{2}}.$$

844. To construct a space angle of two faces equivalent to any polyhedral angle, only involves constructing a lune equivalent to a spherical polygon, as in 731.

845. ***To find the area of a lune.***

RULE. *Multiply its angle in radians by twice its squared radius.*

FORMULA. $L = 2r^2u$.

PROOF. By 703, lunes are as their angles,

$\therefore$ a lune is to a hemisphere as its angle is to a straight angle,

$$\frac{L}{2r^2\pi} = \frac{u}{\pi},$$

$$\therefore \; L = 2r^2u.$$

846. COROLLARY I. A lune measures twice as many steradians as its angle contains radians.

847. COROLLARY II. If two-faced space angles are equal, their lune angles are equal; so a dihedral angle may be measured by the plane angle between two perpendiculars, one in each face, from any point of its edge.

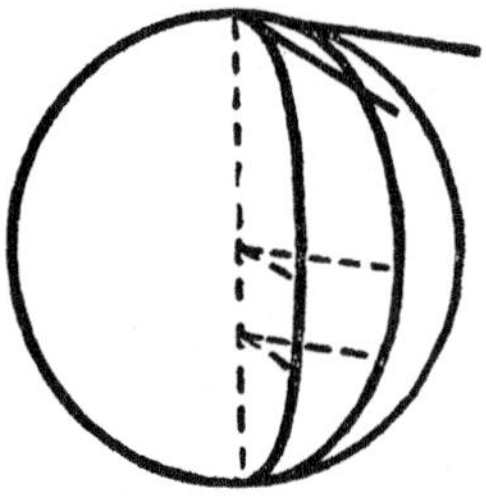

848. Suppose the vertex of a space angle is put at the center of a sphere, then the planes which form the space angle will cut the sphere in arcs of great circles, forming a spherical polygon, whose angles may be taken to measure the dihedral angles of the space angle, and whose sides measure its face angles.

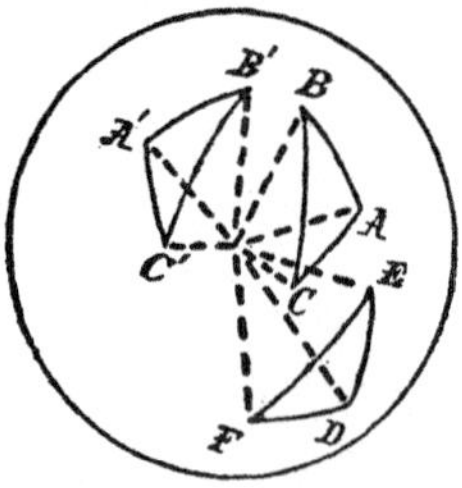

Hence from any property of spherical polygons we may infer an analogous property of space angles.

For example, the following properties of trihedral angles have been proved in our treatment of spherical triangles: —

I. Trihedral angles are either congruent or symmetrical which have the following parts equal: —

(1) Two face angles and the included dihedral angle.

(2) Two dihedral angles and the included face angle.

(3) Three face angles.

(4) Three dihedral angles.

(5) Two dihedral angles and the face angle opposite one of them, provided the edge of the third dihedral angle of neither trihedral makes right angles with any line in the half of the opposite face not adjacent to one of the face angles equal by hypothesis.

(6) Two face angles and the dihedral angle opposite one of them, provided the other pair of opposite dihedral angles are not supplemental.

II. As one of the face angles of a trihedral angle is greater than, equal to, or less than, another, the dihedral angle which it subtends is greater than, equal to, or less than, the dihedral angle subtended by the other.

III. Symmetrical trihedral angles are equivalent.

IV. Space angles having the same number and sum of dihedral angles are equivalent.

V. The face angles of a convex space angle are together less than a perigon.

849. ***To find the area of a spherical triangle.***

RULE. *Multiply its spherical excess in radians by its squared radius.*

FORMULA. $\widehat{\triangle} = er^2$.

PROOF. By 711, a triangle is equal to a lune whose angle is $\frac{1}{2}e$; therefore, by 845, $\widehat{\triangle} = r^2e$.

850. COROLLARY I. A spherical triangle measures as many steradians as its e contains radians.

851. COROLLARY II. By 714, to find the area of a spherical polygon, multiply its spherical excess in radians by its squared radius.

CHAPTER V.

THE MEASUREMENT OF VOLUMES.

852. The *Volume* of a solid is its ratio to an assumed unit.

853. The *Unit for Measurement of Volume* is a cube whose edge is the unit for length.

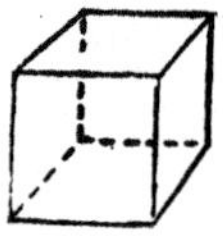

Cubic Centimeter.

854. METRIC UNITS FOR VOLUME.

Solid.		Liquid.		Cubic.		
1 kilostere			=	1 cubic dekameter	=	1000 cubic meters.
1 hectostere					=	100 cubic meters.
1 dekastere					=	10 cubic meters.
1 stere	=	1 kiloliter	=	1 cubic meter	=	1 cubic meter.
1 decistere	=	1 hectoliter			=	.1 cubic meter.
1 centistere	=	1 dekaliter			=	.01 cubic meter.
		1 liter	=	1 cubic decimeter	=	.001 cubic meter.
		1 deciliter			=	.0001 cubic meter.
		1 centiliter			=	.00001 cubic meter.
				1 cubic centimeter	=	.000001 cubic meter.
				1 cubic millimeter	=	.000000001 cubic meter.

The authorized abbreviation for cubic is the index ^{3}, as in 1cm.^{3} for 1 cubic centimeter; that for stere is s., and for liter l..

855. ***To find the volume of a quader.***

RULE. *Multiply together its length, breadth, and thickness.*

PROOF. By 768, two quaders have the ratio compounded of the ratios of their bases and altitudes.

856. COROLLARY. The volume of any cube is the third power of the length of an edge. This is why the third power of a number is called its cube.

857. ***To find the volume of any parallelopiped.***

RULE. *Multiply its altitude by the area of its base.*

PROOF. By 769, any parallelopiped is equivalent to a quader of equivalent base and equal altitude.

858. ***To find the volume of any prism.***

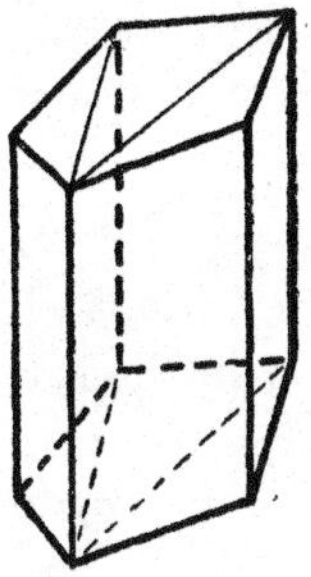

RULE. *Multiply its altitude by its base.*

FORMULA. $V.P = aB.$

PROOF. By 771, any three-sided prism is half a parallelopiped, with base twice the prism's base and the same altitude.

Thus the rule is true for triangular prisms, and consequently for all prisms; since, by passing planes through one lateral edge, and all the other lateral edges excepting the two adjacent to this one, we can divide any prism into triangular prisms of the same altitude, whose triangular bases together make the polygonal base.

859. ***To find the volume of any cylinder.***

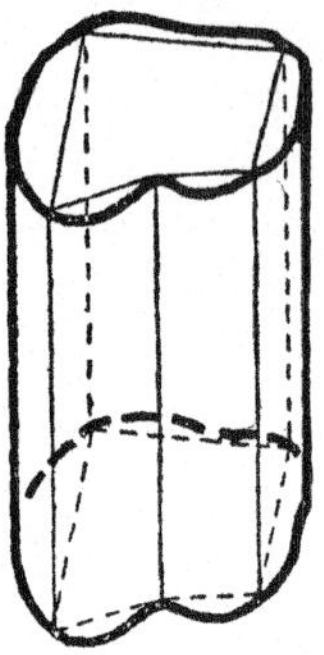

RULE. *Multiply its altitude by its base.*

PROOF. The cylinder is the limit of an inscribed prism when the number of sides of the prism is indefinitely increased, the base of the cylinder being the limit of the base of the prism.

But always the variable prism is to its variable base in the constant ratio a; therefore, by 798, Principle of Limits, their limits will be to one another in the same ratio.

860. ***To find the volume of any pyramid.***

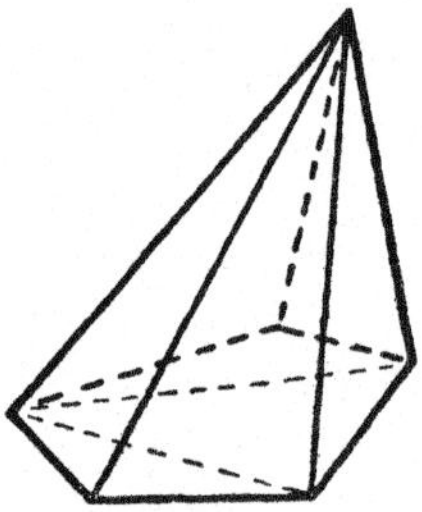

RULE. *Multiply one-third of its altitude by its base.*

FORMULA. $Y = \frac{1}{3}aB.$

PROOF. By 775, any triangular pyramid is one-third of a triangular prism of the same base and altitude.

The rule thus proved for triangular pyramids is true for all pyramids; since by passing planes through one lateral edge, and all the other lateral edges excepting the two adjacent to this one, we can divide any pyramid into triangular pyramids of the same altitude whose bases together make the polygonal base.

861. *To find the volume of any cone.*

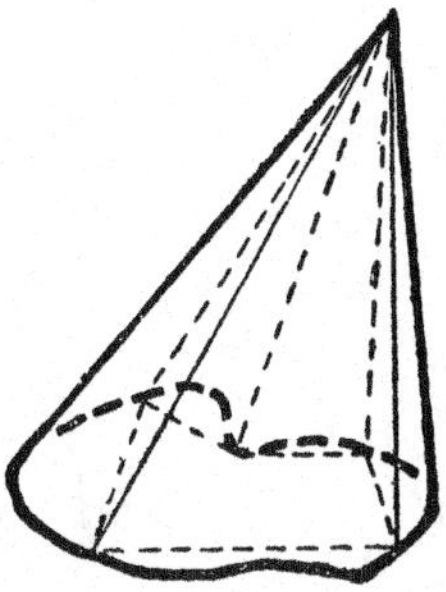

RULE. *Multiply one-third its altitude by its base.*

FORMULA WHEN BASE IS A CIRCLE. $V.K = \frac{1}{3}ar^2\pi$.

PROOF. The base of a cone is the limit of the base of an inscribed pyramid, and the cone is the limit of the pyramid. But always the variable pyramid is to its variable base in the constant ratio $\frac{1}{3}a$.

Therefore their limits are to one another in the same ratio, and $V.K = \frac{1}{3}aB$.

SCHOLIUM. This applies to all solids determined by an elastic string stretching from a fixed point to a point describing any closed plane figure.

862. COROLLARY. The volume of the solid generated by the revolution of any triangle about one of its sides as axis, is one-third the product of the triangle's area into the length of the circle described by its vertex.

$$V = \tfrac{2}{3}\pi r \triangle.$$

PRISMATOID.

863. A *Prismatoid* is a polyhedron whose bases are any two polygons in parallel planes, and whose lateral faces are triangles

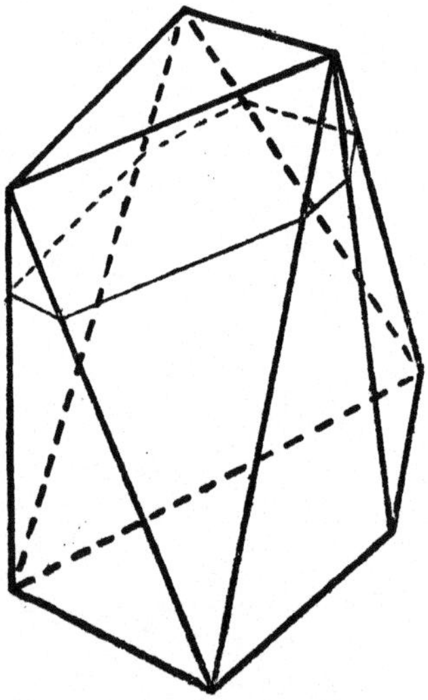

determined by so joining the vertices of these bases that each lateral edge, with the preceding, forms a triangle with one side of either base.

864. A number of different prismatoids thus pertain to the same two bases.

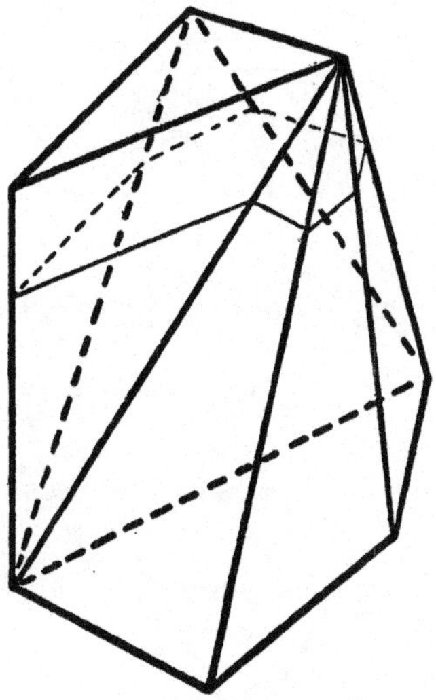

865. If two basal edges which form with the same lateral edge two sides of two adjoining faces are parallel, then these

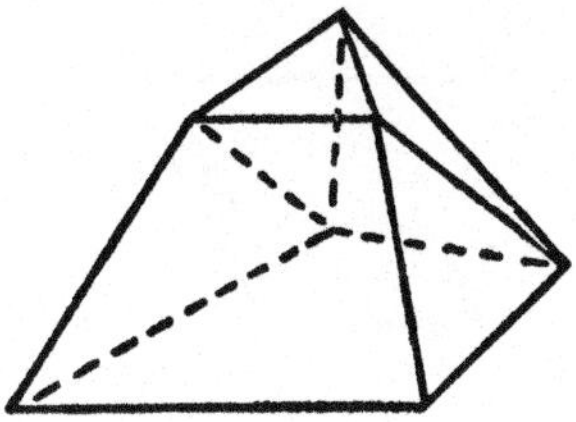

two triangular faces fall in the same plane, and together form a trapezoid.

866. A *Prismoid* is a prismatoid whose bases have the same number of sides, and every corresponding pair parallel.

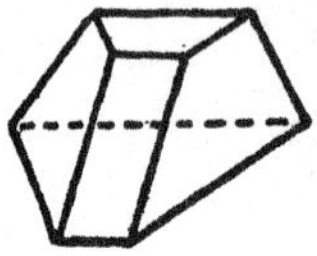

867. A frustum of a pyramid is a prismoid whose two bases are similar.

868. COROLLARY. Every three-sided prismoid is the frustum of a pyramid.

869. If both bases of a prismatoid become sects, it is a tetrahedron.

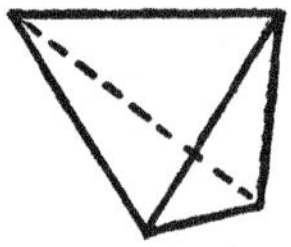

870. A *Wedge* is a prismatoid whose lower base is a rectangle, and upper base a sect parallel to a basal edge.

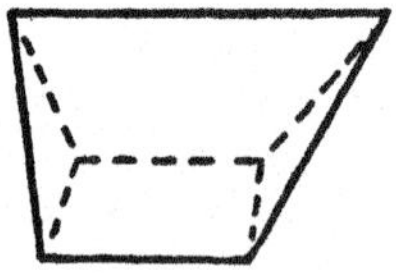

871. The altitude of a prismatoid is any sect perpendicular to both bases.

872. A *Cross-Section* of a prismatoid is a section made by a plane perpendicular to the altitude.

873. ***To find the volume of any prismatoid.***

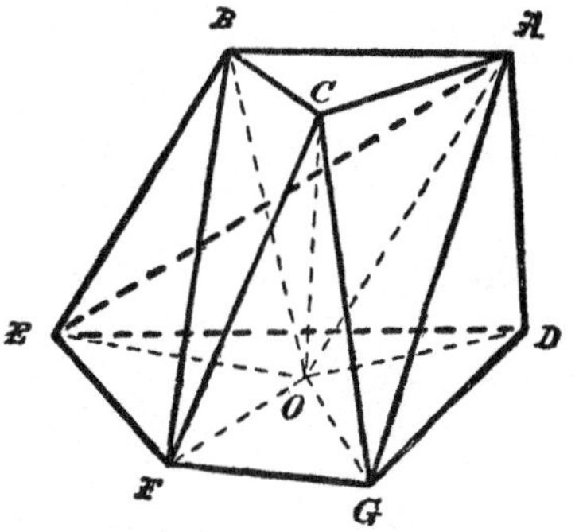

RULE. *Multiply one-fourth its altitude by the sum of one base and three times a cross-section at two-thirds the altitude from that base.*

FORMULA. $D = \frac{a}{4}(B + 3T)$.

PROOF. Any prismatoid may be divided into tetrahedra, all of the same altitude as the prismatoid; some, as $CFGO$, having their apex in the upper base of the prismatoid, and for base a portion of its lower base; some, as $OABC$, having base in the upper, and apex in the lower, base of the prismatoid; and the others, as $ACOG$, having for a pair of opposite edges a sect in the plane of each base of the prismatoid, as AC and OG.

Therefore, if the formula holds good for tetrahedra in these three positions, it holds for the prismatoid, their sum.

In (1), call T_1 the section at two-thirds the altitude from the base B_1; then T_1 is $\frac{1}{3}a$ from the apex. Therefore, by 772,

$$T_1 : B_1 :: (\tfrac{1}{3}a)^2 : a^2, \qquad \therefore \quad T_1 = \tfrac{1}{9}B_1,$$

$$\therefore \quad D_1 = \frac{a}{4}(B_1 + 3T_1) = \frac{a}{4}(B_1 + \frac{1}{3}B_1) = \frac{1}{3}aB,$$

which, by 860, equals Y, the volume of the tetrahedron.

In (2), $B_1 = 0$, being a point, and T_1 is $\frac{2}{3}a$ from the apex;

$$\therefore \quad T_1 : B_2 :: (\tfrac{2}{3}a)^2 : a^2, \qquad \therefore \quad T_1 = \tfrac{4}{9}B_2,$$

$$\therefore \quad D_2 = \frac{a}{4}(B_1 + 3T_1) = \frac{a}{4}(0 + \frac{4}{3}B_2) = \frac{1}{3}aB_2 = Y.$$

In (3), let *KLMN* be the section T_1.

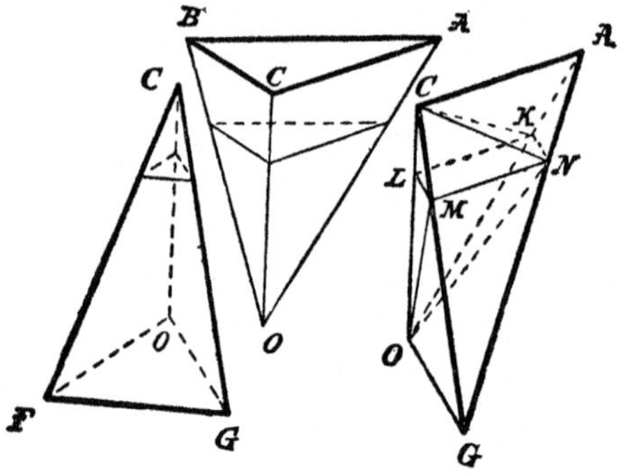

Join *CK*, *CN*, *OM*, *ON*. Now

$$\triangle\, ANK : \triangle\, AGO :: AN^2 : AG^2 :: (\tfrac{1}{3}a)^2 : a^2 :: 1 : 9.$$

$$\triangle\, GNM : \triangle\, GAC :: GN^2 : GA^2 :: (\tfrac{2}{3}a)^2 : a^2 :: 4 : 9.$$

But the whole tetrahedron D_3 and the pyramid *CANK* may be considered as having their bases in the same plane, *AGO*, and the same altitude, a perpendicular from *C*;

$$CANK : D_3 :: \triangle\, ANK : \triangle\, AGO :: 1 : 9,$$

$$\therefore \quad CANK = \tfrac{1}{9}D_3.$$

In same way,

$$O\,GNM : D_3 :: \triangle\,GNM : \triangle\,GAC :: 4 : 9,$$

$$\therefore\quad O\,GNM = \tfrac{4}{9}D_3.$$

$$\therefore\quad C\,ANK + O\,GNM = \tfrac{5}{9}D_3,$$

$$\therefore\quad C\,KLMN + O\,KLMN = \tfrac{4}{9}D_3.$$

But, by 860,

$$C\,KLMN + O\,KLMN = \tfrac{1}{3}\,.\,\tfrac{1}{3}aT_1 + \tfrac{1}{3}\,.\,\tfrac{2}{3}aT_1 = \tfrac{1}{3}aT_1,$$

$$\therefore\quad \frac{4}{9}D_3 = \frac{1}{3}aT_1,\qquad \therefore\quad D_3 = \frac{a}{4}3T_1 = \frac{a}{4}(B_1 + 3T_1),$$

since here

$$B_1 = 0.$$

874. Corollary I. Since, in the frustum of a pyramid, B and T are similar,

$$\therefore\quad V.F = \frac{a}{4}B\left(1 + \frac{3w_2^2}{w_1^2}\right),$$

where w_1 and w_2 are corresponding sides of B and T.

875. Corollary II. For the frustum of a cone of revolution,

$$V.F = \frac{a}{4}\pi r_1^2\left(1 + \frac{3r_2^2}{r_1^2}\right),$$

where r_2 is the radius of T.

876. ***To find the volume of a globe.***

RULE. *Multiply the cube of its radius by* $\frac{4}{3}\pi$.

FORMULA. $G = \frac{4}{3}\pi r^3$.

PROOF. By 644, a tetrahedron on edge, and a globe with the tetrahedron's altitude for diameter, have all their corresponding cross-sections equivalent if any one pair are equivalent.

Hence the volumes may be proved equivalent, as in 774, and

$$G = \tfrac{3}{4}aT.$$

But $a = 2r$, and, by 522, $T = \frac{2}{3}r \,.\, \frac{4}{3}r \,.\, \pi$,

$$\therefore \quad G = \tfrac{3}{4} \,.\, 2r \,.\, \tfrac{2}{3}r \,.\, \tfrac{4}{3}r \,.\, \pi = \tfrac{4}{3}\pi r^3.$$

877. COROLLARY I. Globes are to each other as the cubes of their radii.

878. COROLLARY II. Similar solids are to one another as the cubes of any two corresponding edges or sects.

DIRECTION.

879. If two points starting from a state of coincidence move along two equal sects which do not coincide, that quality of each movement which makes it differ from the other is its *Direction*.

880. If two equal sects are terminated at the same point, but do not coincide, that quality of each which makes it differ from the other is its direction.

881. The part of a line which could be generated by a tracing-point, starting from a given point on that line, and moving on the line without ever turning back, is called a *Ray* from that given point as origin.

882. Two rays from the same origin are said to have the same direction if they coincide; otherwise, they are said to have different directions.

883. Two rays which have no point but the origin in common, and fall into the same line, are said to have *opposite* directions.

884. Two rays lying on parallel lines have parallel-same directions if they are on the same side of the line joining their origins.

885. Two rays lying on parallel lines have parallel-opposite directions if they are on opposite sides of the line joining their origins.

886. A sect is said to have the same direction as the ray of which it is a part.

887. A sect is definitely fixed if we know its initial point, its parallel-same direction, and its length.

888. The operation by which a sect could be traced if we knew its initial point, that is, the operation of carrying a tracing-point in a certain parallel-same direction until it passes over a given number of units for length, is called a *Vector.*

889. The position of B relative to A is indicated by the length and parallel-same direction of the sect $\overline{AB}$ drawn *from* A *to* B. If you start from A, and travel, in the direction indicated by the ray from A through B, and traverse the given number of units, you get to B. This parallel-same direction and length may be indicated equally well by any other sect, such as $\overline{A'B'}$, which is equal to $\overline{AB}$ and in parallel-same direction.

890. As indicating an operation, the vector $\overline{AB}$ is completely defined by the parallel-same direction and length of the transferrence. All vectors which are of the same magnitude and parallel-same direction, and only those, are regarded as equal.

Thus $\overline{AB}$ is not equal to $\overline{BA}$.

PRINCIPLE OF DUALITY.

JOIN OF POINTS AND OF LINES.

891. The line joining two points is called the *Join of the Two Points.* The point common to two intersecting lines is called the *Join of the Two Lines.*

892. PENCIL OF LINES. A fixed point, A, may be joined to all other points in space.

We get thus all the lines which can be drawn through the point A. The aggregate of all these lines is called a *Pencil of Lines.* The fixed point is called the *Base* of the pencil. Any one of these lines is said to be a line in the pencil, and also to be a *line in the fixed point.* In this sense, we say not only that a point may lie in a line, but also that a *line may lie in a point,* meaning that the line passes through the point.

893. In most cases, we can, when one figure is given, construct another such that lines take the place of points in the first, and points the place of lines.

Any theorem concerning the first thus gives rise to a corresponding theorem concerning the second figure. Figures or theorems related in this manner are called *Reciprocal Figures* or *Reciprocal Theorems.*

894. Small letters denote lines, and the join of two elements is denoted by writing the letters indicating the elements, together.

Thus the join of the points A and B is the line AB, while ab denotes the join, or point of intersection, of the lines a and b.

ROW OF POINTS, PENCIL OF LINES.

895. A line contains an infinite number of points, called *a Row of Points,* of which the line is the *Base.*

A row is all points in a line.

The reciprocal figure is *all lines in a point,* or all lines passing through the point.

A *Flat Pencil* is the aggregate of all lines in a plane which pass through a given point. In plane geometry, by a pencil we mean a flat pencil.

RECIPROCAL THEOREMS.

896₁. A point moving along a line describes a row.

897₁. A sect is a part of a row described by a point moving from one position, *A,* to another position, *B.*

896′. A line turning about a point describes a pencil.

897′. An angle is a part of a pencil described by a line turning from one position, *a,* to another position, *b.*

Thus to sect *AB* corresponds ∡ *ab.*

LINKAGE.

898. The *Peancellier Cell* consists of a rhombus movably

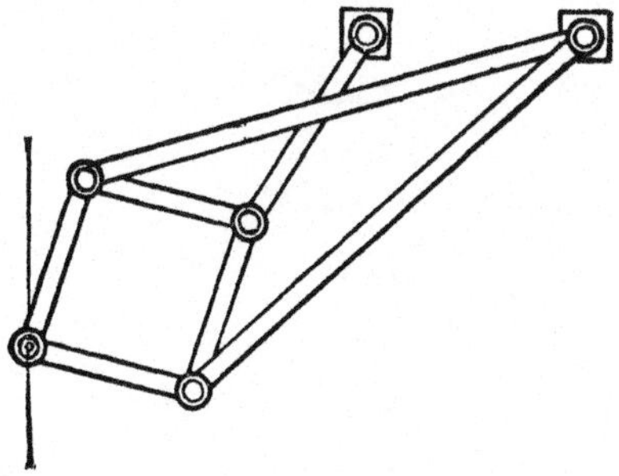

jointed, and two equal links movably pivoted at a fixed point, and at two opposite extremities of the rhombus.

TO DRAW A STRAIGHT LINE.

899. Take an extra link, and, while one extremity is on the fixed point of the cell, pivot the other extremity to a fixed point. Then pivot the first end to one of the free angles of the rhombus. The opposite vertex of the rhombus will now describe a straight line, however the linkage be pushed or moved.

PROOF. By the bar FD, the point D is constrained to move on the circle ADR; therefore $\measuredangle\, ADR$, being the angle in a semicircle, is always right.

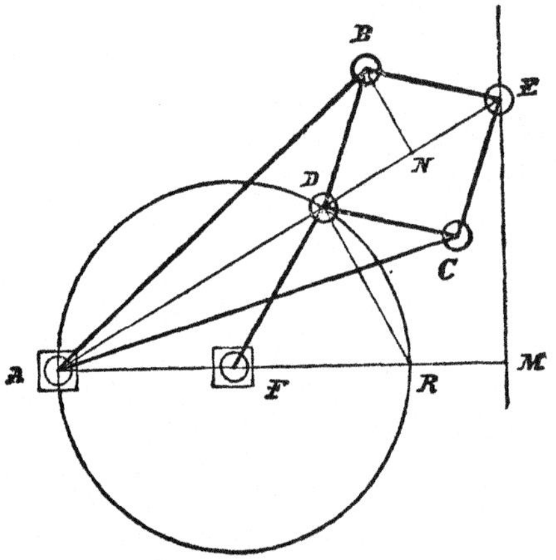

If, now, E moves on $EM \perp AM$,

$$\therefore \quad \triangle ADR \sim \triangle AME,$$

$$\therefore \quad DA : AR :: AM : AE,$$

$$\therefore \quad DA \,.\, AE = RA \,.\, AM.$$

Therefore, if $AE \,.\, AD$ is constant, E moves on the straight line EM. But because $BDCE$ is a rhombus, and $AB = AC$,

$\therefore$ D and N are always on the variable sect AE.

Always

$$AB^2 = AN^2 + NB^2, \quad \text{and} \quad \overline{BE}^2 = EN^2 + \overline{NB}^2,$$

$$\therefore \; AB^2 - \overline{BE}^2 = AN^2 - \overline{NE}^2$$
$$= (AN + NE)(AN - NE) = AE \, . \, AD.$$

900. COROLLARY. The efficacy of our cell depends on its power to keep, however it be deformed, the product of two varying sects a constant.

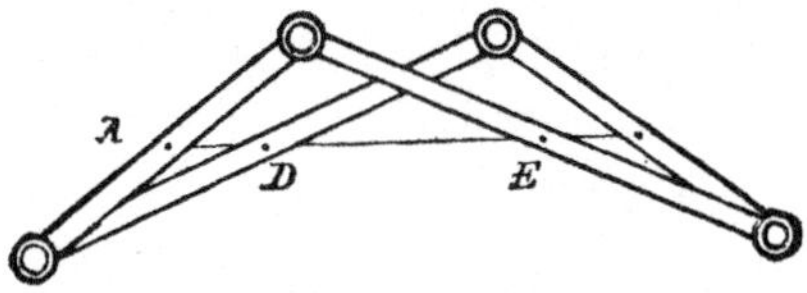

Therefore a Hart four-bar cell, constructed as in the accompanying figure, may be substituted for the Peancellier six-bar cell, since $AE \, . \, AD$ equals a constant.

Also, for the Hart cell may be substituted the quadruplane, four pivoted planes.

CROSS-RATIO.

901. If four points are collinear, two may be taken as the extremities of a sect, which each of the others divides internally or externally in some ratio.

The ratio of these two ratios is called the *Cross-Ratio* of the four points.

The cross-ratio $\dfrac{AC}{CB} : \dfrac{AD}{DB}$ is written $(ABCD)$.

Distinguishing the "step" AB from BA, as of opposite "sense," and taking the points in the two groups of two in a definite order, to write out a cross-ratio, make first the two bars, and put crossing these the letters of the first group of two, thus $\frac{A}{\ \ B} : \frac{A}{\ \ B}$; then fill up, crosswise, the first by the first letter of the second group of two, the second by the second.

If we take the points in a different order, the value of the cross-ratio may change.

We can do this in twenty-four different ways by forming all permutations of the letters.

But of these twenty-four cross-ratios, groups of four are equal, so that there are really only six different ones.

We have the following rules:—

I. *If in a cross-ratio the two groups be interchanged, its value remains unaltered.*

$$(ABCD) = (CDAB).$$

II. *If in a cross-ratio the two points belonging to one of the two groups be interchanged, the cross-ratio changes to its reciprocal.*

$$(ABCD) = \frac{1}{(ABDC)}.$$

I. and II. are proved by writing out their values.

III. *From II. it follows, that, if we interchange the elements in each pair, the cross-ratio remains unaltered.*

$$(ABCD) = (BADC).$$

IV. *If in a cross-ratio the two middle letters be interchanged, the cross-ratio changes into its complement.*

$$(ABCD) = 1 - (ACBD).$$

This is proved by taking the step-equation for any four collinear points,

$$BC \,.\, AD + CA \,.\, BD + AB \,.\, CD = 0,$$

and dividing it by $CB \,.\, AD$.

THEOREM IV.

902. *If any four concurrent lines are cut by a transversal, any cross-ratio of the four points of intersection is constant for all positions of the transversal.*

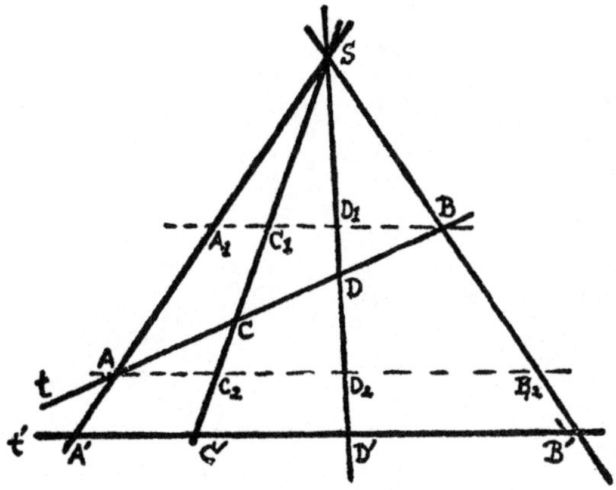

HYPOTHESIS. ***Let** A, B, C, D, **be the intersection points for** t, **the transversal in one position, and let** A', B', C', D', **be the corresponding intersection points for** t', **the transversal in another position.***

CONCLUSION. $(ABCD) = (A'B'C'D')$; that is,

$$\frac{AC}{CB} : \frac{AD}{DB} = \frac{A'C'}{C'B'} : \frac{A'D'}{D'B'}.$$

PROOF. Through the points A and B on t draw parallels to t', which cut the concurrent lines in C_2, D_2, B_2, and A_1, C_1, D_1.

$$\triangle ACC_2 \sim \triangle BCC_1, \quad \text{and} \quad \triangle ADD_2 \sim \triangle BDD_1,$$

$$\therefore \quad \frac{AC}{CB} = \frac{AC_2}{C_1B}, \quad \text{and} \quad \frac{AD}{DB} = \frac{AD_2}{D_1B},$$

where account is taken of "sense."

Hence

$$\frac{AC}{CB} : \frac{AD}{DB} = \frac{AC_2}{C_1B} : \frac{AD_2}{D_1B} = \frac{AC_2}{AD_2} : \frac{C_1B}{D_1B};$$

but

$$\frac{AC_2}{AD_2} = \frac{A'C'}{A'D'} \quad \text{and} \quad \frac{C_1B}{D_1B} = \frac{C'B'}{D'B'},$$

$$\therefore \quad \frac{AC}{CB} : \frac{AD}{DB} = \frac{A'C'}{A'D'} : \frac{C'B'}{D'B'} = \frac{A'C'}{C'B'} : \frac{A'D'}{D'B'}.$$

EXERCISES.

BOOK I.

120. Show how to make a rhombus having one of its diagonals equal to a given sect.

121. If two quadrilaterals have three consecutive sides, and the two contained angles in the one respectively equal to three consecutive sides and the two contained angles in the other, the quadrilaterals are congruent.

122. Two circles cannot cut one another at two points on the same side of the line joining their centers.

123. Prove, by an equilateral triangle, that, if a right-angled triangle have one of the acute angles double of the other, the hypothenuse is double of the side opposite the least angle.

124. Draw a perpendicular to a sect at one extremity.

125. Draw three figures to show that an exterior angle of a triangle may be greater than, equal to, or less than, the interior *adjacent* angle.

126. Any two exterior angles of a triangle are together greater than a straight angle.

127. The perpendicular from any vertex of an acute-angled triangle on the opposite side falls within the triangle.

128. The perpendicular from either of the acute angles of an obtuse-angled triangle on the opposite side falls outside the triangle.

129. The semi-perimeter of a triangle is greater than any one side, and less than any two sides.

130. The perimeter of a quadrilateral is greater than the sum, and less than twice the sum, of the diagonals.

131. If a triangle and a quadrilateral stand on the same base, and the one figure fall within the other, that which has the greater surface shall have the greater perimeter.

132. If one angle of a triangle be equal to the sum of the other two, the triangle can be divided into two isosceles triangles.

133. If any sect joining two parallels be bisected, this point will bisect any other sect drawn through it and terminated by the parallels.

134. Through a given point draw a line such that the part intercepted between two given parallels may equal a given sect.

135. The medial from vertex to base of a triangle bisects the intercept on every parallel to the base.

136. Show that the surface of a quadrilateral equals the surface of a triangle which has two of its sides equal to the diagonals of the quadrilateral, and the included angle equal to either of the angles at which the diagonals intersect.

137. Describe a square, having given a diagonal.

138. ABC is a right-angled triangle; $BCED$ is the square on the hypothenuse; $ACKH$ and $ABFG$ are the squares on the other sides. Find the center of the square $ABFG$ (which may be done by drawing the two diagonals), and through it draw two lines, one parallel to BC, and the other perpendicular to BC. This divides the square $ABFG$ into four congruent quadrilaterals. Through each mid-point of the sides of the square $BCED$ draw a parallel to AB or AC. If each be extended until it meets the second of the other pair, they will cut the square $BCED$ into a square and four quadrilaterals congruent to $ACKH$ and the four quadrilaterals in $ABFG$.

139. The orthocenter, the centroid, and the circumcenter of a triangle are collinear, and the sect between the first two is double of the sect between the last two.

140. The perpendicular from the circumcenter to any side of a triangle is half the sect from the opposite vertex to the orthocenter.

141. Sects drawn from a given point to a given circle are bisected; find the locus of their mid-points.

142. The intersection of the lines joining the mid-points of opposite sides of a quadrilateral is the mid-point of the sect joining the mid-points of the diagonals.

143. A parallelogram has central symmetry.

SYMMETRY.

144. No triangle can have a center of symmetry, and every axis of symmetry is a medial.

145. Of two sides of a triangle, that is the greater which is cut by the perpendicular bisector of the third side.

146. If a right-angled triangle is symmetrical, the axis bisects the *right* angle.

147. An angle in a triangle will be acute, right, or obtuse, according as the medial through its vertex is greater than, equal to, or less than, half the opposite side.

148. If a quadrilateral has axial symmetry, the number of vertices not on the axis must be even; if none, it is a symmetrical trapezoid; if two, it is a *kite*.

149. A *kite* has the following seven properties; from each prove all the others by proving that a quadrilateral possessing it is a kite.

(1) One diagonal, *the axis*, is the perpendicular bisector of the other, which will be called the *transverse axis*.

(2) The axis bisects the angles at the vertices which it joins.

(3) The angles at the end-points of the transverse axis are equal, and equally divided by the latter.

(4) Adjacent sides which meet on the axis are equal.

(5) The axis divides the kite into two triangles which are congruent, with equal sides adjacent.

(6) The transverse axis divides the kite into two triangles, each of which is symmetrical.

(7) The lines joining the mid-points of opposite sides meet on the axis, and are equally inclined to it.

150. A symmetrical trapezoid has the following five properties; from each prove all the others by proving that a quadrilateral possessing it is a symmetrical trapezoid.

(1) Two opposite sides are parallel, and have a common perpendicular bisector.

(2) The other two opposite sides are equal, and equally inclined to either of the other sides.

(3) Each angle is equal to one, and supplemental to the other, of its two adjacent angles.

(4) The diagonals are equal, and divide each other equally.

(5) One median line bisects the angle between those sides produced which it does not bisect, and likewise bisects the angle between the two diagonals.

151. Prove the properties of the parallelogram from its central symmetry.

152. A kite with a center is a rhombus; a symmetrical trapezoid with a center is a rectangle; if both a rhombus and a rectangle, it is a square.

BOOK II.

153. The perpendicular from the centroid to a line outside the triangle equals one-third the sum of the perpendiculars to that line from the vertices.

154. If two sects be each divided internally into any number of parts, the rectangle contained by the two sects is equivalent to the sum of the rectangles contained by all the parts of the one, taken separately, with all the parts of the other.

155. The square on the sum of two sects is equivalent to the sum of the two rectangles contained by the sum and each of the sects.

156. The square on any sect is equivalent to four times the square on half the sect.

157. The rectangle contained by two internal segments of a sect grows less as the point of section moves from the mid-point.

158. The sum of the squares on the two segments of a sect is least when they are equal.

159. If the hypothenuse of an isosceles right-angled triangle be divided into internal or external segments, the sum of their squares is

equivalent to twice the square on the sect joining the point of section to the right angle.

160. Describe a rectangle equivalent to a given square, and having one of its sides equal to a given sect.

161. Find the locus of the vertices of all triangles on the same base, having the sum of the squares of their sides constant.

162. The center of a fixed circle is the point of intersection of the diagonals of a parallelogram; prove that the sum of the squares on the sects drawn from any point on the circle to the four vertices of the parallelogram is constant.

163. Thrice the sum of the squares on the sides of any pentagon is equivalent to the sum of the squares on the diagonals, together with four times the sum of the squares on the five sects joining, in order, the mid-points of those diagonals.

164. The sum of the squares on the sides of a triangle is less than twice the sum of the rectangles contained by every two of the sides.

165. If from the hypothenuse of a right-angled triangle sects be cut off equal to the adjacent sides, the square of the middle sect thus formed is equivalent to twice the rectangle contained by the extreme sects.

BOOK III.

166. From a point, two equal sects are drawn to a circle. Prove that the bisector of their angle contains the center of the circle.

167. Describe a circle of given radius to pass through a given point and have its center in a given line.

168. Equal chords in a circle are all tangent to a concentric circle.

169. Two concentric circles intercept equal sects on any common secant.

170. If two equal chords intersect either within or without a circle, the segments of the one equal the segments of the other.

171. Divide a circle into two segments such that the angle in the one shall be seven times the angle in the other.

172. ABC and ABC' are two triangles such that $AC = AC'$; prove the circle through A, B, C, equal to that through A, B, C'.

173. Pass a circle through four given points. When is the problem possible?

174. If *AC* and *BD* be two equal arcs in a circle *ABCD*, prove chord *AB* parallel to chord *CD*.

175. Circles described on any two sides of a triangle as diameters intersect on the third side, or the third side produced.

176. If one circle be described on the radius of another as diameter, any chord of the larger from the point of contact is bisected by the smaller.

177. If two circles cut, and from one of the points of intersection two diameters be drawn, their extremities and the other point of intersection will be collinear.

178. Find a point inside a triangle at which the three sides shall subtend equal angles.

179. On the produced altitudes of a triangle the sects between the orthocenter and the circumscribed circle are bisected by the sides of the triangle.

180. If on the three sides of any triangle equilateral triangles be described outwardly, the sects joining their circumcenters form an equilateral triangle.

181. If two chords intersect at right angles, the sum of the squares on their four segments is equivalent to the square on the diameter.

182. The opposite sides of an inscribed quadrilateral are produced to meet; prove that the bisectors of the two angles thus formed are at right angles.

183. The feet of the perpendiculars drawn from any point in its circumscribed circle to the sides of a triangle are collinear.

184. From a point *P* outside a circle, two secants, *PAB*, *PDC*, are drawn to the circle *ABCD*; *AC*, *BD*, are joined, and intersect at *O*. Prove that *O* lies on the chord of contact of the tangents drawn from *P* to the circle. Hence devise a method of drawing tangents to a circle from an external point by means of a ruler only.

185. A variable chord of a given circle passes through a fixed point; find the locus of its mid-point.

186. Find the locus of points from which tangents to a given circle contain a given angle.

187. The hypothenuse of a right-angled triangle is given; find the loci of the corners of the squares described outwardly on the sides of the triangle.

BOOK IV.

188. If a quadrilateral be circumscribed about a circle, the sum of two opposite sides is equal to the sum of the other two.

189. Find the center of a circle which shall cut off equal chords from the three sides of a triangle.

190. Draw a line which would bisect the angle between two lines which are not parallel, but which cannot be produced to meet.

191. The angle between the altitude of a triangle and the line through vertex and circumcenter equals half the difference of the angles at the base.

192. The bisector of the angle at the vertex of a triangle also bisects the angle between the altitude and the line through vertex and circumcenter.

193. If an angle bisector contains the circumcenter, the triangle is isosceles.

194. If a circle can be inscribed in a rectangle, it must be a square.

195. If a rhombus can be inscribed in a circle, it must be a square.

196. The square on the side of a regular pentagon is greater than the square on the side of the regular decagon inscribed in the same circle by the square on the radius.

197. The intersections of the diagonals of a regular pentagon are the vertices of another regular pentagon; so also the intersections of the alternate sides.

198. The sum of the five angles formed by producing the alternate sides of a regular pentagon equals a straight angle.

199. The circle through the mid-points of the sides of a triangle, called the medioscribed circle, passes also through the feet of the altitudes, and bisects the sects between the orthocenter and vertices.

HINT. Let ABC be the triangle; H, K, L, the mid-points of its sides; X, Y, Z, the feet of its altitudes; O its orthocenter; U, V, W, the mid-points of AO, BO, CO.

$LHWU$ is a rectangle; so is $HKUV$; and the angles at X, Y, Z, are right.

200. The mediocenter is midway between the circumcenter and the orthocenter.

201. The diameter of the circle inscribed in a right-angled triangle, together with the hypothenuse, equals the sum of the other two sides.

202. An equilateral polygon circumscribed about a circle is equiangular if the number of sides be odd.

203. Given the vertical angle of a triangle and the sum of the sides containing it; find the locus of the circumcenter.

204. Given the vertical angle and base of a triangle; find the locus of

(1) The center of the inscribed circle.

(2) The centers of the escribed circles.

(3) The orthocenter.

(4) The centroid.

(5) The mediocenter.

205. Of all rectangles inscribable in a circle, show that a square is the greatest.

BOOK VI.

206. The four extremities of two sects are concyclic if the sects cut so that the rectangle contained by the segments of one equals the rectangle contained by the segments of the other.

207. If two circles intersect, and through any point in their common chord two other chords be drawn, one in each circle, their four extremities are concyclic.

208. Does the magnitude of the third proportional to two given sects depend on the order in which the sects are taken?

209. Through a given point inside a circle draw a chord so that it shall be divided at the point in a given ratio.

210. If two triangles have two angles supplemental and other two equal, the sides about their third angles are proportional.

211. Find two sects from any two of the six following data: their sum, their difference, the sum of their squares, the difference of their squares, their rectangle, their ratio.

212. If two sides of a triangle be cut proportionally, the lines drawn from the points of section to the opposite vertices will intersect on the medial from the third vertex.

213. Given the base of a triangle and the ratio of the two sides; find the locus of the vertex.

MISCELLANEOUS.

214. Find the locus of a point at which two adjacent sides of a rectangle subtend supplementary angles.

215. Draw through a given point a line making equal angles with two given lines.

216. From a given point place three given sects so that their extremities may be in the same line, and intercept equal sects on that line.

217. Divide a sect into two parts such that the square of one of the parts shall be half the square on the whole sect.

218. Find the locus of the point at which a given sect subtends a given angle.

219. Given a curve, to ascertain whether it is an arc of a circle or not.

220. If two opposite sides of a parallelogram be bisected, and lines be drawn from these two points of bisection to the opposite angles, these lines will be parallel, two and two, and will trisect both diagonals.

221. Through two given points on opposite sides of a line draw lines to meet in it such that the angle they form is bisected.

222. The squares on the diagonals of a quadrilateral are double of the squares on the sides of the parallelogram formed by joining the mid-points of its sides.

223. If a quadrilateral be described about a circle, the angles subtended at the center by two opposite sides are supplemental.

224. Two circles touch at A and have a common tangent BC. Show that BAC is a right angle.

225. Find the locus of the point of intersection of bisectors of the angles at the base of triangles on the same base, and having a given vertical angle.

226. Find the locus of the centers of circles which touch a given circle at a given point.

227. Two equal given circles touch each other, and each touches one side of a right angle; find the locus of their point of contact.

228. In a given line find a point at which a given sect subtends a given angle.

229. Describe a circle of given radius to touch a given line and have its center on another given line.

230. At any point in the circle circumscribing a square, show that one of the sides subtends an angle thrice the others.

231. Divide a given arc of a circle into two parts which have their chords in a given ratio.

232. The sect of a common tangent between its points of contact is a mean proportional between the diameters of two tangent circles.

233. Any regular polygon inscribed in a circle is a mean proportional between the inscribed and circumscribed regular polygons of half the number of sides.

234. The circumcenter, the centroid, the mediocenter, and the orthocenter form a harmonic range.

EXERCISES IN GEOMETRY OF THREE DIMENSIONS AND IN MENSURATION.

The most instructive problems in geometry of three dimensions are made by generalizing those first solved for plane geometry. This way of getting a theorem in solid geometry is often difficult, but a number of the exercises here given are specially adapted for it.

In the author's "Mensuration" (published by Ginn & Co.) are given one hundred and six examples in metrical geometry worked out completely, and five hundred and twenty-four exercises and problems, of which also more than twenty are solved completely, and many others have hints appended.